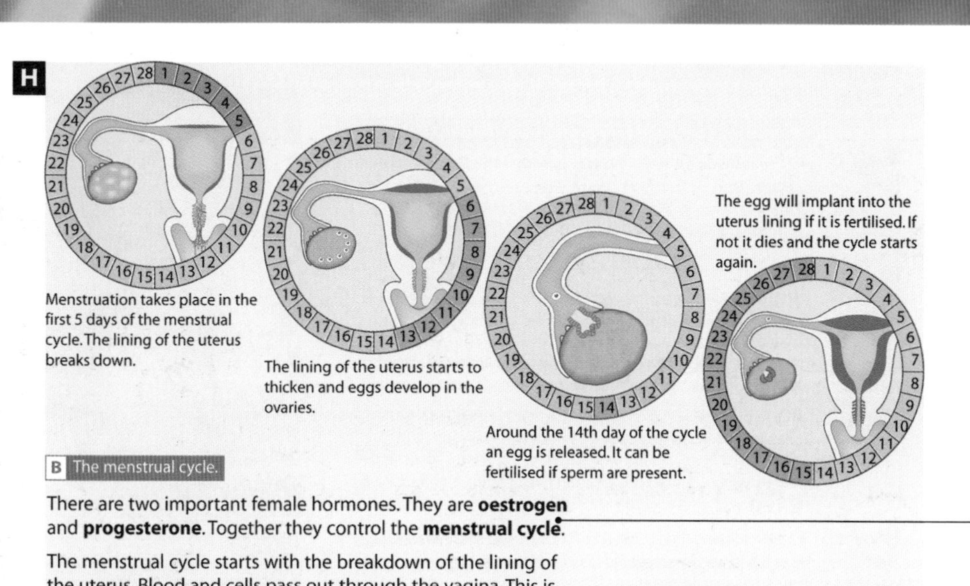

B The menstrual cycle.

There are two important female hormones. They are **oestrogen** and **progesterone**. Together they control the **menstrual cycle**.

The menstrual cycle starts with the breakdown of the lining of the uterus. Blood and cells pass out through the vagina. This is menstruation, but is usually called a period. As soon as the period has finished a new egg starts to develop in the woman's ovary. It grows inside a special fluid-filled ball called a follicle. About 14 days into the cycle, the follicle bursts and releases the egg into the oviduct. This is called ovulation.

If the egg is fertilised by a male's sperm, it implants into the uterus wall. It will develop into a baby. If the egg is not fertilised it dies and passes out through the vagina. The next period will then start, and the whole process begins again.

The process is controlled by hormones. As the egg develops in the follicle the ovary starts to make oestrogen. Oestrogen causes the lining of the uterus to start thickening and stops any more eggs from developing.

After ovulation, the follicle forms a yellow body. This starts to produce progesterone, which makes the uterus lining thicken even more.

So oestrogen and progesterone together make sure that the uterus is ready for a fertilised egg to implant. If a fertilised egg does implant, both hormones continue to be produced. This stops the next period from happening. If the egg is not fertilised, hormone production stops. The next period starts and the whole cycle starts again.

4 Why is it important that periods stop if a woman becomes pregnant?

5 What two things would happen if a woman's ovaries did not produce enough oestrogen?

P

Higher Questions

75

Glossary words
You will need to know the meaning of some key words. These are shown in **bold**. The glossary at the back of each topic gives you a list of all the key words and what they mean.

Questions
There are lots of questions on the page to help you think about the main points in each double-page section.

How to use your ActiveBook

The ActiveBook is an electronic copy of the book, which you can use on a compatible computer. The CD-ROM will only play while the disc is in the computer. The ActiveBook has these features:

Glossary
Click this tab to see all of the key words and what they mean. Click 'play' to listen to someone read them out to help you pronounce them.

DigiList
Click on this tab to see menus which list all the electronic files on the ActiveBook.

ActiveBook tab
Click this tab at the top of the screen to access the electronic version of the book.

Key words
Click on any of the words in **bold** to see a box with the word and what it means. Click 'play' to listen to someone read it out for you to help you pronounce it.

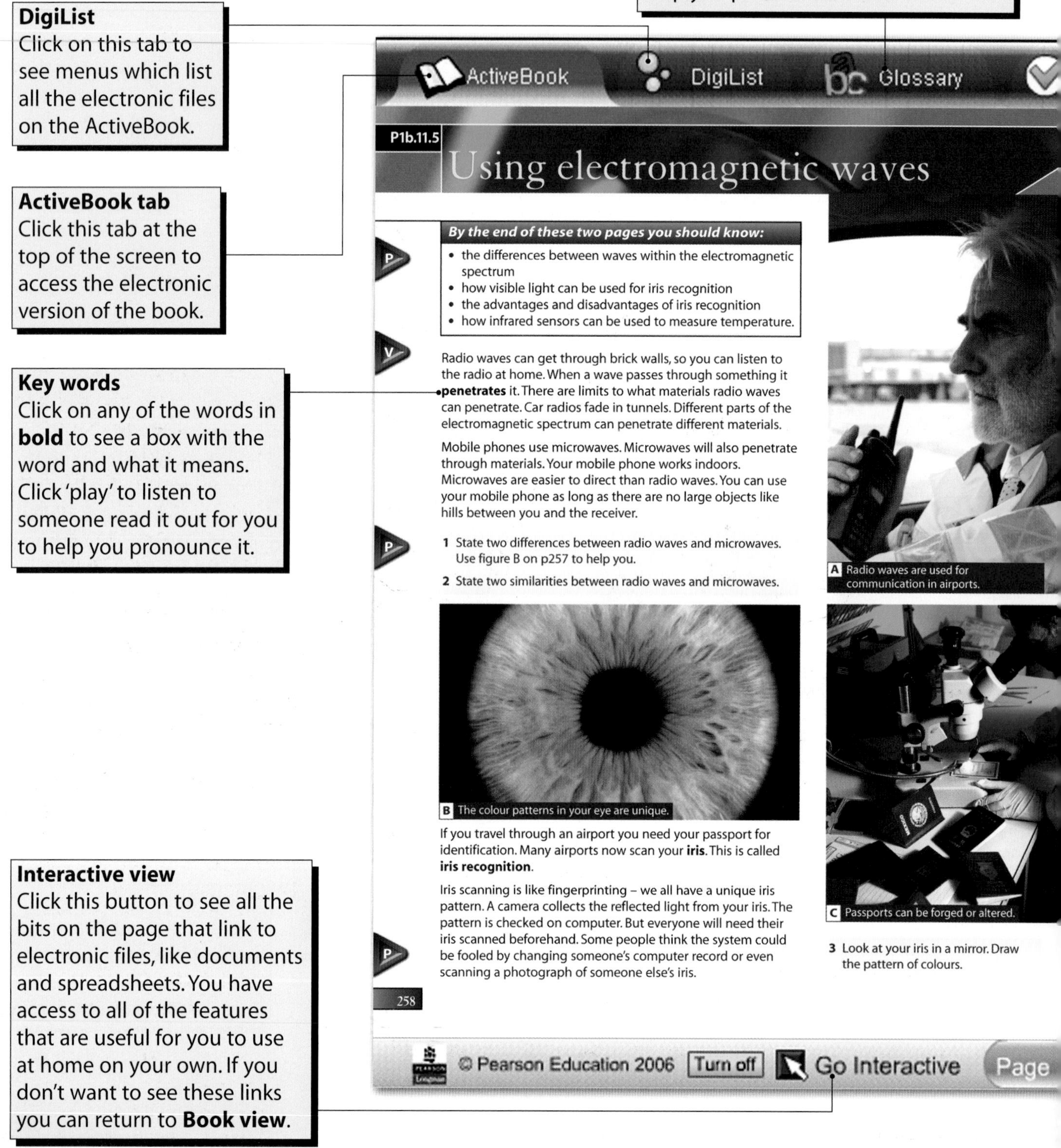

Interactive view
Click this button to see all the bits on the page that link to electronic files, like documents and spreadsheets. You have access to all of the features that are useful for you to use at home on your own. If you don't want to see these links you can return to **Book view**.

advancing learning, changing lives

GCSE Science Students' Book

Edexcel 360Science

Edexcel's own course for the new specification

James de Winter
Richard Laird
Nigel Saunders
Richard Shewry
Paul Spencer
Martin Stirrup

A PEARSON COMPANY

How to use this book

The book is divided into 12 topics. Each topic has a one-page introduction and is then divided into ten double page sections. At the end of each topic there is a set of questions that will help you practise for your exams and a glossary of key words for the unit.

As well as the paper version of the book there is a CD-ROM called an ActiveBook. For more information on the ActiveBook please see the next two pages.

What to look for on the pages of this book:

Numbering
This tells you what unit, topic and lesson you are in.
B1b = unit
3 = topic
9 = lesson

Learning outcomes
These tell you what you should know after you have studied these two pages.

Higher material
If you are hoping to get a grade between A* and C you need to make sure that you understand the bits with this symbol **H** next to it (as well as everything else in the book). Even if you don't think you will get a grade above a C have a go and look at these bits – you might surprise yourself (and your teacher!).

'Have You Ever Wondered?'
These questions are there to help you think about the way science works in your life. Your teacher might ask you what you think.

B1b.3.9

Hormones and contraception

By the end of these two pages you should be able to:

- explain how manufactured sex hormones can be used for contraception
- H describe how reproduction is controlled by hormones
- explain how the menstrual cycle is controlled by hormones.

During the Olympic Games in 1988, Olympic sprinter Ben Johnson won gold in the 100 m. He was sent home a few days later in disgrace and stripped of his gold medal after testing positive for drugs.

He was using a drug which is a synthetic version of the male sex hormone **testosterone**. It works by increasing muscle growth. This is the same as happens when boys go through puberty. Some women athletes have been known to take the drug. There can be pretty unpleasant side effects. Some athletes must be desperate to win if they are prepared to risk this. They also risk being banned from their sport if they are caught by one of the random drug tests. An athlete never knows if they are going to be tested.

1 Why might increased muscle growth be useful to an athlete?

2 Why do most people think that using such drugs is cheating?

How contraceptive pills work?

Sex hormones can be used to stop **pregnancy**. The pill contains sex hormones. It has to be taken by a woman over the course of a month and works by stopping the ovaries from releasing a new egg every month. This is called **contraception**.

3 How does the contraceptive pill work?

A Ben Johnson used banned substances to help him win.

74

Target sheets
Click on this tab at the top of the screen to see a target sheet for each topic. Save the target sheet on your computer and you can fill it in on screen. At the end of the topic you can update the sheet to see how much you have learned.

Help
Click on this tab at any time to search for help on how to use the ActiveBook.

t sheets Help

Have you ever wondered:
Do night vision goggles you see in the movies really work?

Objects emit **radiation** due to their temperature. You may see the **emission** of visible light if objects are very hot. Hot objects actually emit a range of different types of electromagnetic radiation. A light bulb emits visible light because the filament is very hot.

Hot objects emit infrared radiation. Infrared radiation carries most of the heat from the Sun to Earth. The surface temperature of an object can be measured by detecting the amount and type of infrared radiation. This is called infrared thermography.

6 In the film *Predator*, an alien can only detect infrared light. The hero makes himself invisible to the alien by covering himself with wet mud. Why might this make the hero invisible to the alien?

4 List three things that emit visible light when they are hot.

5 Give two examples of electromagnetic radiation which have a lower frequency than visible light.

D An infrared image of a human face.

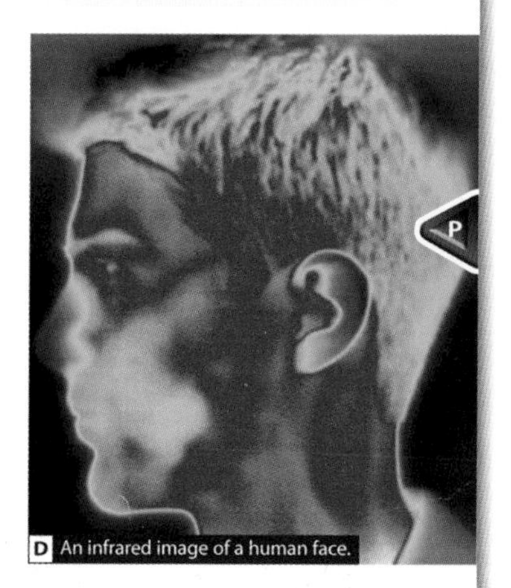

E

F An aircraft wing in infrared light showing cracks that have let in water, which has frozen.

This wing is made of new material which is light and strong. But if water gets into the wing it could freeze and cause damage. An infrared detector will show water in the wing if it is inspected soon after landing. The water will have frozen because of the cold temperature at high altitude. Ice will emit less infrared than the rest of the wing.

7 State one use of microwave radiation.

8 Gamma-rays are used to detect cracks in aircraft wings. Describe two differences between gamma-rays and visible light. Use figure B on p257.

Summary Exercise

Higher Questions

259

1 of 255

Summary exercise
Your teacher may ask you to complete the summary exercise found on the ActiveBook. This is a fill-in-the-gaps activity that will make sure you understand the science covered. The exercise will also be useful for revision.

Higher questions
The ActiveBook also contains higher-level questions. If you are hoping to get a grade between A* and C you should answer these questions. They include a summary question. Completing this will give you a summary of the whole double page and will be useful for revision.

Zoom feature
Just click on a section of the page and it will magnify so that you can read it easily on screen. This also means that you can look closely at photos and diagrams.

Page navigation
You can turn one page at time, or you can type in the number of the page you want and go straight to that page.

Contents

How to use this book 2
How to use your ActiveBook 4

B1a.1	**Environment**	**9**
B1a.1.1	Competition for resources	10
B1a.1.2	Sampling and estimating populations	12
B1a.1.3	Our influence on the environment	14
B1a.1.4	Chains, webs and pyramids	16
B1a.1.5	Wheat versus meat	18
B1a.1.6	Going organic	20
B1a.1.7	Changing organisms	22
B1a.1.8	Natural selection	24
B1a.1.9	The evidence for evolution	26
B1a.1.10	Variation	28
B1a.1.11	Questions	30
B1a.1.12	Glossary	32
B1a.2	**Genes**	**33**
B1a.2.1	Asexual reproduction	34
B1a.2.2	Genes and DNA	36
B1a.2.3	Variation	38
B1a.2.4	Dominant and recessive alleles	40
B1a.2.5	Nature or nurture?	42
B1a.2.6	It's in the genes	44
B1a.2.7	The Human Genome Project	46
B1a.2.8	Gene therapy	48
B1a.2.9	Clones and transgenic organisms	50
B1a.2.10	Designer babies	52
B1a.2.11	Questions	54
B1a.2.12	Glossary	56
B1b.3	**Electrical and chemical signals**	**57**
B1b.3.1	Reaction times	58
B1b.3.2	The nervous system	60
B1b.3.3	The human brain	62
B1b.3.4	Sense organs	64
B1b.3.5	The eye	66
B1b.3.6	Reflexes	68
B1b.3.7	The reflex arc	70
B1b.3.8	Hormones	72
B1b.3.9	Hormones and contraception	74
B1b.3.10	Hormones and fertility	76
B1b.3.11	Questions	78
B1b.3.12	Glossary	80
B1b.4	**Use, misuse and abuse**	**81**
B1b.4.1	TB in London	82
B1b.4.2	Controlling tuberculosis	84
B1b.4.3	Microorganisms and disease	86
B1b.4.4	The first line of defence	88
B1b.4.5	The second line of defence	90
B1b.4.6	The third line of defence	92
B1b.4.7	Types of drugs	94
B1b.4.8	Pain-relief	96
B1b.4.9	Drug misuse and abuse	98

B1b.4.10 Tobacco 100
B1b.4.11 Questions 102
B1b.4.12 Glossary 104

C1a.5 Patterns in properties 105

C1a.5.1 A map of the elements 106
C1a.5.2 Conducting heat 108
C1a.5.3 Colourful chemistry 110
C1a.5.4 Atomic structure 112
C1a.5.5 Group 1 – the alkali metals 114
C1a.5.6 Group 0 – the noble gases 116
C1a.5.7 Group 7 – the halogens 118
C1a.5.8 The salt formers 120
C1a.5.9 Using the periodic table 122
C1a.5.10 Forensic science 124
C1a.5.11 Questions 126
C1a.5.12 Glossary 128

C1a.6 Making changes 129

C1a.6.1 Oxygen and oxidation 130
C1a.6.2 The reactivity series 132
C1a.6.3 Ores 134
C1a.6.4 Neutralisation 136
C1a.6.5 Making useful salts 138
C1a.6.6 Baking with bubbles 140
C1a.6.7 Breaking down in the heat 142
C1a.6.8 Chemicals in food 144
C1a.6.9 Liquids and gases in the home 146
C1a.6.10 Solids in the home 148
C1a.6.11 Questions 150
C1a.6.12 Glossary 152

C1b.7 There's only one Earth 153

C1b.7.1 Getting energy from fuels 154
C1b.7.2 The dangers of incomplete combustion 156
C1b.7.3 Global warming and fossil fuels 158
C1b.7.4 The Earth's changing climate 160
C1b.7.5 Tackling the problem of fossil fuels 162
C1b.7.6 Is there an alternative to oil? 164
C1b.7.7 Waste not, want not… [illegible]
C1b.7.8 What do we get from crude oil? 168
C1b.7.9 What can we get from air? 170
C1b.7.10 What can we get from sea water? 172
C1b.7.11 Questions 174
C1b.7.12 Glossary 176

C1b.8 Designer products 177

C1b.8.1 Getting the right materials 178
C1b.8.2 Making new materials 180
C1b.8.3 The future could be very small… 182
C1b.8.4 The nanoparticle balance sheet 184
C1b.8.5 Making beer and wine 186
C1b.8.6 How alcohol affects the human body 188
C1b.8.7 What alcohol does to society 190
C1b.8.8 Intelligent packaging 192
C1b.8.9 How does mayonnaise work? 194
C1b.8.10 Design and test 196
C1b.8.11 Questions 198
C1b.8.12 Glossary 200

P1a.9 Producing and measuring electricity 201

P1a.9.1	Telecommunications	202
P1a.9.2	Direct current	204
P1a.9.3	Producing electric current	206
P1a.9.4	Current, voltage and resistance	208
P1a.9.5	Resistance, lamps and computers	210
P1a.9.6	Cells, time and recharging	212
P1a.9.7	Controlling the flow of electricity	214
P1a.9.8	Changing resistance	216
P1a.9.9	Smaller and more powerful	218
P1a.9.10	Science leading to technology	220
P1a.9.11	Questions	222
P1a.9.12	Glossary	224

P1a.10 You're in charge 225

P1a.10.1	Oil, coal and gas won't last forever	226
P1a.10.2	The National Grid	228
P1a.10.3	Motors	230
P1a.10.4	Wires, fuses and safety	232
P1a.10.5	Power	234
P1a.10.6	Efficiency	236
P1a.10.7	Electricity, units and money	238
P1a.10.8	Energy-efficiency and saving money	240
P1a.10.9	Solar cells	242
P1a.10.10	Technology, electricity and medicine	244
P1a.10.11	Questions	246
P1a.10.12	Glossary	248

P1b.11 Now you see it, now you don't 249

P1b.11.1	Wave basics	250
P1b.11.2	Reflecting waves	252
P1b.11.3	Earthquake waves	254
P1b.11.4	The wave equation	256
P1b.11.5	Using electromagnetic waves	258
P1b.11.6	More uses of electromagnetic waves	260
P1b.11.7	Microwaves and mobile phones	262
P1b.11.8	Harmful electromagnetic waves	264
P1b.11.9	Digital information	266
P1b.11.10	The digital revolution	268
P1b.11.11	Questions	270
P1b.11.12	Glossary	272

P1b.12 Space and its mysteries 273

P1b.12.1	A weighty problem	274
P1b.12.2	Our place in the universe	276
P1b.12.3	Comets and asteroids	278
P1b.12.4	Space travel	280
P1b.12.5	Launching into space	282
P1b.12.6	Observations of the universe	284
P1b.12.7	Searching for life	286
P1b.12.8	Life of a star	288
P1b.12.9	The universe: past, present and future	290
P1b.12.10	Endless discoveries	292
P1b.12.11	Questions	294
P1b.12.12	Glossary	296

Periodic Table 297
Index 298
Acknowledgements 302

Environment

A A Masai tribesman in the Serengeti.

In the 1850s, Charles Darwin described the struggle for life. The plains of the Serengeti in Africa provide an excellent example of this. Living things interact and compete together. The fittest ones survive. Over time a species gradually changes and improves its chances of survival. The fiercest predators eat well but the fastest prey escape and live to reproduce. If a species fails to adapt and improve it becomes extinct, like the dinosaurs.

All species interact with each other and with their environment. Humans are upsetting the Earth's balance but we can now reverse some of these changes. **Sustainable development** and **conservation** programmes show that we need not destroy the planet. Advances in science and technology hold the key to restoring this delicate balance.

In this topic you will learn that:

- animals and plants depend upon each other
- all organisms are adapted to their environment
- there is often competition between organisms for resources
- natural selection is a long process over many generations.

Look at these statements and sort them into the following categories:

I agree, I disagree, I want to find out more

- The Sun provides all of the energy needed to support life on Earth.
- There is no limit to the length of a food chain.
- Organisms in an environment have different adaptations that enable them to survive.
- The number of foxes controls the size of a population of rabbits.
- Humans evolved from apes.
- Evolution is something that only occurred millions of years ago.

Competition for resources

By the end of these two pages you should know:

- that living things compete with each other for the things they need
- that predators compete with each other for prey
- that animal populations are affected by numbers of predators and by competition for resources.

A Lunch-time on the savannah.

Charles Darwin described life as a struggle for survival. Animals compete with each other for resources. The savannah is a hot, grassy plain in Africa. There are very few trees and many of them have sharp thorns. The grass provides food for huge herds of zebras, wildebeest, gazelles and many other **species**. Because they all eat the same food, these animals compete for the best grass. **Competition** between different species, such as zebra and wildebeest, is called **inter-species** competition. Competition within a single species, say between two zebras, is called **intra-species** competition.

As a general rule the strongest, best-adapted animals get to eat most of the food. These animals survive and breed so their **population** increases at the expense of other weaker, less-adapted species. Sometimes species adapt so that they can live comfortably side by side. In addition to food, animals compete for water, shelter, a suitable mate and places to breed.

Have you ever wondered:

How do different organisms make different changes to solve the same environmental problem?

1. Suggest why there are few trees growing on the Serengeti savannah in Tanzania.
2. Most of the trees have long thorns. How might this help the trees to survive?
3. A large herd of wildebeest are feeding on the savannah. Grass is in short supply. Name the type of competition which would be taking place in these conditions.
4. Name two other resources which wildebeest would be competing for.

B 'Nature, red in tooth and claw'. A predator eats its prey on the savannah.

Predators are hunters which kill and eat other animals. The animal which is hunted is called the **prey**. **Scavengers** are animals which eat the 'left-overs' of animals killed by predators. The large herbivores on the savannah are prey to animals like lions and cheetahs. Other predators and scavengers eat smaller herbivores such as gazelles and dik-diks. The plants and animals on the savannah are **interdependent**. Changes in the numbers of one **organism** affect the numbers of other organisms.

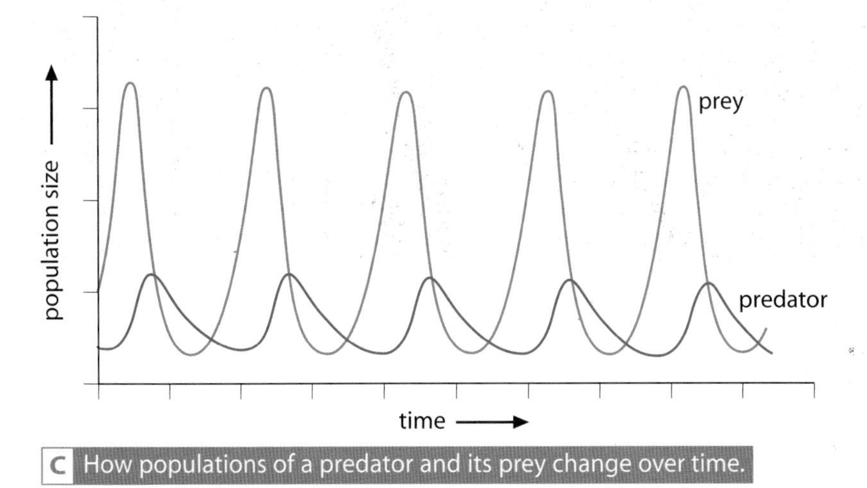

C How populations of a predator and its prey change over time.

Have you ever wondered:

Does the number of foxes control the number of rabbits or does the number of rabbits control the number of foxes?

In years when there is plenty of grass, competition between the herbivores is reduced. The number of herbivores will increase. Young herbivores are easy prey for predators. This extra source of meat allows predators to successfully raise more young. When the grass fails to grow, or if there are too many predators, the herbivore population will drop. Figure C shows how the populations change over a long period of time.

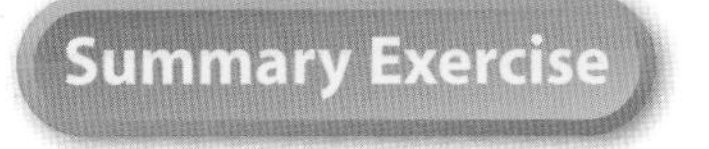

5 Suggest why young herbivores are easy prey for predators.

6 Suggest what will happen to the size of the predator population if the herbivore population falls. Give a reason for your answer.

Higher Questions

Sampling and estimating populations

By the end of these two pages you should know:

- why it is important to sample populations
- how data from samples can be used to estimate population sizes
- some pros and cons of using computer models to predict how changing one factor might affect a population.

It is important for us to know how many of each species are living in an area. For example, the Tanzanian authorities may need to know how many lions there are in the Serengeti. The problem is that it is sometimes difficult to know exactly how many of each species there are. Animals move around and if there are lots it is difficult to count them. Plants can be tricky too because there are so many of them and they are so small.

Pupils using a quadrat to sample plants found growing in the school field.

How do scientists estimate the population sizes of different species in an area such as a large field? A population is the total number of a species, such as dandelions or spiders, in a given area. It is called an estimate because it is not really possible to count every individual. A better way is to sample a small area, and scale up the results. A **quadrat** is a square frame of known size, say 1 m^2. Quadrats are used to sample plants (and animals that don't move much). To sample a field, several quadrats are placed randomly in the field. Every species found in each quadrat is identified and counted. This gradually builds up a picture of the kinds and numbers of organisms living in the field.

Example: Suppose a field has an area of 10,000 m^2. The pupils study 25 quadrats and count a total of 80 dandelion plants. Each quadrat has an area of 1 m^2, so they sample 25 m^2, which is one four-hundredth of the field area (25/10,000). Estimate the size of the dandelion population.
Solution: The dandelion plant population in the field is estimated to be about $80 \times 400 = 32{,}000$.

1 Scientists estimate the size of a population. What is a population?

2 Another group of pupils used 20 quadrats to sample the same 10,000 m^2 field. Their sample contained a total of 120 daisy plants. Estimate the size of the daisy population in the field.

3 Suggest why the size of the daisy population is only a rough estimate.

P

Populations change over time and it is difficult to keep track of them. Temperature, the availability of food and water, predation and disease can all affect the numbers of organisms. We cannot keep counting populations – it takes too long and is expensive to do. **Computer models** can be used to simulate the effect a change in conditions might have on populations. So, it would be possible to model what will happen to the zebra population of the Serengeti if the lion population soared. Computer models are useful because they allow us to predict what could happen to a population and then do something about it if possible. The problem is that there are so many interactions in the natural world that it is difficult to be sure that computer models are reliable.

Summary Exercise

4 Explain the advantages and disadvantages of computer models.

P

5 To estimate the size of animal populations, scientists sometimes mark the animals with ear-tags or rings on their legs. How can this help?

6 Animals can also be tracked by fitting them with radio transmitters and using satellites to map their movements. Suggest how this information could be useful for scientists.

Higher Questions

Our influence on the environment

H ***By the end of these two pages you should know:***

- how economic, industrial and population changes affect the environment.

The human population has dramatically increased in size over the last 100 years (see graph A). All of these extra people need food, water and somewhere to live. Everyone wants a good standard of living which means higher demand for things like electricity and cars. Generating more electricity, burning forest to clear land for farming and burning fuel in cars all produces more carbon dioxide. All of this puts pressure on the **environment**. The environment is the place where an organism lives and is made up of many different features such as air, water, soil and other living things.

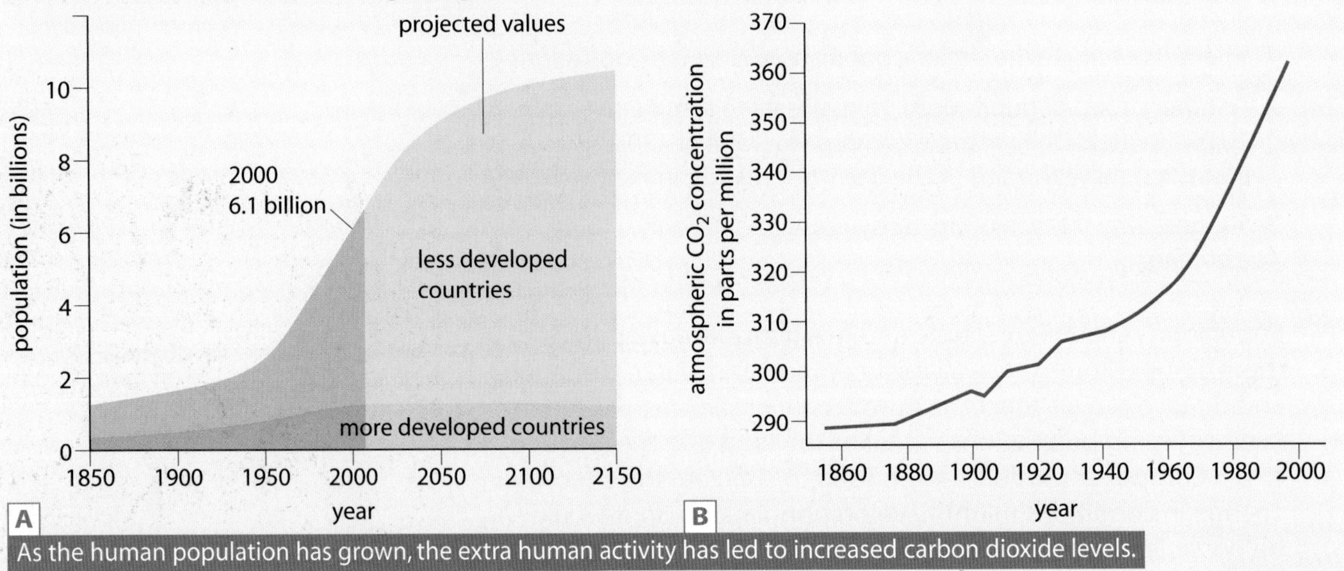

As the human population has grown, the extra human activity has led to increased carbon dioxide levels.

1 Look at the graph of human population growth.
   a What was the human population in the years 1860 and 2000?
   b What could the population be in 2050?

2 Look at the graph of carbon dioxide levels. What were the levels of carbon dioxide in 1860 and 2000?

3 Explain what the link is between increased population and more carbon dioxide in the atmosphere.

Carbon dioxide is a 'greenhouse gas'. This means that it traps heat in the Earth's atmosphere. If there is more heat trapped in the atmosphere then the Earth could become warmer. This could lead to melting of polar ice and rising sea levels. Many places in Britain and across the world would be at risk from flooding if sea levels rose.

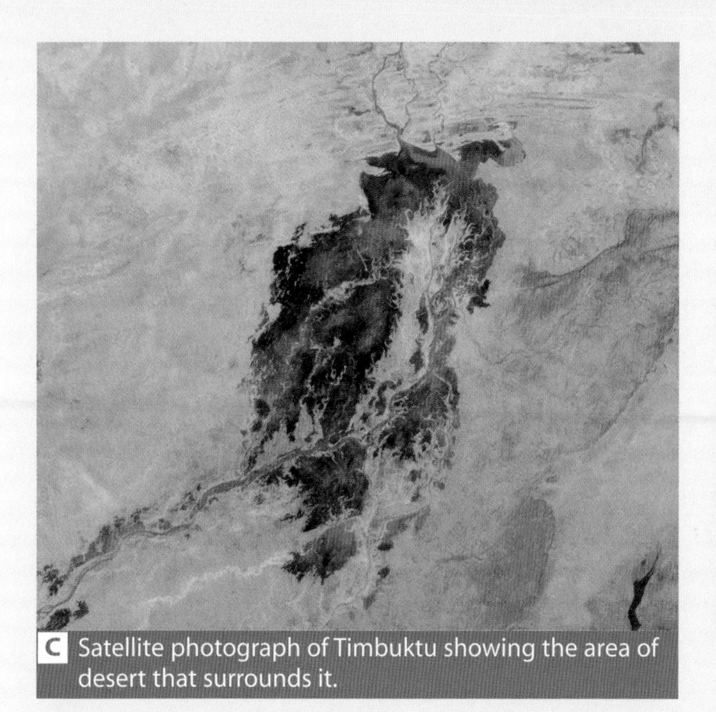

C Satellite photograph of Timbuktu showing the area of desert that surrounds it.

D A sandstorm as the desert advances.

The satellite photograph shows Timbuktu in Mali, West Africa. The area around Timbuktu is the Sahel, which means 'the edge of the desert'. Nomads used to graze animals in this area until farmers began planting peanuts to sell. This change in land use, together with local climate changes, has resulted in the loss of soil fertility. This is called **desertification**.

Recently an area of South American rainforest the size of Belgium was cleared to grow crops and to graze animals – the trees were sold as timber. This is **deforestation**. This can lead to soil washing away and increased flooding.

Here are some of the things that you can do to help the environment. Use less energy – make sure you turn off lights when you leave a room. Don't leave your TV or computer on standby. Try not to use the car for every journey – walk to school if you can, get the bus or ride your bike. Recycle whenever you can. Glass, aluminium and paper are just a few of the things that you can recycle.

E Deforestation is a global concern, not just a local issue.

4 Look at the satellite image of Timbuktu.
  a Explain how changes in land use could reduce soil fertility.
  b Suggest why sandstorms are more common in overgrazed areas.

5 What is deforestation? Give three reasons why deforestation is still taking place.

6 List three things that you can do to reduce your impact on the environment.

Higher Questions

Chains, webs and pyramids

By the end of these two pages you should know:

- what a food chain is
- that many food chains link together to make a food web
- that pyramids can show the numbers, biomass or energy at different stages in a food chain.

A A food chain in the Serengeti savannah.

The diagram shows a **food chain** in the East African Serengeti. Food chains are examples of energy-transfer chains. The arrows show the flow of energy through the food chain. Because green plants use photosynthesis to make their own food they are called **producers**. Almost every food chain in the world begins with a green plant.

Have you ever wondered:

How can the Sun's energy support all life on Earth?

1 Suggest another food chain that you would find in the Serengeti.

2 Give two life processes, in addition to growth, that need energy.

The Serengeti-Mara is the 26,000 km^2 home to more than 3 million large animals. After the rains, over 2 million grazers move into the area, feeding on grasses and acacia bushes. Large predators use the area as a hunting ground to find food. The savannah **ecosystem** has lots of different food chains. When food chains are linked together, they make a **food web**. The top carnivores have few predators. When top carnivores die, scavengers eat their bodies.

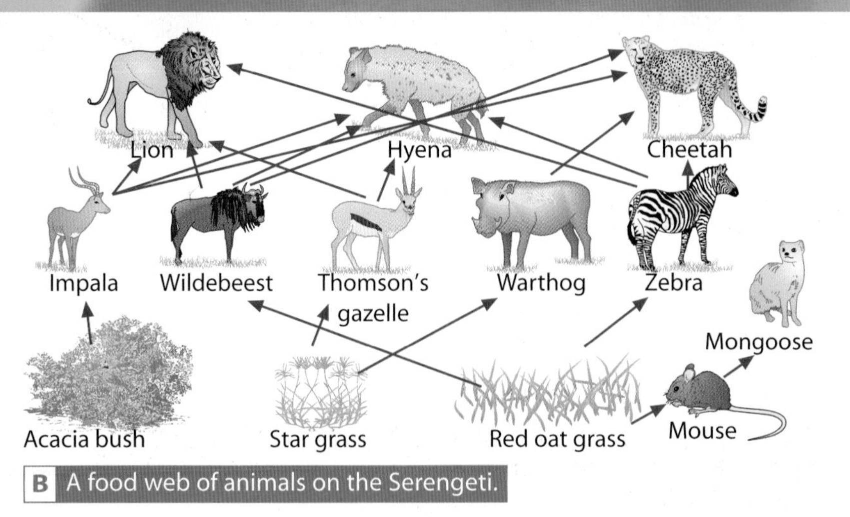

B A food web of animals on the Serengeti.

The Serengeti-Mara plains are not endless. Every lion needs about 350 grazers to feed it. Each grazer needs enough grass to survive. About 2000 lions live in the Serengeti-Mara. This gives each lion about 13 km^2 of hunting ground. The same area contains about 350 grazers, mainly Thomson's gazelle, wildebeest and zebra.

Food chains tell you how energy is transferred from one organism to another but not much more. A food pyramid is a numerical (**quantitative**) way of showing the organisms in a food chain. There are two kinds of food pyramid: **pyramids of number** and **pyramids of biomass. Biomass** is the dry mass of living matter. Pyramids of number show how many of each organism are found at each level of a food chain. Pyramids of biomass show the amount of living matter at each level of the pyramid.

Organism	Number of individuals	Average biomass of one individual (kg)	Total biomass of all individuals (kg)
Lion (secondary consumer)	1	200	200
Grazers (primary consumers)	350	157	54,950
Grass (primary producer)	4,000,000	0.17	680,000

C Table of biomass for the hunting area of one lion (13 km^2).

The Masai people use part of the area to raise cattle. They use pens to protect their cattle from hungry lions. The Masai cattle have to compete with other grazers for grass. The whole area is in a delicate balance and has become a National Park. Scientists monitor the numbers of animals and there are strict rules about hunting. This helps to conserve the wide variety of life in the Serengeti.

P

3 Use the food web to construct a food chain that includes impala.

4 In the first lesson you looked at predator–prey relationships. Use your knowledge to explain what could happen if you removed all the impala from the Serengeti.

A

P

P

5 What does a pyramid of number show?

6 Why is a pyramid of biomass more useful than a pyramid of number?

Higher Questions

Wheat versus meat

By the end of these two pages you should know:

- why producing a field of wheat provides more food energy than producing a field of cows
- what happens to the 'lost energy' in a food chain.

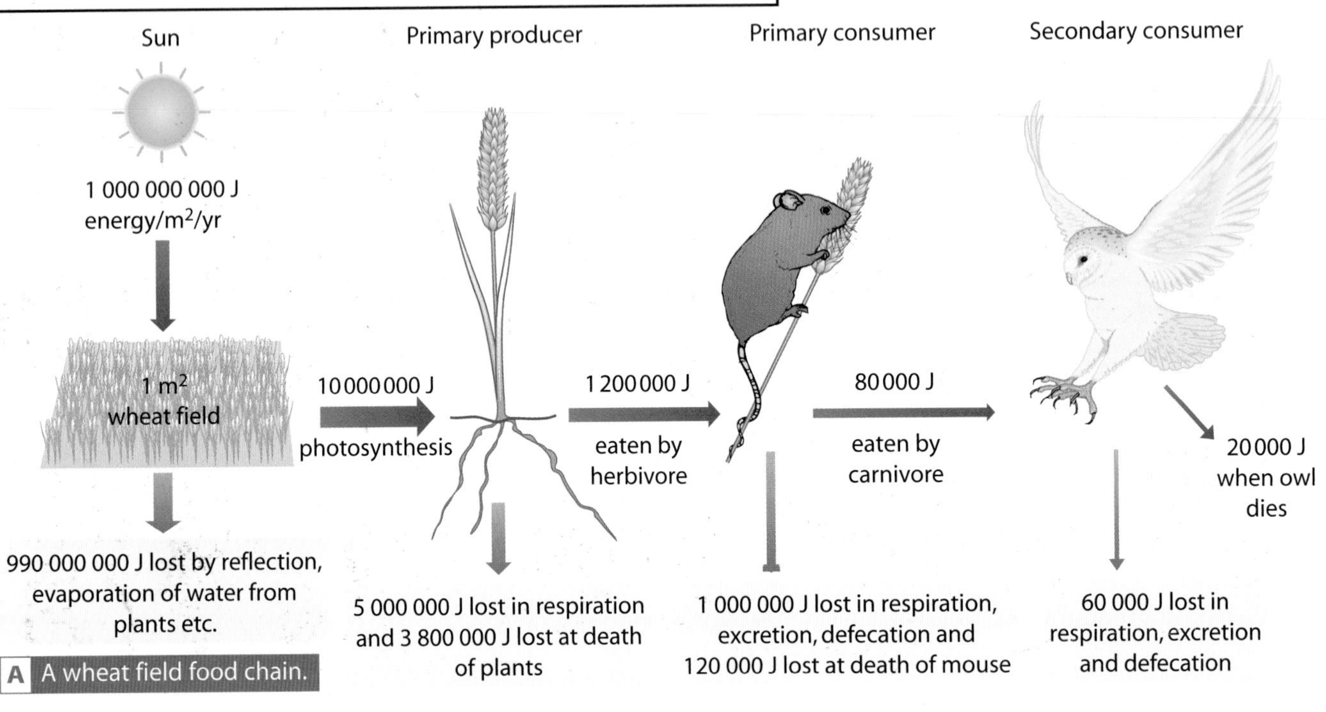

A A wheat field food chain.

Food chains are made up of no more than three or four links. Here you will find out why.

Each square metre of a wheat field receives about 1000 megajoules (MJ; 1MJ is 1,000,000 joules or J) of light energy from the Sun each year. Just over 1% of this energy is absorbed by the wheat and used for photosynthesis. This transfers light energy into chemical energy in the form of carbohydrates.

1 How much light energy is transferred to chemical energy by 1 m^2 of wheat plants? Give your answer in megajoules.

Have you ever wondered:

Why don't food chains go on forever?

The next link in the chain is the harvest mice and they eat plants. Only about 10% of the energy captured by the plant is transferred to mice. The owl then eats the mice. Once again only about 10% of the energy is transferred to the owl. Of the 10 MJ captured by photosynthesis only 0.08 MJ is finally transferred to the owl. These figures are fairly standard for food chains.

2 Use the information in figure A to calculate the total amount of energy lost from this food chain

a in respiration, defecation and respiration

b when the organisms die.

The human population is increasing at a dramatic rate. All these people need to eat. Producing enough food for the world's population is an increasingly hard task.

Here are three ways of producing our food. Which method is the most efficient?

Energy is lost at every step in a food chain. One solution to food shortages might be to cut down the number of steps in human food chains. Almost half of the world's cereal production is used to feed livestock. It is a very inefficient way of making protein. Using grain to produce meat in developed countries pushes up the price of cereals. Developing countries cannot afford enough grain to feed themselves. They face famine. Producing 1000 J of intensively reared chicken needs 12,000 J of grain. Producing 1000 J of beef needs 10,000 J of grain.

Some domestic animals (sheep, goats, rabbits, free-range chickens and ducks) are 'rough grazers'. Sheep and goats thrive in areas where it is impossible to grow cereal crops. They convert plant food, which would otherwise not be used, into meat to feed people. If you eat free-range eggs or drink goats' milk produced by rough grazers, you can have a relatively clear conscience so far as energy is concerned.

Have you ever wondered:

Which grows more quickly – grass or cow?

Summary Exercise

3 What is a 'rough grazer'?

4 Why does cutting down the number of steps in the food chain make it more efficient?

5 Explain why using goats, sheep and free-range chickens to produce protein is preferable to rearing cattle.

6 Keeping animals in confined spaces makes them produce more meat and eggs for the same amount of food. Explain why.

?

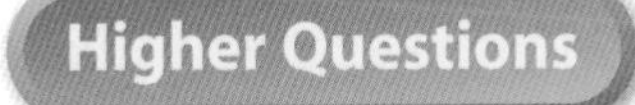

Going organic

By the end of these two pages you should know:

- what is meant by the term organic farming
- about some of the methods used by organic farmers.
- why organic products are more expensive than non-organic products.

In recent years people have started to ask questions about the food we eat. Modern **factory farming** is an intensive method of farming. It uses chemicals such as artificial fertilisers, herbicides and pesticides. It relies heavily on non-renewable resources. Crops are grown in huge fields and are harvested by large, expensive machinery. Animals are fed special diets and are raised intensively. These farms are highly mechanised and automated and produce food efficiently and cheaply. However, some people disapprove of these methods. **Organic farming** is a more natural way of producing food. It avoids the use of artificial fertilisers and chemical pesticides. Organic methods are often better for the environment. Growing legumes, such as beans and peas, improves soil fertility by adding nitrates. Some people also think the food is better for our health. However, it is usually more expensive to buy. This is because fields are small, crops may take longer to grow and the yield may vary. Organic produce may also be more fragile and difficult to transport.

A Organic food is becoming more popular.

B A comparison of organic and intensive farming.

1. What are legumes and how do they help the soil?
2. During decay a lot of heat is produced. Well-rotted manure and compost is relatively free from weed seeds and pests. Explain why.

People do not want caterpillars or maggots in their food. Pesticides prevent this but poisons can build up in food chains by **bioaccumulation**. **Organic** farmers sometimes use **biological control**. Using predators to kill pests is an example of this. For example, ladybirds eat aphids and other sap-sucking pests. Slugs and snails can be controlled by tiny worms which burrow into their bodies and kill them. Caterpillars can be controlled using special wasps. The wasps lay their eggs inside the young caterpillar. As the caterpillar grows, the eggs hatch. A small maggot grows inside the caterpillar. It feeds very selectively, avoiding the caterpillar's vital organs and nervous system. Only when the caterpillar is almost full size does the maggot finally kill it. The maggot then turns into a wasp.

One advantage of these methods is that they do not harm beneficial insects like bees. Biological control leaves no poisonous residues, although there may be a small amount of pest damage to the crop. Biological control works best in closed environments such as glass-houses. Companion planting also helps to keep pests at bay. For example, growing rows of onions or garlic between rows of carrots can prevent root damage by carrot fly. Other plants like marigolds and lavender also deter pests and help to attract pollinating insects.

C The Eden Project in Cornwall uses biological control.

3 Describe one example of biological control.

4 Explain why using a predator to control pests cannot prevent all of the damage to crops.

5 Suggest why biological control is more effective in a glass-house than in an open field.

6 What is companion planting? Suggest why it is not widely used in intensive farming situations.

Summary Exercise

Higher Questions

Changing organisms

By the end of these two pages you should know:

- what selective breeding and genetic engineering are and how they are used to change organisms
- H that crops can be genetically modified
- why growers use these techniques to develop new crop varieties.

Before anyone knew anything about genetics, **selective breeding** of animals and plants was taking place. Selective breeding means choosing organisms with the features you want and deliberately mating them together. Wild animals were caught, tamed and domesticated. Dogs and cats were kept to help with hunting or for keeping down vermin. Cattle were bred for meat and milk. By choosing the best and **breeding** from them, humans developed animals that suited their needs. Breeding plants was more accidental at first. Early farmers grew a wild grass called spelt and wild wheat together. Some of the seeds would have been collected and planted the following year. This produced a slow but gradual improvement in the crop.

A Pollinating a flower in a selective breeding programme.

B a Friesian cow

C a Hereford bull

Friesian–Hereford crosses produce good milk and beef.

D a Friesian–Hereford cross

1 Suggest why early farmers found it easier to breed new varieties of animals than to breed new varieties of plants.

2 What features did early farmers select when breeding domesticated animals such as cattle or sheep?

3 Why has there been a rapid increase in the number of new crop varieties in the last 100 years?

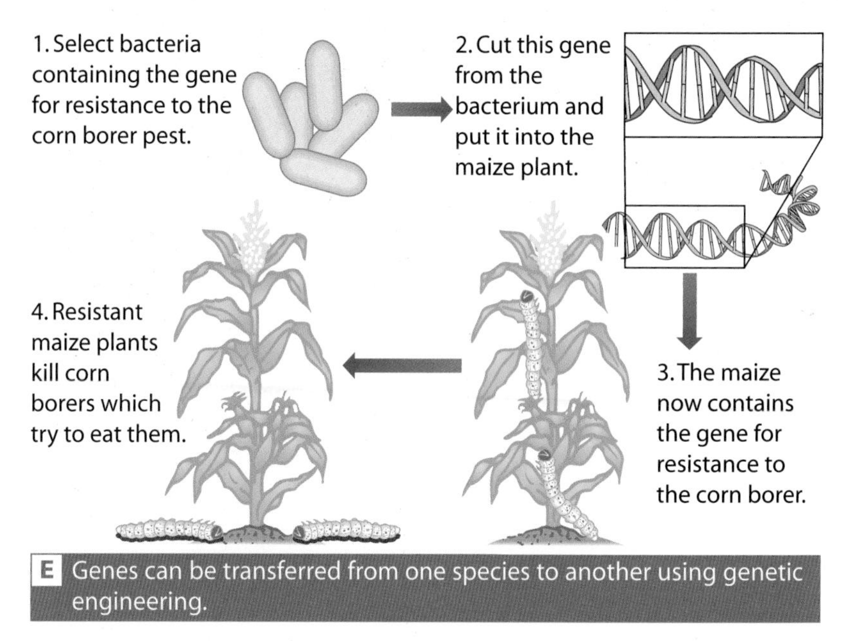

E Genes can be transferred from one species to another using genetic engineering.

It is now possible to use **genetic engineering** to transfer single genes from one organism to another. This can even be done between different species. **Genetically modified** organisms, or GM organisms, contain genes that have been altered or taken from other species. Growers and retailers use GM organisms because they have improved yields and better storage. Genetic modification is quick and it enables single **characteristics** to be altered in just one generation.

4 Genetic modification allows a breeder to make more precise changes to an organism than selective breeding. Explain why.

Have you ever wondered:

Why are so many people worried about GM technology?

H In some GM tomatoes, the gene which makes ripening tomatoes turn soft has part of its DNA reversed. These tomatoes remain firm for longer and do not get squashed in transit. Another GM tomato variety contains genes from an Antarctic fish. This gene makes an 'anti-freeze' which reduces damage to tomatoes caused by cold conditions. There is a variety of GM maize which makes its own insecticide and another variety of GM soya bean which is resistant to weedkillers.

One concern is that putting genes from one species into another species may have unforeseen consequences. If the genes for making insecticide or weedkiller resistance 'escape' into other plants, these plants may become troublesome pests. We cannot be sure of the long-term effects of eating GM foods containing natural insecticides or what they will do to food chains.

5 Maize is wind-pollinated. The pollen of one variety of GM maize contains insecticide. Suggest why growing this variety of GM maize might have unforeseen environmental effects.

6 Suggest why it is useful to have a variety of soya bean that is resistant to weedkiller.

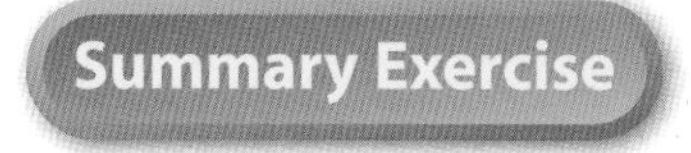

Higher Questions

Natural selection

By the end of these two pages you should know:

- that natural selection can lead to the formation of new species
- that individuals can have characteristics that help them to reproduce more successfully
- that less well adapted species can become extinct.
- some differences between natural selection, selective breeding and genetic engineering.

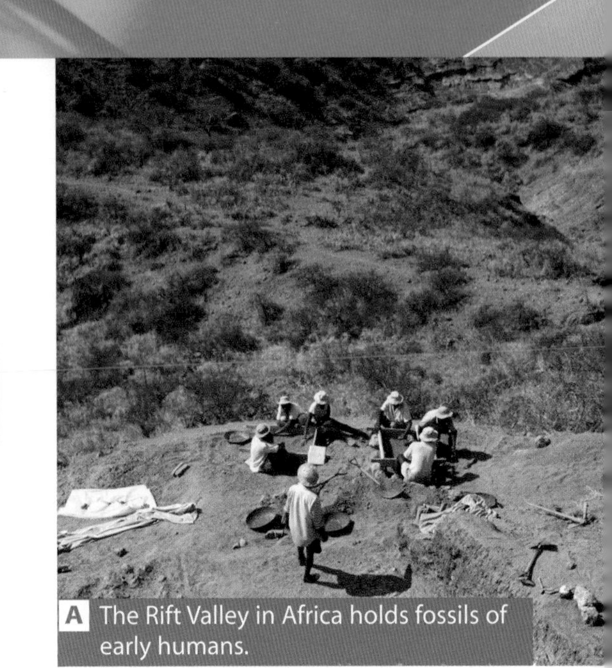

A The Rift Valley in Africa holds fossils of early humans.

Like every other living thing humans have evolved over time. Charles Darwin tried to explain why changes like these occurred. In the nineteenth century, Darwin developed the theory of **evolution** through **natural selection**. He made four observations and three deductions that changed the way biology developed.

B How Darwin's four observations and three deductions lead to the theory of evolution through natural selection.

Darwin called this process natural selection. A population would gradually change over many generations. This gradual change is called evolution. Variation in a species means that individuals have different characteristics. Some characteristics, such as ones that make the individual unattractive to predators and attractive to potential mates, help successful **reproduction**. Individuals without those characteristics will be less likely to reproduce. This means that more offspring in the population will have the favourable characteristics, and over generations the whole species will change and become better adapted to its environment.

Natural selection happens over very long periods of time, and is not controlled by humans. In artificial selection, such as selective breeding or genetic engineering, humans deliberately breed plants or animals to get the characteristics they want. Using genetic engineering to manipulate genes means an organism with new characteristics can be produced very quickly.

Have you ever wondered:

How does natural selection 'know' how to create a new species?

Have you ever wondered:

Is evolution still taking place?

Adaptation to the environment is important for organisms to evolve. If there is a big change in the environment then an organism may not be able to change quickly enough. About 65 million years ago a large meteor crashed into the Earth. This may have caused a change in the climate. Perhaps dinosaurs could not change fast enough and they died out. When a population dies out completely we say that it has become **extinct**.

Have you ever wondered:

What would happen to the human race if we were all the same?

4 Rats are troublesome pests. A poison called warfarin was developed to kill rats. At first the poison was successful but now populations of warfarin-resistant 'super-rats' have developed. Suggest how.

5 Bacteria can divide into two every 20 minutes. This enables them to evolve very rapidly. Explain why.

Summary Exercise

1 Explain why the number of wild rabbits stays nearly the same year after year despite the fact that many baby rabbits are born.

2 Explain why some rabbits are more likely to survive than others.

3 Darwin said that the characteristics of a population would gradually change after many generations. What was his explanation for this gradual change?

C Did dinosaurs become extinct because they could not evolve fast enough?

Higher Questions

The evidence for evolution

By the end of these two pages you should know:

- that fossils provide evidence for evolution
- why Darwin had difficulty in getting his ideas about evolution and natural selection accepted.

The nineteenth century was the age of the **fossil** hunters. It also saw the birth of a new science, geology (the study of rocks). Darwin was interested in both fossils and geology and realised that they provided evidence for evolution. The early geologists found that certain rocks held certain fossils. Deeper, older rocks contain different fossils to younger rocks. Fossils found in younger rocks are more like the animals and plants that still exist today. The animals and plants change over time. This is evidence that organisms have evolved.

1 In order for organisms to become fossilised, sediments must cover their body parts. Explain why.

2 Explain why, as a general rule, deeper rock layers are older than ones nearer to the surface.

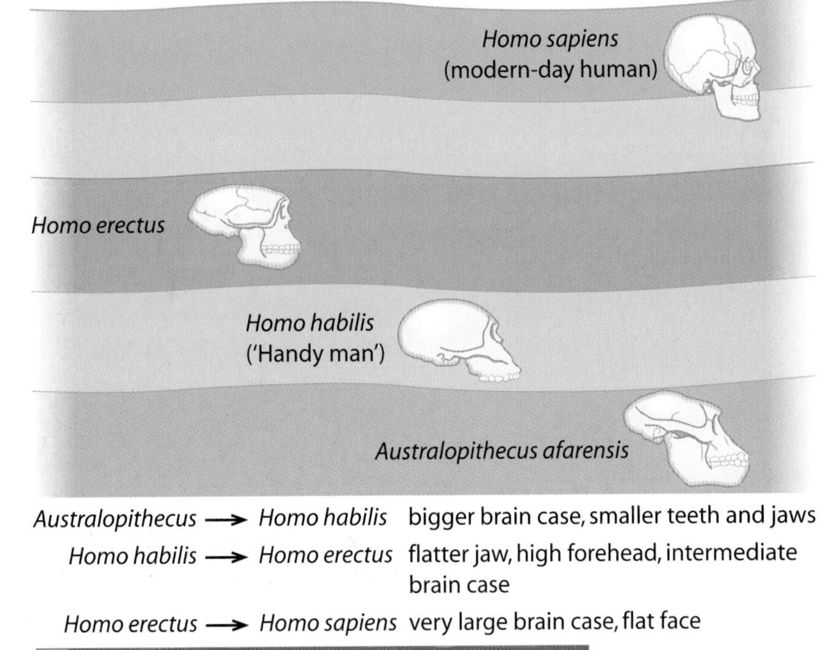

Australopithecus →	*Homo habilis*	bigger brain case, smaller teeth and jaws
Homo habilis →	*Homo erectus*	flatter jaw, high forehead, intermediate brain case
Homo erectus →	*Homo sapiens*	very large brain case, flat face

B Human evolution can be traced through fossils.

Darwin wrote about his ideas in a book called *On the Origin of Species*. He explained how natural selection could change a species. He used fossils found in **sedimentary rocks** as evidence. He said that animals and plants which did not adapt and change had become extinct. He also said that apes and humans probably evolved from the same ancestor.

1. When a plant or animal dies, it falls to the bottom of the sea bed and is buried in mud or sand.

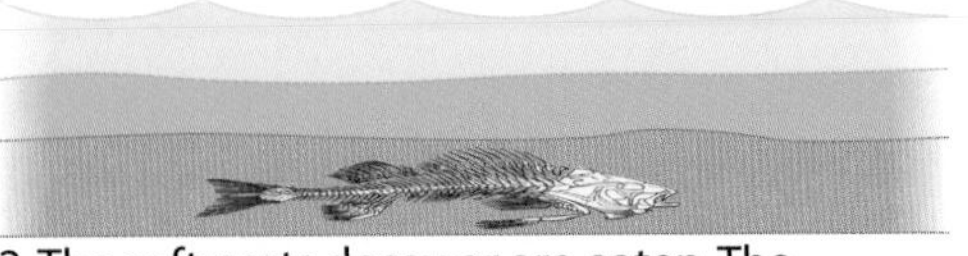

2. The soft parts decay or are eaten. The organism is then covered by layers of sediment. Hard parts such as shells, teeth or bones are preserved.

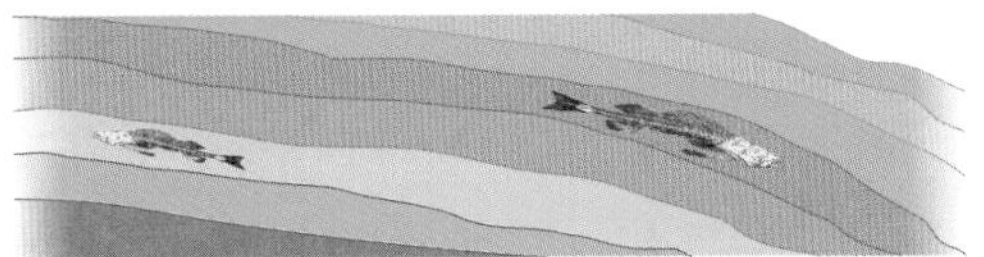

3. The sediments are squashed to form rocks.

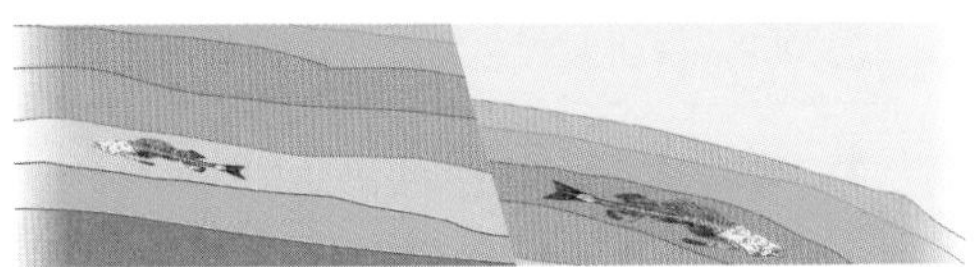

4. Later, earth movements raise the rocks above sea level. The rock erodes, exposing the fossil.

A Fossils form when organisms die and become covered in sediment.

3 Suggest why younger rocks contain fossils of organisms which are similar to those still living today.

4 Look at figure B. Describe what has happened to the skulls over time as humans have evolved.

C This cartoon appeared in 1871.

Have you ever wondered:

Why did a cartoon of Charles Darwin drawn as an ape appear in a national newspaper when he proposed his theory of evolution?

H Darwin lived at a time when the Church strongly influenced peoples' beliefs. Darwin's ideas conflicted with the Bible. They did not fit with the story of Adam and Eve. There were also other established theories about evolution. The press made fun of Darwin. They accused him of saying that men evolved from monkeys, not that we might have shared a common ancestor.

Although Darwin realised that some characteristics were inherited, he could not explain the mechanism of inheritance. Darwin was unaware of Mendel's ideas about genes. He was not able to explain how characteristics were passed from generation to generation. The fossil record at that time also had large gaps in it. Since Darwin's time, new discoveries and improved techniques have added greatly to our knowledge. With modern technology, we can tell the age of rocks and the fossils found in them. We have pieced together stages in our own evolution and, using modern genetics, we have filled in even more of the evolution jigsaw.

We can now look at DNA from different species and compare it. The haemoglobin in our blood is identical to the haemoglobin in ape blood. We share many of our genes with chimpanzees!

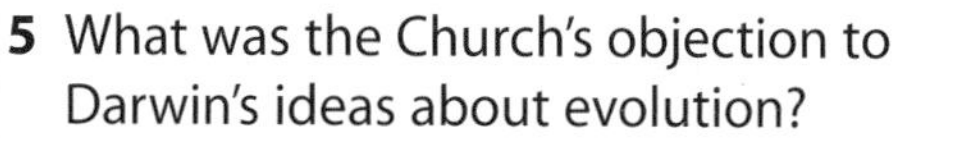

5 What was the Church's objection to Darwin's ideas about evolution?

6 Why did the public wrongly believe that humans evolved from monkeys?

7 Describe how modern evidence helps to support Darwin's ideas about evolution.

Summary Exercise

Higher Questions

Variation

By the end of these two pages you should know:

- why classification is necessary and the difficulties when trying to classify organisms
- how scientists classify living things in a systematic way.

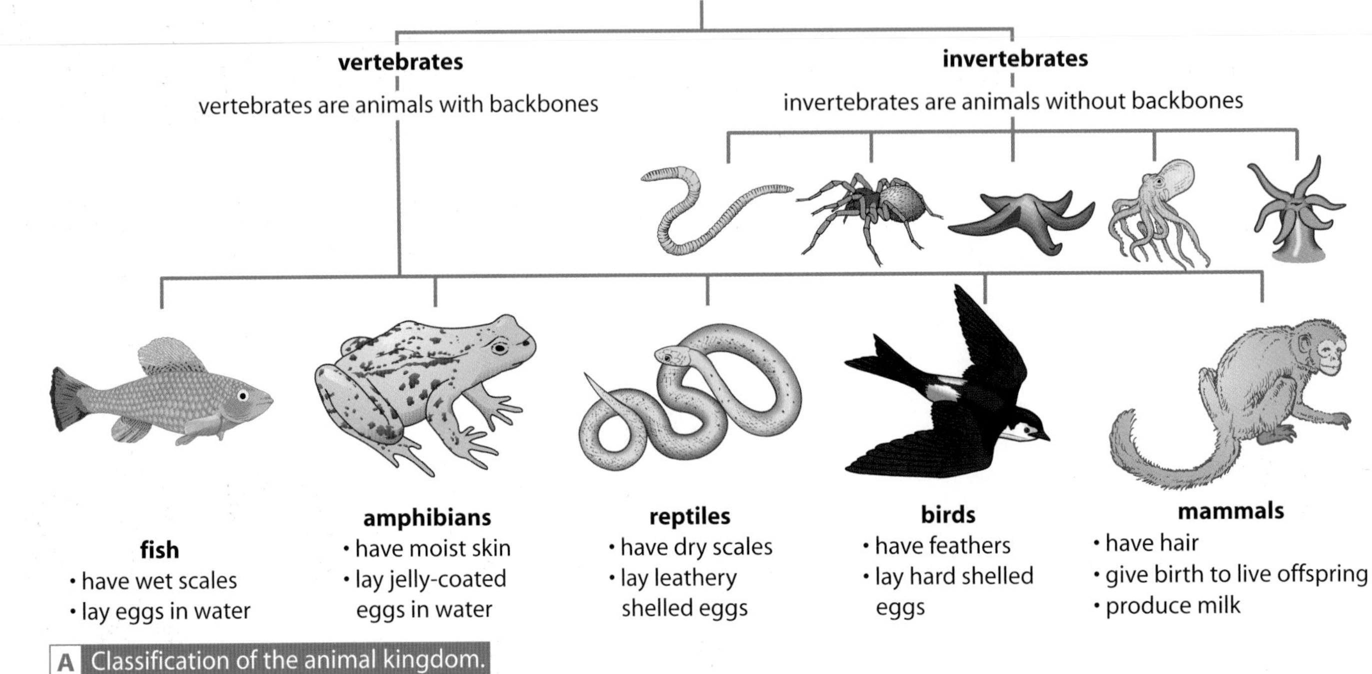

A Classification of the animal kingdom.

The living world contains millions of different species. In the nineteenth century, naturalists and explorers travelled the world looking for new species. It became necessary to group things and record them in a logical way. Grouping organisms began to show the relationships between different plants and animals. To begin with this **classification** was haphazard but gradually it became more organised to follow a systematic pattern. This made it easier to identify, name and classify new species when they were discovered.

It was Darwin's ideas on evolution which affected the way scientists group things. They tried to put related organisms together in groups which showed how they had evolved. Humans were grouped with apes in a group called the **primates**. Primates were placed with other mammals. The mammals were grouped with the other vertebrates like birds, reptiles, amphibians and fish. These were part of a larger group, including animals without backbones. This group, which includes all animals, is the animal **kingdom**. Eventually it was decided to have five kingdoms: bacteria, single-celled organisms (the **Protoctista**), fungi, plants and animals.

1 Why is it important to have an agreed way of grouping things?

2 Why did increased travel make classification more important?

3 Why did scientists start grouping organisms with other, more closely related, species?

Classifying things is not always easy. Take *Euglena* for instance. *Euglena* has chloroplasts and can make its own food by photosynthesis, like a plant. It swims towards the light and has a flexible membrane with no cell wall, like an animal. Is *Euglena* a plant or an animal? Because of these problems, scientists created a kingdom called Protoctista for all single-celled organisms like *Euglena* to go into.

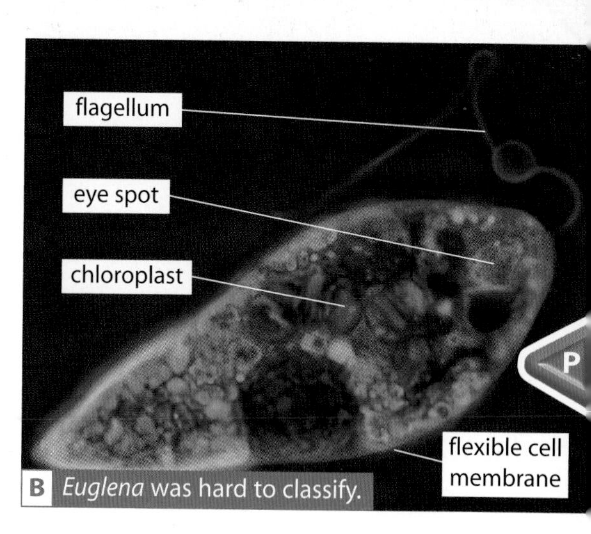

B *Euglena* was hard to classify.

4 What features of *Euglena* make it appear to be
 a like a plant?
 b like an animal?

5 Which kingdom does *Euglena* belong to? Give a reason for your answer.

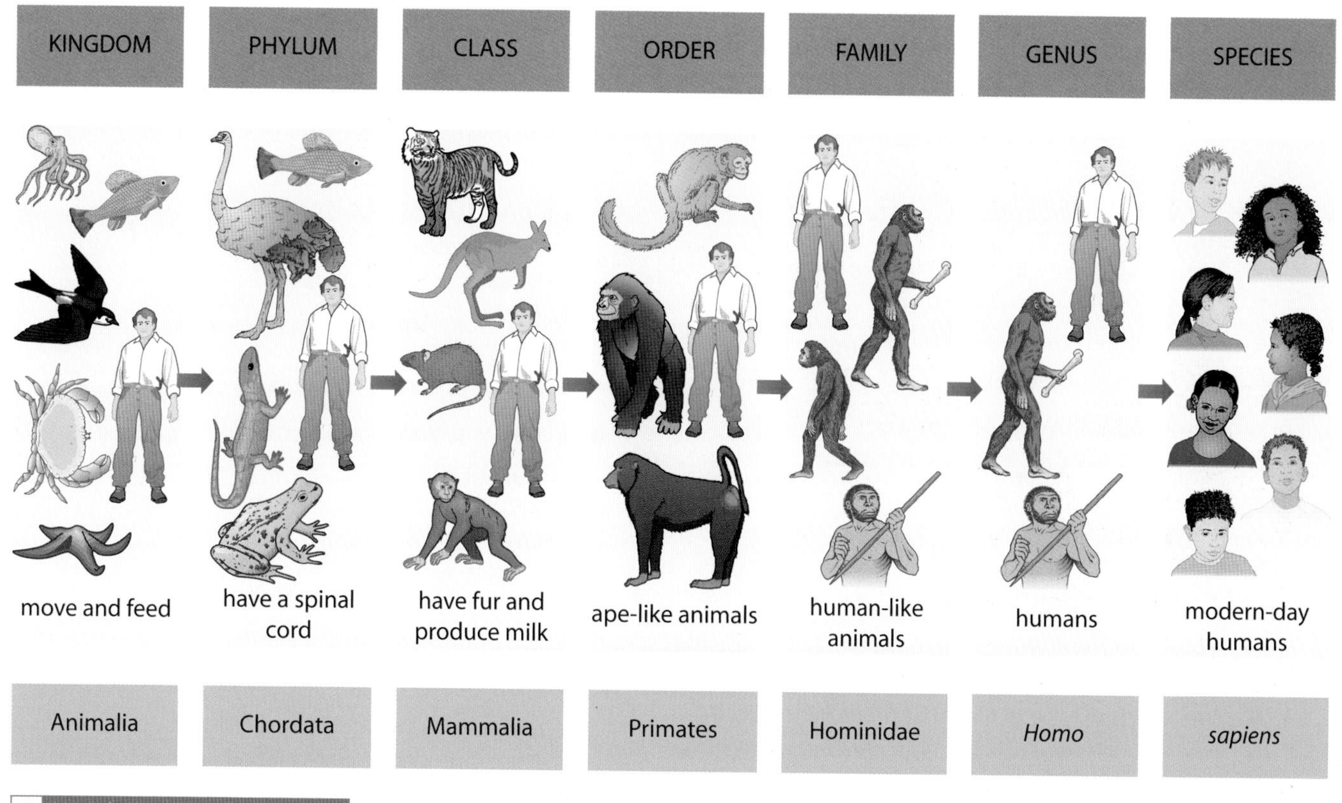

C How humans are classified.

A Swedish botanist called Carl Linnaeus had the idea of using two Latin names for each living thing. Linnaeus' way of naming things is now used all over the world. This avoids things being given more than one name in different parts of the world.

6 What is the species name for humans?

7 Why do we give every different species a two-part Latin name?

P

Summary Exercise

Higher Questions

Questions

D

D

Multiple choice questions

1 If the number of prey animals in an ecosystem falls what will happen to the predators?
- **A** Their numbers will increase.
- **B** Their numbers will remain steady.
- **C** Their numbers will decrease too.
- **D** Their numbers will increase and then decrease.

2 Which of the following is an example of intra-specific competition?
- **A** Zebra and wildebeest competing for grass.
- **B** Trees and grass competing for water and minerals.
- **C** Pond weed and fish competing for space in an aquarium.
- **D** Foxes competing for rabbits.

3 A group of pupils use quadrats to sample plants growing in the school grounds. How should the quadrats be used?
- **A** Placed randomly in the school grounds.
- **B** Arranged in a square pattern.
- **C** Placed side by side in a long strip.
- **D** Placed carefully to avoid areas under trees.

4 Which of the following summarises all of the feeding relationships in an ecosystem?
- **A** food chain
- **B** food web
- **C** pyramid of number
- **D** pyramid of biomass

5 Which of the following statements about computer models of populations is **false**.
- **A** They do not take births and deaths into account.
- **B** They can *exactly* predict changes in population.
- **C** They do not take the effects of seasonal changes into account.
- **D** They do not take the supply of food into account.

6 What fraction of energy is transferred from one consumer to the next in a food chain?
- **A** About 1%.
- **B** About 10%.
- **C** About 90%.
- **D** 100%.

7 Which of the following is an example of biological control?
- **A** Planting clover and cereals together in the same field.
- **B** Using hives of bees to pollinate fruit trees.
- **C** Adding composted vegetable matter to soil to make it more fertile.
- **D** Using ladybirds to eat aphids.

8 Which of the following is an example of genetic engineering?
- **A** Choosing two organisms and breeding them together to produce a new variety with selected genes from both parents.
- **B** Deliberately cross-pollinating two flowers to mix their genes.
- **C** Transferring genes from one species to another.
- **D** Reproducing an organism by cloning it.

9 Who first put forward the idea of evolution by natural selection?
- **A** Francis Crick
- **B** Charles Darwin
- **C** Carl Linnaeus
- **D** Gregor Mendel

10 Which of the following statements about evolution is untrue?
- **A** Generations of organisms adapt and change over long periods of time.
- **B** Fossils are evidence that evolution has taken place.
- **C** Organisms are still evolving.
- **D** Humans evolved from apes.

11 Modern humans are called *Homo sapiens*. Fossils of early humans include *Homo erectus* and *Homo habilis*. This means that they all belong to the same
- **A** class.
- **B** genus.
- **C** order.
- **D** species.

12 Which two of the following pairs of animals are most closely related to each other?
- **A** Two members of the same genus.
- **B** Two members of the same class.
- **C** Two members of the same phylum.
- **D** Two members of the same order.

13 Which of the following does not help the environment?
- **A** Turning off lights when you leave the room.
- **B** Using the bus or your bike.
- **C** Leaving your television on standby.
- **D** Recycling.

14 In dense woodland, what resource are the plants likely to be competing most strongly for?
- **A** air
- **B** light
- **C** minerals
- **D** water

15 Which of the following is true?
- **A** Carbon dioxide levels are falling.
- **B** As the human population of the Earth rises so does the amount of carbon dioxide in the atmosphere.
- **C** The human population of the Earth is falling.
- **D** As the human population of the Earth rises carbon dioxide levels fall.

Short-answer questions

1 A scientist calculated the mass of leaves eaten by 100 caterpillars in 1 day. He collected all of their waste droppings and weighed them. Finally he weighed the caterpillars to see how much weight they had gained. The diagram shows his results.

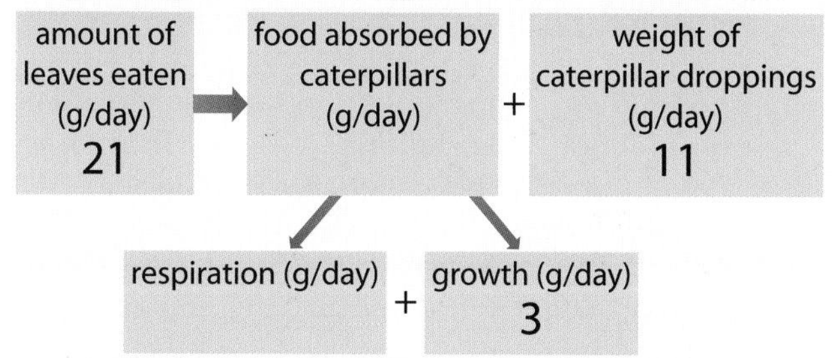

a Complete the diagram by calculating the mass of food
- **(i)** absorbed by the caterpillars each day.
- **(ii)** used in respiration each day.

b How much of the food eaten by the caterpillars is actually used for growth?

Caterpillars are eaten by birds. A bird eats 15 g of caterpillars each day and increases its mass by 2 g.

c What mass of leaves would have been eaten to provide this mass of caterpillars each day?

d **(i)** Use this information to draw a pyramid of biomass for leaves, caterpillars and birds.
(ii) Explain why this pyramid of biomass is the shape it is.

2 **a** What is deforestation and what are its causes?
b Deforestation leads to the loss of many species, not just trees. This is called loss of biodiversity. Explain why this happens.

3 Charles Darwin made four observations and three deductions to explain his theory of evolution by natural selection.
- **a** What is natural selection?
- **b** Outline Darwin's four observations and three deductions.

Darwin used fossils as evidence that evolution had taken place.
- **c** What is a fossil?
- **d** How do fossils show that evolution has taken place?

People made fun of Darwin. A newspaper drew Darwin looking like an ape.
- **e** How did the cartoon show a lack of understanding of Darwin's ideas?

4 Read the following passage and then answer the questions.

D

P

> Gregor Mendel used selective breeding to make new pea varieties. His method showed how genes are inherited. Until about 30 years ago, selective breeding was the standard way of improving crops. Two parent plants are chosen for their characteristics. Pollen is taken from one parent and placed on the stigma of the other parent. When the fruits are ripe, the seeds are collected and planted. The seeds grow into the next generation of plants. The breeder then chooses the offspring that have inherited the correct combination of characteristics from the original parents. This is repeated until pure-breeding plants are obtained.

- **a** Describe how you would produce a new breed of purple-flowering peas from a population of plants that had a mixture of flower colours.
- **b** Why is selective breeding a slow process?
- **c** How is selective breeding different to natural selection?

Glossary

D

adaptation How organisms change in order to be more suited to their environment.
bioaccumulation The build up of toxic substances in a food chain (or in the body of an organism).
biological control Natural methods used to prevent pest damage to crops.
biomass This is the mass of an organism (or a population of organisms) when all of the water has been removed from their tissues.
breeding Mating organisms in order to produce an increase in population size.
characteristic A feature of an organism which can either be inherited or modified by the environment. For example, skin colour, number of petals or blood group.
classification A systematic way of grouping things.
competition Where two organisms both want the same thing, for example food, space or light.
computer model Uses mathematical formulae to work out what might happen if certain changes are made.
conservation Trying to maintain habitats and prevent the loss of species.
deforestation Clearing trees for timber or to make space for raising crops, grazing animals or putting up buildings.
desertification Turning land into desert. It happens because animals have been allowed to graze the land until it is bare.
ecosystem A group of plants and animals which live and interact together.
environment The place where an organism lives. It is made up of many different factors such as air, water, soil and other living things.
evolution A gradual change in the characteristics of a species over many generations.
extinct When a group of organisms dies out because it cannot adapt itself to a new situation.
factory farming Intensive farming which depends on artificial fertilisers and chemical pesticides.
food chain A chain to show how energy moves between plants and animals.
food web A group of inter-linked food chains.
fossil An imprint in the rock left by the body of an organism which lived millions of years ago.
genetic engineering Transferring genes from one organism to another, or changing genes in an organism by altering their DNA.
genetically modified An organism that has had its characteristics altered by the use of genetic engineering.
interdependence When a change to one organism brings about a change in another organism.
inter-species Between two different species.
intra-species Between members of the same species.
kingdom A major category of classification. For example, animals, plants and fungi.
natural selection How environmental factors such as disease or predation alter the characteristics of a species.
organic (when applied to food) Food produced without artificial fertilisers, pesticides or herbicides.
organic farming Farming which uses natural methods of maintaining soil fertility and controlling pests.
organism Any living thing.
population The total number of one species of an organism living in an area.
predator An animal which eats other animals. For example, a cheetah.
prey An animal which is hunted and killed by a predator. For example, a zebra.
primates The order to which humans and the apes belong.
producer Organisms which make their own food by photosynthesis. For example, a green plant.
Protoctista The kingdom to which *Euglena* (and all single-celled organisms) belongs.
pyramid of biomass A diagram which shows how much living material is found at each link in a food chain.
pyramid of number A diagram which shows the number of living organisms at each link in a food chain.
quadrat A square frame which is placed on the ground to get a sample of the organisms living in a small area.
quantitative Using numbers to describe something.
reproduction Propagation of a species. The process of producing the next generation of individuals.
scavenger An animal which feeds on dead remains or food left by another animal. For example, a hyena.
sedimentary rock A rock formed when particles settle together (usually under water) and become compressed to form layers. Examples are limestone, sandstone and shale (mudstones).
selective breeding Deliberately mating together organisms to produce a plant or animal that has useful combinations of characteristics.
species A population of organisms which breed together and produce fertile young. The name given to the finest level of grouping in classification. Several species make up a genus.
sustainable development Using methods which do not use up limited resources and so do no harm to the environment. For example, replanting trees to replace ones which are cut down.

Genes

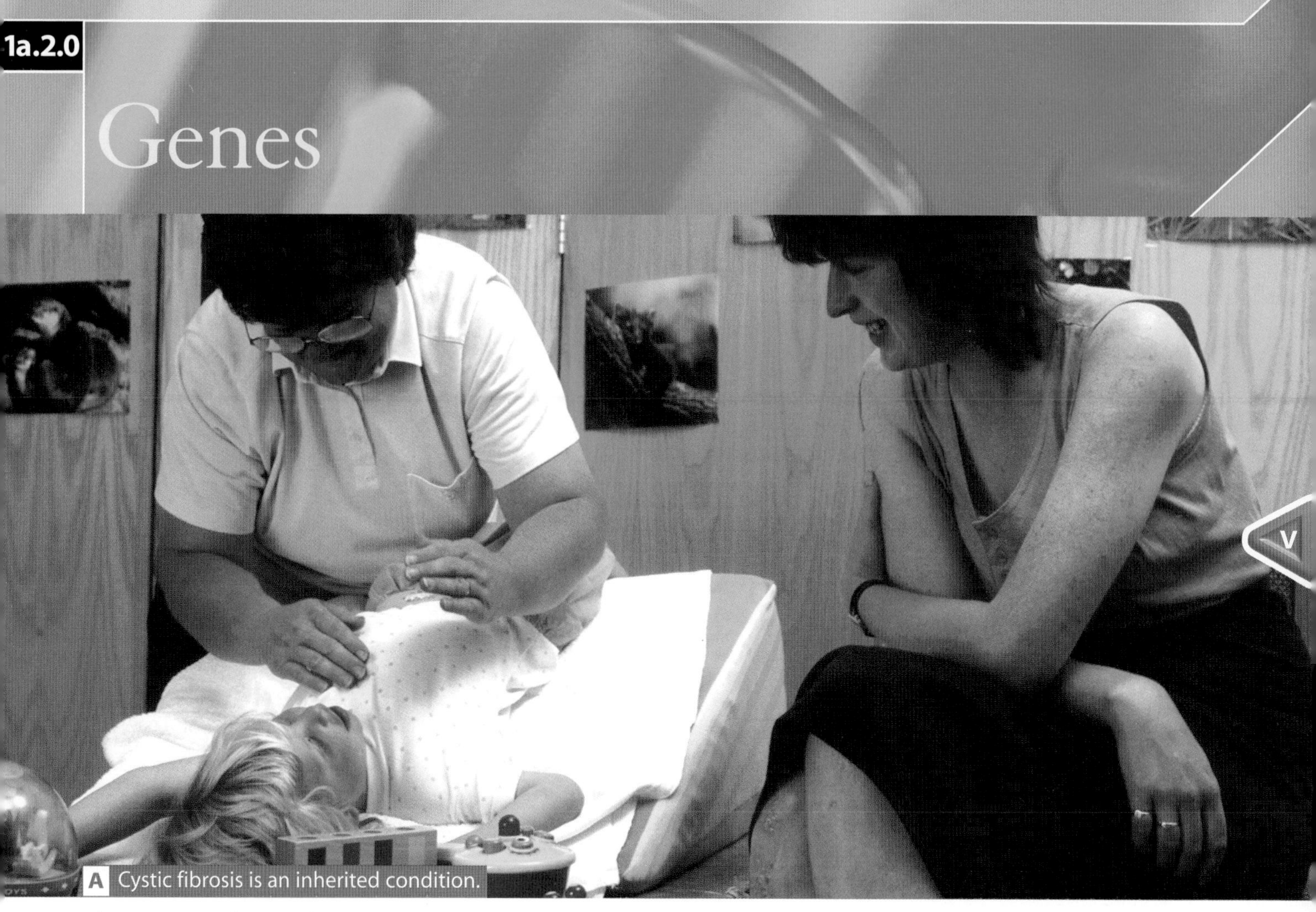
A Cystic fibrosis is an inherited condition.

The science of **genetics** is only 150 years old. Our knowledge of genes has developed rapidly. DNA is the molecule which makes genes. Its shape was only solved in 1953. By 2001 we knew the whole of the human genetic code. The complete human code, or human genome, takes up about 600,000 pages of A4. In book form it would take up 90 metres of shelving in your library!

Knowledge of genetics has allowed us to develop tests for genetic disorders like cystic fibrosis. It offers hope for new treatments and cures for conditions that have had no cure until now. For example, bacteria with human genes now make the insulin used to treat diabetes. Genetics lets us find out why genes go wrong and how to correct them. We can screen embryos to check that they are free from serious genetic errors.

In this topic you will learn that:

- characteristics of organisms are dependent on their genes
- sexual reproduction leads to variation
- genetic modifications are used for a range of purposes
- there are many ethical considerations associated with advances in genetic modification.

Look at these statements and sort them into the following categories:

I agree, I disagree, I want to find out more

- Screening embryos for genetic errors is unnatural. We should not play with nature.
- Putting human genes into bacteria is dangerous because we might produce harmful 'super-bugs'.
- People should have their DNA tested so they can change their lifestyle to avoid developing genetic disorders.

V

D

Asexual reproduction

By the end of these two pages you should know:

- that asexual reproduction produces organisms, called clones, with identical genes
- that many plants produce natural clones by asexual reproduction.

Have you ever wondered:

Are clones really like they are in the movies?

When a baby is born, people often comment that it has its father's eyes or its mother's nose. As humans we all look like a blend of our parents because we inherit equal numbers of **genes** from each of them. Genes are little bits of a chemical called **DNA**. They are found in the **nucleus** of a **cell** and carry instructions. The instructions for making any living thing are carried in its genes.

However, some plants and animals use **asexual reproduction**. The offspring have only one parent, so all of their genes come from one source. They have the same genes as the parent and as each other. They are called **clones** and, provided they develop in similar **environments**, they will all look alike. Anything that uses asexual reproduction can reproduce very quickly. There is another advantage as well. If the parent has survived to reproduce then the offspring will probably survive too.

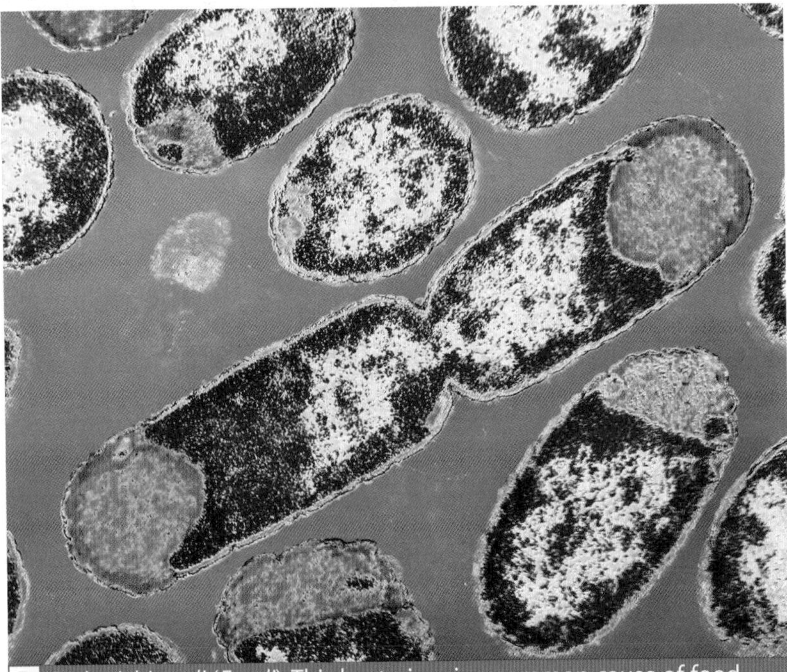

B *Escherichia coli* (*E. coli*). This bacterium is a common cause of food poisoning.

A Spider plants produce plantlets with the same genes as the parent plant. The ones nearest the window get more light so they grow faster.

1 Plantlets on a spider plant (photo A) all have the same genes but they are not identical to each other. Suggest why they might be slightly different.

2 Plant growers reproduce plants asexually by splitting them or by taking cuttings. What are the advantages of this type of reproduction?

Bacteria reproduce asexually. In ideal conditions a bacterium can divide into two every 20 minutes. Some bacteria, such as some varieties of *E. coli*, cause food poisoning. When these varieties of *E. coli* reproduce, the ability to cause food poisoning passes to the offspring. This makes it important to control *E. coli* in places where food is prepared.

3 An *E. coli* bacterium divides into two every 20 minutes. Calculate the number of *E. coli* bacteria there would be after
a 2 hours **b** 6 hours.

C Many important food plants reproduce asexually. The offspring are naturally occurring clones.

Garlic, shallots and other members of the onion family produce bulbs. This is a form of asexual reproduction. Potatoes produce underground tubers. Potatoes originally grew in the high Andes in Peru. Bulbs and tubers allow plants to survive the winter. Strawberry plants can also reproduce asexually. They produce 'runners' with small plantlets on the end, which will produce a 'strawberry patch' that spreads across any bare ground.

All of the new plants are exactly like their parents. Once growers have a plant that produces a good crop and is resistant to disease they can produce lots of identical plants in a short time. There is a disadvantage to all the new plants being identical. If one plant gets an infection or disease they may all get it.

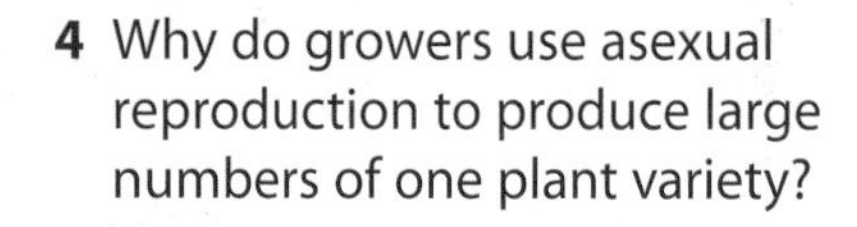

4 Why do growers use asexual reproduction to produce large numbers of one plant variety?

5 Apart from producing a lot of potatoes and being resistant to disease, what other characteristics might a grower want a potato to have?

6 Offspring that are produced by asexual reproduction show very little variation. Explain why.

Summary Exercise

Higher Questions

Genes and DNA

By the end of these two pages you should know:

- that genes are the instructions that decide what living things look like
- that genes are made of DNA
- that genes are parts of chromosomes, which are found in the nucleus
- that DNA fingerprinting can be used to help solve crimes.

P

Your **characteristics** depend on the genes you inherit from your parents. A characteristic can mean what you look like or how easily you do things such as play a musical instrument. Each gene is an instruction for one of your characteristics.

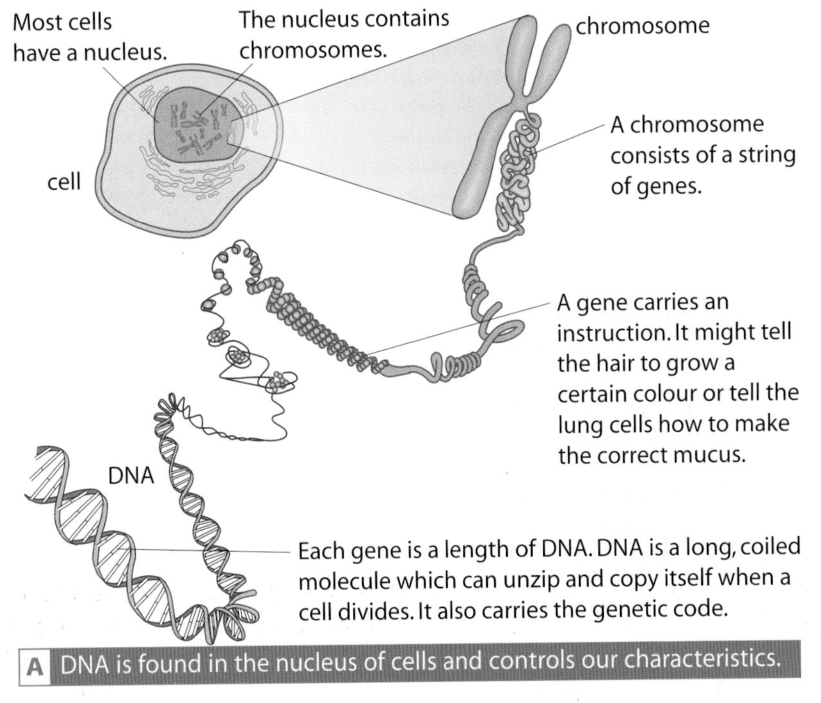

A DNA is found in the nucleus of cells and controls our characteristics.

A

As the diagram shows, in a cell's nucleus are X-shaped **chromosomes** made up of strings of genes. You have two versions of each chromosome – one from your mother and one from your father. You inherit equal numbers of genes from each parent. But the *mixture* of genes is different in every single person (apart from identical twins). This is why you look a little bit like both of your parents but are not exactly the same as your brother or sister.

V

Genes are made of a chemical called DNA. DNA carries a chemical code which tells a cell how it should work. Every time your cells divide the DNA makes a copy of itself. Sometimes this copying goes wrong. Altering just a tiny part of the DNA molecule can cause a big change in a characteristic.

B Oli has cystic fibrosis due to a faulty gene.

We all make mucus in our airways to trap dust and bacteria. The mucus is quite runny and easily removed from the lungs. People with **cystic fibrosis** (CF) have a faulty gene which alters the mucus and makes it thick and sticky. Build up of the sticky mucus leads to breathing difficulties and chest infections.

P

1 Children have some characteristics from their mother and some from their father. Explain why.

2 In cystic fibrosis (CF), part of the DNA code is altered.
   a How does this affect the way that the cells in the airways work?
   b How does this affect the lungs of a person with CF?

One step in identifying a genetic disorder is to examine the DNA in a sample of blood or saliva. The DNA is copied and cut into small pieces using chemicals called enzymes, which act like chemical scissors. The enzymes cut the DNA when they find a particular code. The pieces of different lengths are separated out and they form bands which can be studied. This is called a **DNA fingerprint** because it is like the fingerprints on your fingers. Your fingerprints can help identify you. DNA fingerprints are even better at identifying people, using their unique genetic code. They are often used by **forensic** scientists to help solve crimes. They can also be used to find out whether people are related to each other.

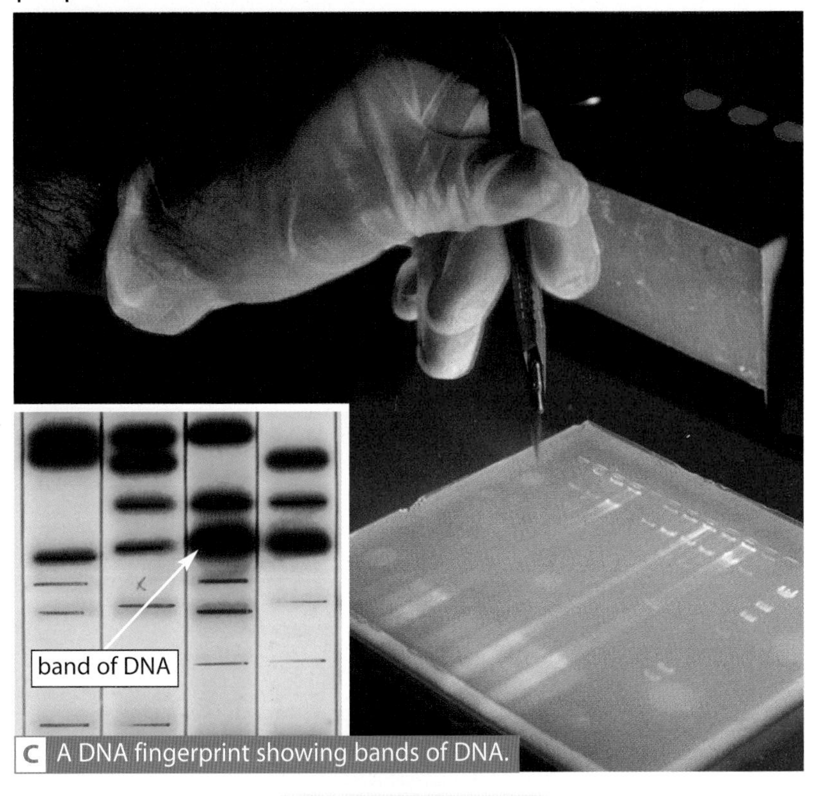

C A DNA fingerprint showing bands of DNA.

3 Look closely at the passage above diagram C. Pretend that the text is a piece of DNA and that an enzyme is moving along the DNA making a cut just after every instance of the word 'the'.
   a How many cuts will the enzyme make?
   b How many words are there in the longest and shortest pieces of the chopped-up passage? Explain why this is like a DNA fingerprint.
   c How is a person's DNA fingerprint unique?

P

Summary Exercise

Higher Questions

Variation

By the end of these two pages you should know:

- what sexual reproduction is
- why sexual reproduction leads to mixing up of genes.

When sperm and egg cells (both called **gametes**) are made, a special kind of cell division takes place. The cells formed have half the usual number of chromosomes and therefore half the genes. The gametes join together during **fertilisation**. The genes come equally from both parents. The mixture of genes passed on by each parent is different every time. **Sexual reproduction** mixes up the genes of two parents in a new, unique individual.

1 What is a gamete?

2 Explain why organisms produced by sexual reproduction have characteristics of both parents.

3 New cells can be produced in two ways: (a) a cell divides to produce two body cells, as in asexual reproduction or (b) two sex cells join, such as when a sperm cell fertilises an egg cell. Which of these ways produces the greatest amount of genetic variation? Explain your answer.

When gametes form, the parents' chromosomes – containing their genetic information – are dealt out at random. This is rather like playing cards. This makes every gamete different. When fertilisation occurs it is pure chance which sperm fertilises which egg. This means that in a population there are lots of different gene combinations. So, sexual reproduction introduces **variation** into the population. This is an advantage. If the environment changes, individuals that have good gene combinations will survive. The population can adapt to change, which leads to evolution. Evolution is what happens when environmental changes cause some individuals to survive and reproduce but cause other individuals to die. Only those with good gene combinations will be lucky. The disadvantage of sexual reproduction is that sometimes you get individuals who have weak combinations of genes and do not survive.

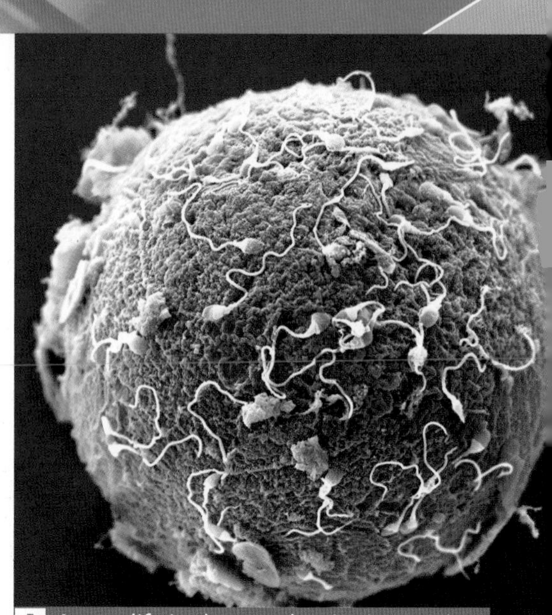

A A new life is about to begin. An egg is surrounded by sperm cells: the egg and sperm contain genetic information which combines to make a new individual.

B Two peppered moths, one light and one dark, on a tree.

These moths are an example of evolution. The lighter moth was once the most common. But, in 1840s England, trees were turned black by pollution in the Industrial Revolution. The darker moth was part of the same population but had a different gene combination. It was better at being camouflaged on blackened trees and avoided being eaten by birds. Within 50 years the darker variant was the commonest form. This is known as natural selection. The moths inherited the best combination of genes to survive.

Have you ever wondered:

Is it possible that Old English Sheepdogs and Yorkshire Terriers both came originally from wolves?

4 Sexual reproduction leads to variation.
- **a** Explain why variation is important.
- **b** Explain two ways in which sexual reproduction leads to variation.

5 What proportion of your genes did you inherit from
- **a** your mother?
- **b** your paternal grandmother (your father's mother)?

6
- **a** Suggest why it took 50 years for the black variant of the moths shown in photo B to become the most common.
- **b** Since the 1950s the amount of smoke in the atmosphere has fallen. The dark-coloured moths are no longer so common. Explain why.

Summary Exercise

Higher Questions

Dominant and recessive alleles

H ***By the end of these two pages you should know:***

- what dominant and recessive alleles are
- that dominant and recessive alleles lead to variation.

A The characteristics of the parents get mixed together and can be seen in the variety of offspring.

The litter of mice in diagram A show many different characteristics. Having lots of different characteristics in a population is called variation. The same kind of variation can be seen in litters of puppies or kittens. In fact, the offspring of most plants and animals show variation.

The litter of mice was produced by sexual reproduction. The eggs and sperm involved in sexual reproduction carry instructions in the form of genes. The gene for fur colour comes in at least two different forms. One form carries the instruction for brown fur. Another form of the gene tells the mouse's cells to produce black fur. Different forms of a gene are called **alleles**. Each baby mouse received a different mixture of alleles from each parent by chance. This is what caused the variation in this litter of mice.

2 Explain why it is not possible to predict *exactly* what all the mice in a litter will look like.

Genetic variation was explained by Gregor Mendel in 1866. Mendel bred different varieties of pea plants together. One of his experiments is shown in diagrams B and C.

In this experiment, Mendel bred pure-breeding white-flowered peas with pure-breeding yellow-flowered peas. (Pure-breeding means the organism has two identical alleles.) He collected the seeds from the new plants and grew them. All of the next **generation** of pea plants, called the F1, had

1 Look carefully at diagram A. The baby mice in this litter are not all alike.

a How many differences can you see? Make a list of these differences.

b Suggest which parent each of these differences might have come from.

yellow flowers. Mendel explained this by saying that the yellow characteristic must be overpowering the white characteristic: the allele for yellow colour is **dominant**.

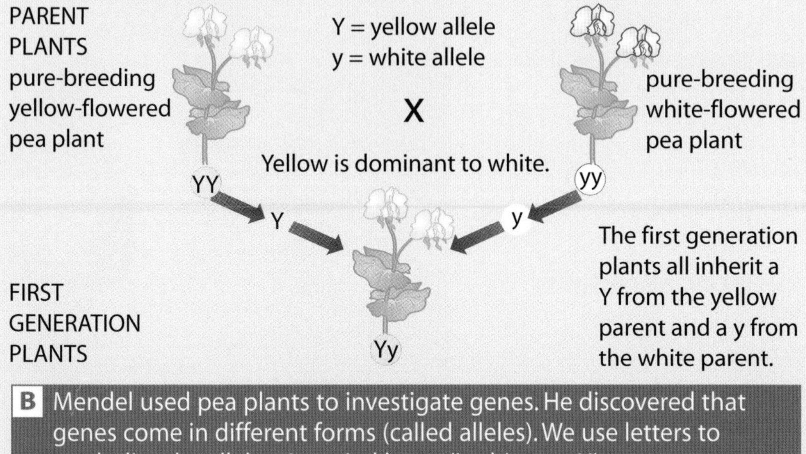

B Mendel used pea plants to investigate genes. He discovered that genes come in different forms (called alleles). We use letters to symbolise the alleles. A capital letter (in this case Y) means a dominant allele and a lower-case letter (y) means a recessive allele.

Next Mendel bred the first generation of yellow-flowered plants with each other. Again he collected their seeds and planted them. The next generation of pea plants (the F2) contained both yellow-flowered and white-flowered plants. Mendel counted the plants and found that there were roughly three yellow-flowered plants for each white-flowered plant. Mendel explained this by saying that the first-generation plants, although yellow flowered, still had information for white flowers in them. He said that the white characteristic must be **recessive** (hidden). When the first-generation plants are bred together they can each pass on either yellow or white alleles in their sex cells. This explains the 1 in 4 chance of getting a plant that inherits white information from both parents.

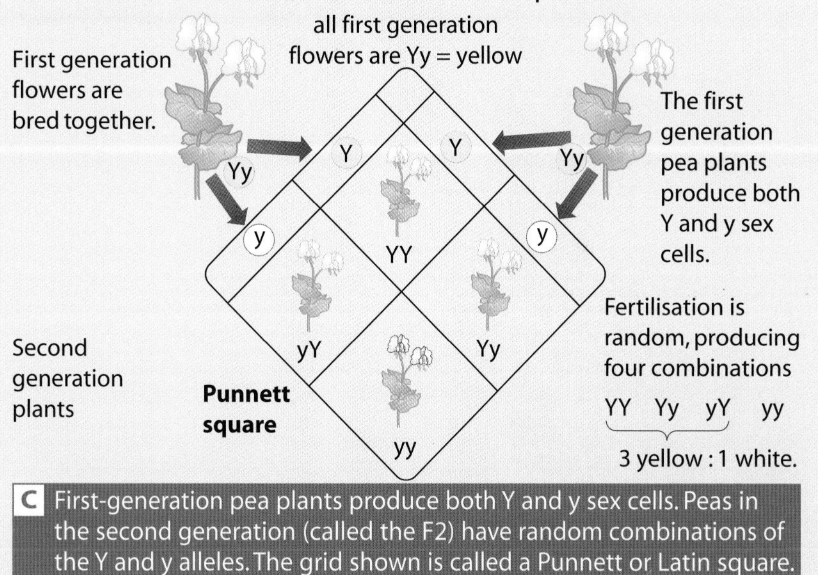

C First-generation pea plants produce both Y and y sex cells. Peas in the second generation (called the F2) have random combinations of the Y and y alleles. The grid shown is called a Punnett or Latin square.

3 Mendel repeated his experiment many times and averaged his results. Why is it important not to rely on the results of just one experiment?

4 What is an allele?

5 What is meant by the terms **a** recessive and **b** dominant?

Summary Exercise

Higher Questions

P

Nature or nurture?

A England's heaviest ever man, Daniel ('Giant') Lambert (1770–1809), weighed 355 kg (almost 54 stone).

H ***By the end of these two pages you should know:***

- that characteristics may be affected by changes in the environment
- that human growth is affected by what we eat
- that plant growth is affected by soil minerals.

People vary in size and shape. Variation in size is caused partly by our genes and partly by what we eat. People with the genetic disorder cystic fibrosis (CF) produce a sticky mucus in their digestive system. This interferes with the digestion and absorption of nutrients. People with CF grow more slowly and must eat a high-calorie diet to have enough energy to fight infections.

Daniel Lambert (picture A) weighed more than a third of a tonne! He was a lightweight compared to American John Minnoch, who probably weighed over 635 kg at his heaviest. What you eat affects how much you weigh. Lambert and Minnoch probably both came from fairly 'normal' families. Their genes gave them the potential to grow to an enormous size but they also had good appetites!

If too little food is eaten and if the diet is not balanced, the body will lack nutrients and will fail to reach its potential size. Whatever genes you have, without enough food your growth rate will slow down. You will lack energy and be prone to disease. Your life span will also be shortened. Lack of food is a problem in developing countries.

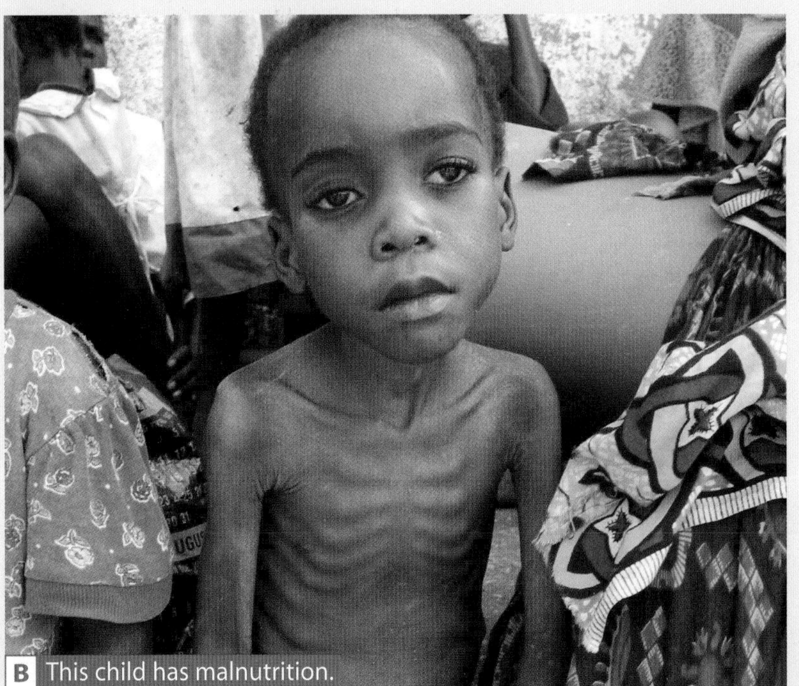

B This child has malnutrition.

There is a different problem in developed countries. Food is plentiful and cheap. People can eat too much of the wrong foods. A third of the population of the USA is seriously overweight. There are no 'genes for being slim'. If you eat too many high-energy foods such as fat and sugar and don't exercise you will get heavier. Heart disease or diabetes can shorten life span. Extra weight also causes problems with movement and it becomes difficult to take exercise.

It's not just animals that need the correct balance of nutrients: plants do too. Soil can contain pests and diseases, and may lack important nutrients. Tomatoes are grown commercially in glasshouses without soil. The roots are bathed in fertiliser solution containing mineral nutrients. Controlling the environment ensures that the plants have ideal growing conditions. It also reduces pest damage so plants grow to their full potential.

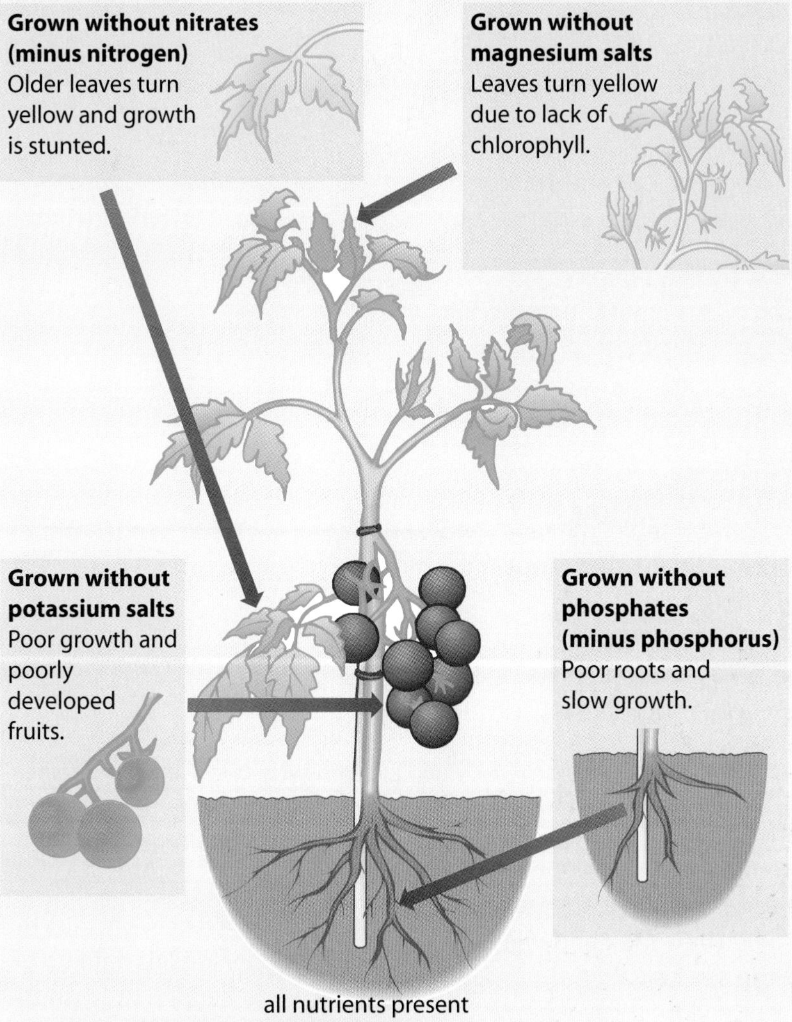

Grown without nitrates (minus nitrogen)
Older leaves turn yellow and growth is stunted.

Grown without magnesium salts
Leaves turn yellow due to lack of chlorophyll.

Grown without potassium salts
Poor growth and poorly developed fruits.

Grown without phosphates (minus phosphorus)
Poor roots and slow growth.

all nutrients present

C How plants use mineral nutrients.

Summary Exercise

1 Giant Lambert and John Minnoch both died quite young. Did their genes or their lifestyles cause their early deaths? Explain your answer.

2 A person's weight depends partly on their genes and partly on their lifestyle. Explain why identical twins, who have the same DNA, can end up being different weights.

Higher Questions

3 A grower notices that leaves on his tomato plants are turning yellow. Use diagram C to decide which mineral might be lacking in the plants.

4 Nitrogen is used in lawn fertiliser to improve the growth and colour of grass. Use diagram C to explain why nitrogen has these two effects on plants.

5 Animal and plant growth is affected by environmental conditions. A poultry farmer grows fields of barley. This is fed to hens which produce eggs. Suggest what the farmer can do to improve the yield of barley and eggs.

It's in the genes

By the end of these two pages you should know:

- that some disorders, such as cystic fibrosis, can be inherited
- that a genetic disorder is caused by faulty alleles (altered forms of a gene)
- how cystic fibrosis is inherited.

A Oli inherited cystic fibrosis because of a faulty gene.

People with cystic fibrosis (CF) produce sticky mucus that clogs the airways in their lungs. This mucus is difficult to clear by coughing and may get infected. This makes it difficult to breathe. People with CF take strong antibiotics. Antibiotics help to fight infection caused by bacteria. The sticky mucus also affects how well their digestive system works.

Alleles come in pairs. Cystic fibrosis is caused by the **inheritance** of two faulty alleles, one from each parent. The parents may show no signs of CF because they each have a working allele in addition to a faulty allele. If, by chance, the egg and sperm both contain the faulty allele then the child will have CF. If, on the other hand, just one of the sex cells has a working allele the child will not have CF but will be a **carrier**. Carriers do not suffer from the disorder but, because they 'carry' a faulty allele, they may pass it on to their offspring. Roughly one person in 25 in Britain is a carrier of a faulty CF allele and CF affects about one child in every 2000.

1 Describe the symptoms of cystic fibrosis.

2 Explain why two carriers of the CF allele have a chance of producing a child with cystic fibrosis.

3 People with cystic fibrosis have physiotherapy to help remove the mucus. The person lies flat and coughs while being patted on the back or on the chest. They also take strong antibiotics. Suggest how these two treatments help to improve their breathing.

Haemophilia	a disorder in which blood fails to clot normally
Glaucoma	a disorder which affects a person's eyesight
Colour-blindness	the person is unable to distinguish red and green colours
Sickle-cell disease	a disorder in which red blood cells change shape, causing great pain
Polydactyly	the person has extra toes or fingers
Huntington's disease	a disorder that affects the nervous system and the brain

B Here is a list of some other human genetic disorders and their effects.

Brian (Annabel's father)

Annabel

Clare (Annabel's mother)

Annabel's parents are both carriers of the CF allele. Because each parent has only one copy of the CF allele neither of them had any symptoms of cystic fibrosis.

carrier (Cc) sperm

carrier (Cc) eggs

Annabel's parents both produce two kinds of sperm and egg.

gametes

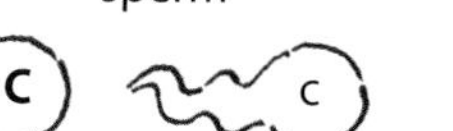

C c

Half of Brian's sperm carry a CF allele (c), the other half carry a normal allele (C). Likewise Clare's eggs carry either a CF allele (c) or a normal allele (C).

Brian's sperm and Clare's eggs join at random.

Gametes	C egg	c egg
c sperm	cC	cc (Annabel)
C sperm	CC	Cc

This produces four possible combinations in their children, one 'normal' (CC), two 'carriers' (cC and Cc) and one with cystic fibrosis (cc).

To have cystic fibrosis a person must have two copies of the faulty allele, so a person with cystic fibrosis must be cc.

C Family tree showing how two carriers can have a 1 in 4 chance of producing a child with CF.

Sickle-cell disease (SCD) is inherited in a similar way to CF. Normal red blood cells are round discs. In SCD the red blood cells become deformed when they release their oxygen. The deformed cells block capillaries and cause pain. Where malaria is common, people with a single sickle-cell allele have resistance to malaria parasites and show only mild symptoms of SCD. This gives them an advantage. People with normal red blood cells may die from malaria and people with two sickle-cell alleles may die from SCD.

4 If Annabel's parents decided to have another child, what would be the chances of it being
a a carrier of cystic fibrosis?
b a person with cystic fibrosis?

5 A carrier of cystic fibrosis has children with someone who has no family history of cystic fibrosis. What are the chances of their children being carriers?

6 Freckled skin is inherited in the same way as cystic fibrosis. Show how two people without freckles could produce a child with freckles.

Summary Exercise

Higher Questions

The Human Genome Project

By the end of these two pages you should know:

- about the work involved in the Human Genome Project (HGP)
- how information gained from the HGP might be used
- why some people are concerned about some uses of this information.

A Scientists on the Human Genome Project have mapped the entire human DNA code. It is the world's longest 'book'.

The instructions that make you look like you do are present in your DNA. DNA is a chemical code. The code is made of units called **bases**. There are only four different bases: A, C, G and T. You can think of a base as a letter in the DNA code. In 1990, scientists began to work out this code. It is about 3 billion bases long. Many people thought it was an impossible task but it was completed in 2001, 4 years ahead of schedule. This work was called the **Human Genome Project**, or HGP for short. Many scientists from all over the world worked on the project, sharing their results on the Internet.

1 Suggest two reasons why the project was finished ahead of schedule.

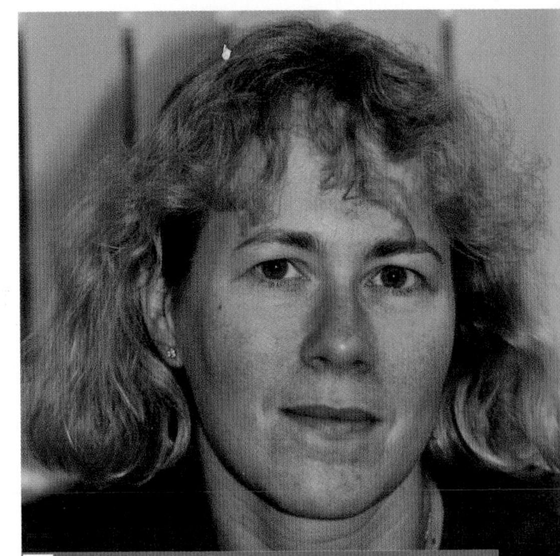

B Lisa Priestley is a Cystic Fibrosis Clinical Nurse specialist. The people she treats could benefit from gene therapy.

Scientists looked at the genes on different chromosomes. The form of the gene which causes cystic fibrosis was found on chromosome 7. The DNA code causing cystic fibrosis has just three bases missing. If we could replace these missing bases, the person would not have cystic fibrosis. Repairing or replacing faulty genes is called **gene therapy**. The HGP has identified other harmful genes. All have DNA which has been altered in some way. Some genes have extra bases, some have missing bases and others have the wrong bases.

Information from the HGP is being used to develop tests to see if you are likely to develop **cancer**. Cancer is caused when normal cells start to grow and divide uncontrollably. Genes discovered by the HGP could be used to make new treatments for people whose genes do not work properly. The HGP has provided us with lots of information but we need to decide what we will do with it.

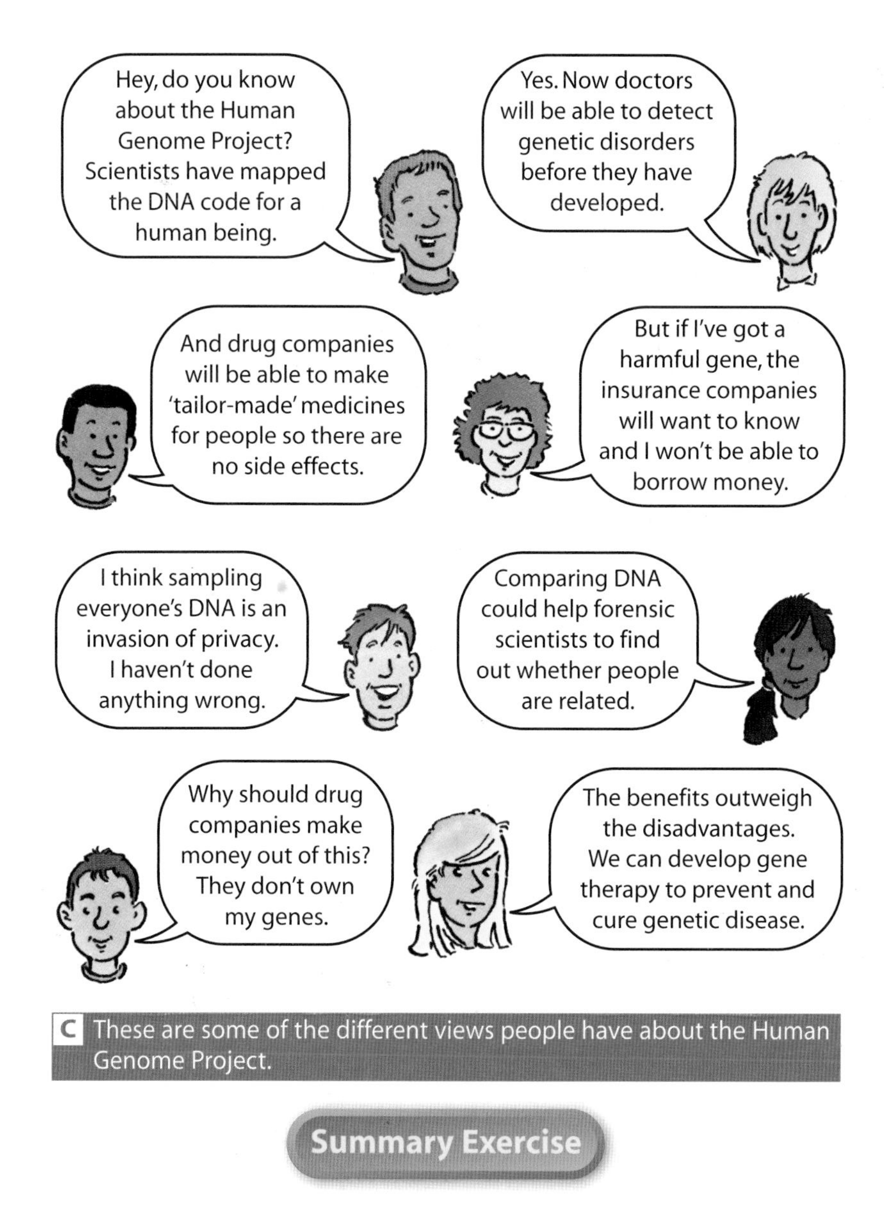

C These are some of the different views people have about the Human Genome Project.

Summary Exercise

2 On which chromosome is the gene for cystic fibrosis found?

3 How does the faulty CF allele differ from the working allele?

4 What is gene therapy?

Gene therapy

By the end of these two pages you should know:

- how gene therapy works and how it might be used to treat CF and breast cancer.

Have you ever wondered:

How can genetics be used to cure diseases?

People with cystic fibrosis are treated with physiotherapy and regular antibiotics. But as scientists discover more about genetic disorders and faulty genes, they may be able to improve treatments using gene therapy.

As well as those for CF, other faulty genes, including ones for heart disease and breast cancer, have been discovered. It is sometimes possible to treat genetic disorders by replacing the faulty alleles with healthy ones. This is difficult because the healthy alleles have to get into the nuclei of millions of cells, but it does work sometimes.

If gene therapy became common it could change lives. For example, someone with CF would not need regular physiotherapy and antibiotics. For someone with breast cancer, gene therapy could mean not having surgery or chemotherapy.

In 1990, two children with a serious genetic disorder had copies of correctly working alleles put into their white blood cells. This was the world's first gene therapy treatment. The children had suffered from a disease called ADA, which affects blood cells. Five years later the children had almost normal lives. Because white blood cells eventually die, this treatment must be repeated to keep working.

Stem cells are special cells in the bone marrow that keep dividing and producing new cells. Putting healthy alleles into stem cells may be a permanent answer to ADA. Unfortunately, finding and purifying enough stem cells to treat is proving difficult.

1 It was necessary to repeat the gene therapy treatment for ADA. Explain why. Why would stem cells avoid the need for repeated treatments?

2 Why are scientists working to find ways of growing large numbers of stem cells?

A Oli visits his physiotherapist Alison for regular treatment for CF.

It might be possible to treat CF using gene therapy. Working copies of the CF gene can be put inside tiny droplets called **liposomes**. A nasal spray can be used to deliver liposomes containing healthy alleles to lung cells. The liposomes dissolve into the cell membranes and the healthy alleles end up in the cells. The healthy allele causes the lungs to make healthy mucus. This improves breathing, makes physiotherapy easier and reduces the risk of chest infections.

A permanent cure needs to put healthy alleles into the lung cells and get them to stay there. **Viruses** can put copies of working alleles into cells. The adenovirus, which causes the common cold, is being used to try and put healthy alleles into lung cells for a permanent cure.

Some forms of breast cancer also have genetic causes. Breast-screening programmes use genetic tests to identify women at greatest risk of breast cancer. These women can be closely watched so any sign of breast cancer is spotted and treated immediately. In the long term, gene therapy might be used to prevent cancer.

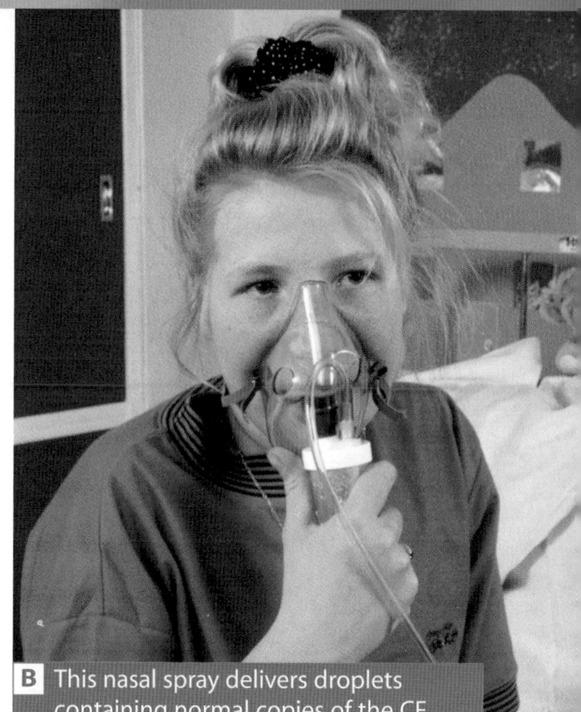

B This nasal spray delivers droplets containing normal copies of the CF allele.

Have you ever wondered:

When will I be able to get medicines made just for me?

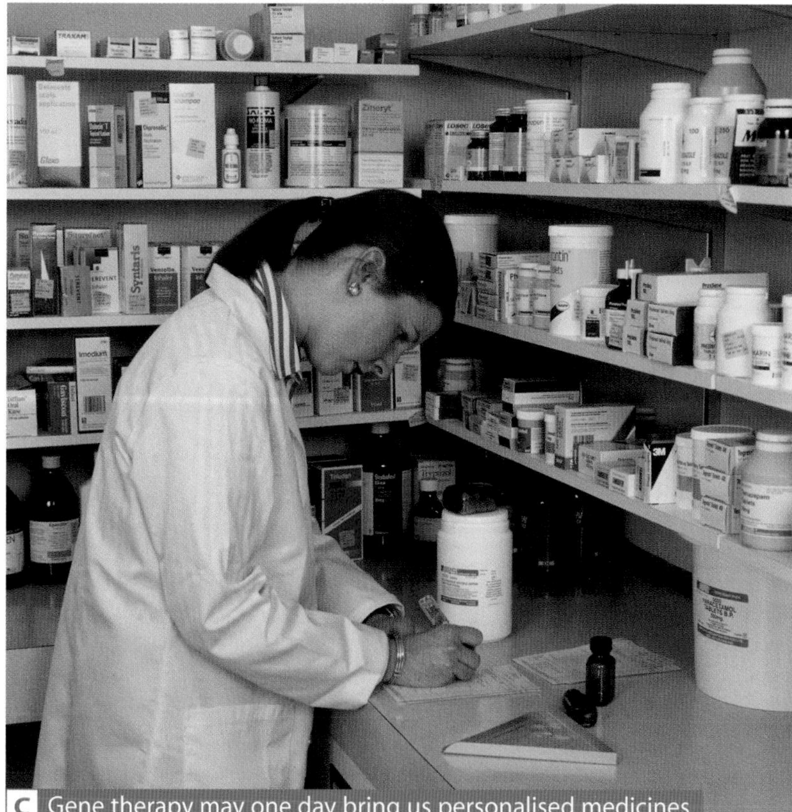

C Gene therapy may one day bring us personalised medicines.

3 What is a liposome and how does it work?

4 Why is this not a permanent cure?

Summary Exercise

Higher Questions

Clones and transgenic organisms

By the end of these two pages you should know:

- about cloning mammals and why it is controversial
- about growing body parts for transplant surgery
- **H** about transgenic animals and their uses.

Have you ever wondered:

How can cows make drugs in their milk?

Have you ever wondered:

Why can we not just breed a racehorse that will win every race?

A clone is a group of identical plants or animals which get all their genes from one parent. Mammals reproduce sexually and have two parents. Dolly the sheep had no father. Her 'egg donor' mother was a Scottish Blackface sheep. One of her unfertilised eggs had its nucleus removed. This was replaced by a nucleus from the body cell of her 'gene mother' a Finn Dorset sheep. Dolly was a Finn Dorset clone with exactly the same genes as her 'gene mother'.

Finn Dorset ewe

Scottish Blackface ewe

donor cell

egg with no nucleus

Donor cells were taken from the udder tissue of a Finn Dorset ewe and cultured for a week in low-nutrient medium to stop them dividing.

An **unfertilised** egg was removed from a Scottish Blackface ewe and the nucleus containing DNA was removed with a micropipette, leaving an egg cell with no nucleus.

The dormant donor cell and the recipient egg cell were placed close together and caused to fuse together using a very gentle electric pulse.

A second electrical pulse triggered cell division, producing a ball of cells after about 6 days.

The developing embryo was implanted into the uterus of another Scottish Blackface ewe.

After 148 days the pregnant Scottish Blackface ewe gave birth to Dolly, a Finn Dorset lamb that was genetically identical to the original Finn Dorset donor.

C How Dolly was produced.

A Dolly the sheep was made in a laboratory.

B Dolly the sheep.

1 Why was it necessary to replace the egg nucleus with a nucleus from a body cell?

2 Explain why Dolly was a clone of her 'gene mother' and not her 'egg donor' mother.

Human organs might be grown from cloned human cells. It is now possible to replace organs such as livers, kidneys, faces and even hearts. **Transplant** patients often wait a long time for an organ to become available. The organ must be closely matched to the patient's tissues to prevent rejection. Human organs cloned from the patient's own cells would match exactly and the patient would not have to wait as long.

To clone human cells you need stem cells. At the moment most stem cells are obtained from embryos. Is it right to make human embryos just for the purpose of extracting cells from them?

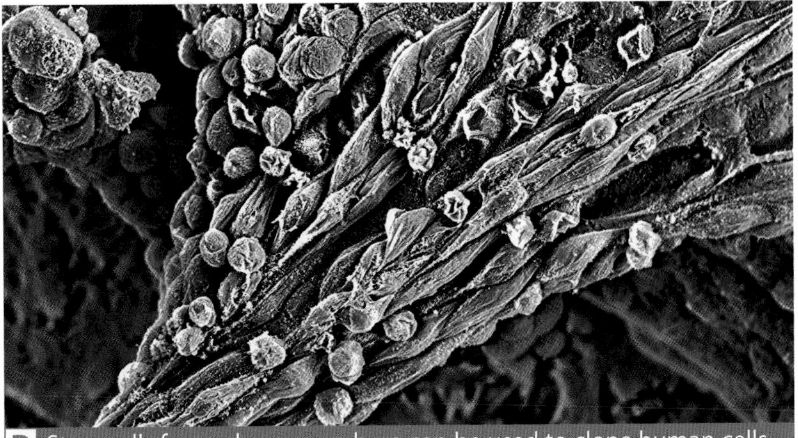

D Stem cells from a human embryo can be used to clone human cells, but there are ethical concerns about this.

3 Explain the advantages and disadvantages of being able to clone body parts.

H Genes from animals or plants can be put into other organisms by **genetic engineering**. These are called **transgenic** organisms. Tracey, a transgenic sheep, had a human gene put into her cells. This gene instructed Tracey's cells to make an enzyme in her milk. This enzyme is used to treat cystic fibrosis.

Transgenic cattle could produce milk with low cholesterol. High levels of cholesterol in your blood increases risk of heart disease. Milk could be produced which contains human **antibodies**. Antibodies help us to fight disease. This milk could be used to treat people with low disease resistance and help them fight infection.

Some people object to making clones and transgenic organisms. They say it is unnatural and cruel and might be dangerous. In fact humans have bred animals and plants for centuries. Even clones are not new. What are the new techniques that scientists use today? Soon it may be possible to clone human body parts for transplant surgery. Society has to decide what should be allowed and what shouldn't. This is called an ethical code. To make a judgement, you must understand the science and weigh up the potential advantages and disadvantages of each treatment.

4 Many people are concerned about cloning animals, although they probably eat cloned vegetables.
   a Why is there a concern about cloning animals?
   b Suggest why there are not similar concerns about cloning plants.
   c What benefits might cloning animals produce?

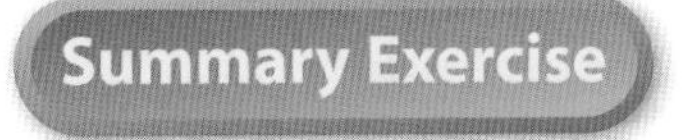

Higher Questions

Designer babies

H ***By the end of these two pages you should know:***

- what is meant by 'designer babies'
- why the idea of designer babies is so controversial.

P

V

'**Designer babies**' develop from embryos *selected* for particular genes. We each have well over 100,000 genes so it is impossible to choose them all. Embryos are tested for severe genetic disorders and only healthy embryos are transplanted into the mother and allowed to develop. **Embryo screening** would mean that parents who were carriers of cystic fibrosis could ensure that only healthy embryos developed into babies.

Embryo selection has been used to create 'saviour siblings'. This happens when parents have a child with a disorder like leukaemia (a blood cancer). Embryos are made in a laboratory and those that are the closest match to the sick child are implanted into the mother. Stem cells are taken from the new baby's umbilical cord and used to treat their brother or sister. There is much debate about this. It is allowed in the USA but not in the UK. Taking stem cells poses no risk to the new baby but any unused embryos are destroyed.

A As someone with CF, Oli will have moral and ethical issues to consider if he wants children.

B These four couples all have different decisions to make about having children.

Anita has three boys. She and her husband Archie want a daughter to complete their family. Brian has haemophilia. His blood fails to clot normally. He does not want to pass the haemophilia allele to his children. Brian's daughters would be carriers of haemophilia although his sons would be unaffected. Callista is a supermodel. She and her partner want a daughter with long legs, blonde hair and blue eyes. Diedre has colour-blindness. Her husband Donald is a pilot. They want a son who can become a pilot too. If their son has colour-blindness, this would be impossible.

1 Anita, Brian, Callista and Diedre all have ideas about what their 'designer babies' should be like.
  a Which families do you think should get the go-ahead for a designer baby?
  b Who do you think should make this decision?
Give reasons for your answers.

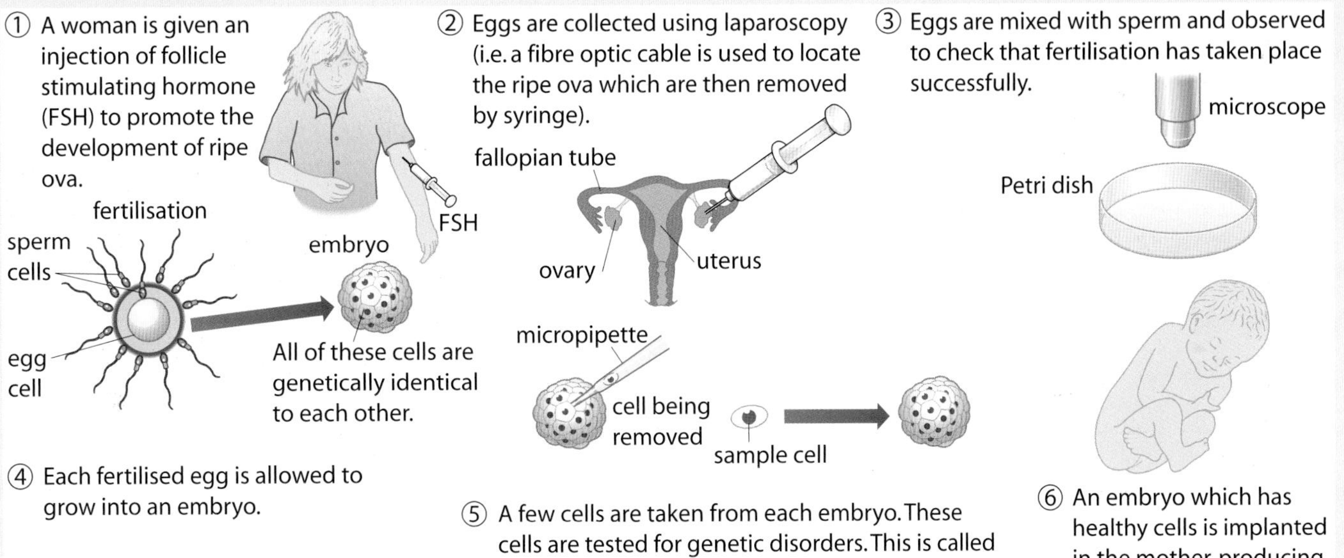

C How a designer baby is made.

2 Babies produced by embryo screening should be free from serious genetic disorders. Why can't we predict exactly what the babies will look like? Which couples in question 1 could embryo screening help? Explain your answer.

3 Do you think making a saviour sibling should be allowed in this country? Give your reasons.

One day, designer babies might be produced by cloning, like Dolly the sheep. The babies would be exactly the same as either their mother or father. It might also be possible to put genes from other animals into human embryos, making transgenic humans.

There is much debate about embryo screening and designer babies. Many people oppose these ideas on ethical and religious grounds. They say that only God had the right to create life and that we should not create life to destroy it. Some say we do not understand the risks involved. Are we in danger of playing God? Science is neither good nor bad but how we use science affects everyone in society.

4 What do you think about designer babies? Should they be allowed or not? Give your reasons.

Higher Questions

Questions

D

D

Multiple choice questions

1 Which of the following is a definition of a clone?
- **A** Organisms which share the same parents.
- **B** Organisms which share the same genes.
- **C** Organisms produced by sexual reproduction.
- **D** Organisms produced by selective breeding.

2 A small section of DNA is called
- **A** the nucleus.
- **B** a gene.
- **C** DNA.
- **D** a chromosome.

3 Look at the following statements and put them in the correct order. To make a DNA fingerprint the following steps are taken:
- **1** DNA is cut using enzymes
- **2** DNA is taken from blood or saliva
- **3** the sections of DNA are separated out into bands
- **4** DNA is copied.
- **A** 1, 2, 3, 4
- **B** 3, 1, 4, 2
- **C** 4, 3, 2, 1
- **D** 2, 4, 1, 3

4 The DNA fingerprint is unique for each individual because
- **A** the DNA is cut in different places each time a fingerprint is run.
- **B** our genes get mixed up when we are babies.
- **C** we get a mixture of genes from our parents.
- **D** genes are made of DNA.

5 Which of the following ways of growing plants such as geraniums would produce the least amount of variation?
- **A** Grown from seeds in a garden.
- **B** Grown from cuttings in a garden.
- **C** Grown from seeds in a glasshouse.
- **D** Grown from cuttings in a glasshouse.

6 A farmer notices that the leaves of her crop are losing their green colour. Which of the following minerals is likely to be missing?
- **A** calcium salts
- **B** magnesium salts
- **C** phosphates
- **D** potassium salts

7 Sexual reproduction results in
- **A** offspring who all have the same genes.
- **B** offspring who cannot adapt easily to the environment.
- **C** offspring who have a mixture of genes.
- **D** offspring who are all likely to get the same disease.

8 Jack and Jill are 'carriers' of a recessive genetic disorder. What is the chance of their first child being affected by this disorder?
- **A** 0% (nil)
- **B** 25% (1 in 4)
- **C** 50% (1 in 2)
- **D** 100% (certainty)

9 Replacing a faulty allele with a healthy working copy is called
- **A** DNA fingerprinting.
- **B** embryo testing.
- **C** gene therapy.
- **D** genetic engineering.

10 What was the main purpose of the Human Genome Project? (HGP)
- **A** To develop embryo testing.
- **B** To develop genetic engineering.
- **C** To produce genetic fingerprints of human DNA.
- **D** To sequence the bases in human DNA.

11 Which of these terms describes an organism containing genes from another species?
- **A** transgenic
- **B** transferred
- **C** transformed
- **D** transposed

12 When Mendel crossed peas with yellow flowers (YY) with peas with white flowers (yy) what happened in the next generation?
- **A** All the pea plants had yellow flowers because the yellow allele is recessive.
- **B** The pea plants had a mixture of white and yellow flowers.
- **C** The flowers were pale yellow.
- **D** All the pea plants had yellow flowers because the yellow allele is dominant.

13 Statements 1–6 are stages in the process of selective breeding a new variety of pea plants. They are not in the correct order.

1 Pollen from the 'pollen parent' is placed on the stigma of the 'seed parent'.
2 Selection of suitable 'pollen parent' and 'seed parent' plants.
3 Removal of anthers from the 'seed parent'.
4 Collection and germination of seeds.
5 Covering flower of 'seed parent' to prevent cross-pollination.
6 Selecting offspring with the desired combination of characteristics.

Which letter shows the order in which the stages are carried out?

A 2→5→3→4→1→6
B 6→1→4→3→5→2
C 2→3→1→5→4→6
D 6→2→3→1→5→4

14 The allele for freckled skin is recessive (f) to non-freckled skin (F). Which of the following describes the likely offspring of the cross Ff×ff?

A All have freckles.
B One-quarter of the offspring have freckles.
C Half of the offspring have freckles.
D None of the offspring have freckles.

15 'Designer baby' means

A a group of embryos has been screened and one has been selected to be free of genetic diseases.
B a baby has been designed to have exactly the features the parents want.
C the baby is a clone of one of the parents.
D a baby which has blue eyes and blonde hair.

Short-answer questions

H **1** Gregor Mendel did a plant-breeding experiment using pea plants. He crossed a pure-breeding, round-seeded pea plant with a pure-breeding, wrinkled-seeded pea plant. All the first-generation plants produced round seeds.

a **i** Which of the alleles is dominant: round seeded or wrinkled seeded? Explain your answer.
ii Choose suitable symbols you could use to represent the two alleles.
round-seeded allele = ______
wrinkled-seeded allele = ______

Mendel then bred the round-seeded, first-generation plants with each other. Approximately three-quarters of their offspring produced round seeds and one-quarter produced wrinkled seeds.

b Show, by means of a genetic diagram, why Mendel obtained this result. Use the symbols you chose for part (a, ii) in your answer.

2 Roses can be reproduced sexually from seeds or asexually by taking cuttings. A rose grower takes pollen from a scented variety of rose and places it on the stigma of an unscented rose variety which has large colourful petals.

a **i** Explain why the rose grower uses two varieties of rose.
ii Suggest what the rose grower is hoping to achieve by using this method of reproduction.

The rose grower then collects the seeds and plants them. When the plants flower, he chooses the ones with the correct characteristics and takes cuttings.

b **i** The plants produced by taking cuttings are all *very similar* to each other. Explain why.
ii Why are the plants *not all identical* to each other?

3 Many farmers in America grow genetically modified (GM) crops such as soya and maize.

a What advantages do farmers hope to gain by growing GM crops?

One variety of GM maize actually manufactures its own insecticide. Some people are concerned about the effects that GM maize might have on the environment, on animals and on people.

b Suggest why people might have these concerns about GM crops.

Genetic engineers are hoping to produce varieties of cereal that can grow in dry soils and need less fertiliser.

c Why might these varieties of cereal become increasingly important in the future?

D

P

Glossary

D

allele Alternative forms of the same gene. For example, the gene controlling flower colour in peas has white and yellow alleles.

antibody A substance produced by white blood cells in response to infection. It finds disease-causing organisms so they can be destroyed by special cells called phagocytes.

asexual reproduction Reproduction in which genes are passed on from only one parent.

base The four bases – adenine (A), cytosine (C), guanine (G) and thymine (T) – make up the 'rungs' of the DNA 'ladder'. In DNA the bases pair with each other: A always pairs with T and C always pairs with G.

cancer Cancer is a result of the rapid uncontrolled growth of cells. These cells form tumours which may spread and invade other organs.

carrier A person whose cells have both a normal and a faulty allele. Symptomless carriers show no signs of the genetic disorder they are carrying.

cell The basic 'unit of life', consisting of a nucleus and cytoplasm surrounded by a membrane.

characteristic A feature of an organism. A result of a gene's instructions to a cell, for example, flower colour or blood group.

chromosome Thread-like structures in the nucleus made up of strings of genes.

clone A group of genetically identical plants or animals produced asexually from one parent.

cystic fibrosis A genetic disorder in which the person produces abnormal, sticky mucus in their lungs.

designer babies Offspring produced with particular characteristics using either genetic modification or embryo screening.

DNA A chemical containing the code which tells a cell how to develop. DNA stands for deoxyribonucleic acid.

DNA fingerprint A pattern of bands produced when short lengths of DNA are separated using an electric current.

dominant An allele which overrides other alleles of a gene so that their effects are hidden.

embryo screening Selecting embryos with particular characteristics, usually to eliminate inherited disorders.

environment The surrounding conditions in which an organism develops.

fertilisation When the nuclei of two sex cells – such as an egg and sperm – join together.

forensics The medical knowledge used in the detection of crime.

gamete A sex cell; for example, an egg or a sperm. Gametes carry only one copy of an allele.

gene A piece of DNA that contains the instructions needed for a particular characteristic, such as eye colour.

gene therapy Replacing faulty alleles with working copies of the affected genes.

generation A term used to describe the descendents of a pair of individuals. For example, the first generation is their children and the second generation is their grandchildren.

genetics The science concerning the inheritance of characteristics.

genetic engineering Techniques used to remove genes from one organism and transfer them to another organism.

Human Genome Project A project, begun in 1990, which set out to map the human genetic code by working out the sequence of the genes in human DNA.

inheritance A term used to describe the passing of genes from parents to offspring.

liposome A small fat-like droplet which can be used to deliver working genes to cells in gene therapy.

nucleus The part of a cell which contains genetic material in the form of chromosomes.

recessive An allele whose effects are hidden by the presence of a dominant allele. A recessive allele must be inherited from both parents in order to show its effect in the offspring.

sexual reproduction Reproduction in which half the genes are inherited from each of two parents.

stem cell Cells which continue to divide and which have the ability to replace cells in all types of tissue.

transgenic A term applied to an organism containing genes taken from another species.

transplant A term used to describe an organ which has been donated by one organism and inserted into the body of another, such as a kidney transplant.

variation This describes the differences shown in a group of organisms; for example, fur colour and flower colour.

virus A simple particle that enters and infects cells. Mild viruses like the adenovirus, which causes the common cold, may be modified and used to put working copies of genes into faulty cells.

b.3.0

Electrical and chemical signals

A Iwan Thomas – an athlete whose body depends on electrical and chemical signals so that he can win.

Things are happening around and inside Iwan all the time. His body detects and responds to these changes. It uses electrical and chemical signals to do this.

The electrical and chemical signals are used to control what his body does. This often means several parts of the body working together. They are coordinated by the electrical and chemical signals.

In this topic you will learn that:

- the body needs to be maintained in an optimum state
- the central nervous system lets your body respond to changes in its surroundings
- hormones regulate the functions of cells and organs
- artificial hormones can be used to control reproduction and alter body functions.

Andrew has made a grid of things he knows and things that he would like to know. Add four more things that you think you know, and four that you want to know. You can change any you do not agree with.

Five things *I think I know* about electrical and chemical signals	Five things *I want to know* about electrical and chemical signals
Insulin is used to control diabetes for my Grandma.	Why do my pupils get bigger in dim light?

Reaction times

By the end of these two pages you should know:

- what a reaction time is and why it is important
- how reaction time can be measured.

To be the fastest person on the planet all you have to do is run 100 m in 9.76 s or less. Everything could depend on how quickly you can react to the starting gun. **Reaction time** is the amount of time between hearing or seeing something and the first movement of your body. For most of us reaction time is about 0.25 s, but for top athletes it is between 0.11 and 0.18 s.

A Reaction time can be pretty important if you want to win. An athlete needs really fast reactions to be first off the blocks.

NORWICH UNION **Super Grand Prix** **Sheffield** 21 August 2005 **MEN 100 Metres** wind 1.1

1	Kim Collins	SKN	10.01
2	Leonard Scott	USA	10.10
3	Michael Frater	JAM	10.11
4	Jason Gardener	UK	10.12
5	Marlon Devonish	UK	10.13
6	Darrel Brown	TRI	10.14
7	Marc Burns	TRI	10.15
8	Ainsley Waugh	JAM	10.31
9	Daniel Plummer	UK	10.46
DQ	Maurice Greene	USA	DQ

B Look at the results of this race. The times between second and fifth are just 0.03 s (three hundredths of a second). Having a faster reaction time could make a difference!

1 What situation might you be in where a quick reaction time would be vital?

2 An athlete will keep practising race starts. Suggest a reason why they do this.

In a race, the time between the gun and the first kick on the starting block is measured electronically. The shortest time possible for the sound of the gun to reach the ear and for that information to be processed by the **brain** is 0.12 s. So any athlete who moves before that time has beaten their own reaction time – they were moving before their brain could have known the gun had fired. This is a false start.

The sound of the starting gun is the **stimulus** for the athlete to move. Stimulus is the scientific word for something you react to. It may be a sudden movement that you see, a hot object that you have touched or the point of a pin pricking your skin.

A stimulus is detected by **sense organs**. Eyes and ears are sense organs. They contain special cells called **receptors**. Some receptors in your eyes detect how much light enters the eye. Others can detect different colours of light. In your ears there are receptors that detect sound. They can detect how loud and what pitch a sound is.

What happens when a sound reaches your ear? Receptors in your ear convert the sound into **electrical impulses**. These **impulses** travel along a **pathway** in your body until they reach a **muscle**. Your muscle starts to **contract** when the impulse reaches it, and you start to move. So there is a pathway between your ear and your muscles. Your reaction time is the time taken for the impulses to travel along this pathway.

C Reaction time is important to an athlete like Iwan.

Top

Reaction time (s)	Distance from bottom of ruler (cm)
0.20	17.5
0.18	12.3
0.16	9.6
0.14	7.6
0.12	6.0
0.10	4.5
0.08	2.8
0.06	1.1
0.04	0.4

Bottom

D A simple way to measure reaction times using a ruler.

There are several ways of measuring reaction time. The simplest is a special ruler like the one above. A friend holds the ruler at the top. Line up your fingers with the bottom edge. When your friend lets go, catch it and record your reaction time. This is not a very precise technique. There are electronic and online reaction timers that can measure much more precisely.

3 Why is the ruler reaction timer not a very precise way of measuring reaction time?

4 What is meant by the term reaction time?

5 Why is reaction time important if you are driving a car?

V

P

Summary Exercise

Higher Questions

The nervous system

By the end of these two pages you should be able to:

- describe the structure of the nervous system
- explain how the nervous system controls the body
- explain that the nervous system can detect changes both inside and outside the body.

P

?

Have you ever wondered:

How does my brain tell my body what to do?

V

Think about how the body deals with everything that is going on around it and inside it. It has to cope with all sorts of changes. This is the job of the **nervous system**. The nervous system carries electrical signals to and from all the different parts of your body. These messages are called impulses.

The main part of the nervous system is the brain and the **spinal cord**. Together they make up the **central nervous system (CNS)**. This is connected to the rest of the body by nerves. Each nerve is made up of a bundle of special cells called **neurones**.

P

1. Give an example of an athletic activity where it is important that senses and movements are coordinated.
2. What are the changes to your surroundings that can affect your body?

A This athlete must respond quickly to changes in his surroundings.

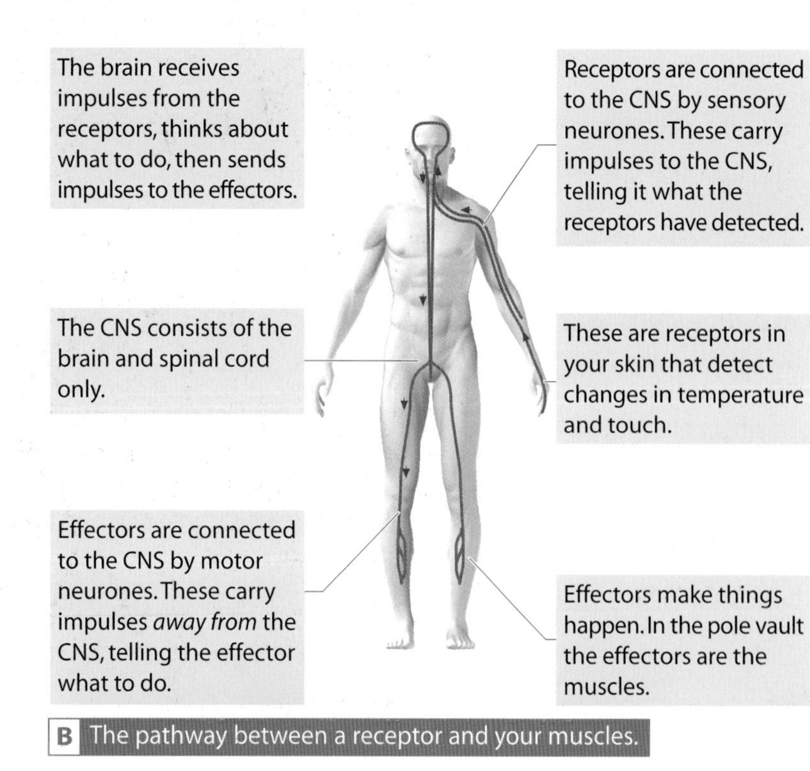

B The pathway between a receptor and your muscles.

Everything the nervous system does follows the same pattern. In and around the body are stimuli. These are changes that can be detected, like heat, sound, light and pain. There are receptors to detect these changes. The receptors are special cells. There are receptors in your skin that detect changes in temperature and touch. **Effectors** make things happen. In the pole vault the effectors are the muscles. Receptors are connected to the central nervous system by **sensory neurones**. These carry impulses *to the CNS*, telling it what the receptors have detected. Effectors are connected to the central nervous system by **motor neurones**. These carry impulses *away from the CNS*, telling the effector what to do.

3 In what ways are sensory and motor neurones different?

P

The impulses carried by the neurones are electrical signals. They pass very quickly along the neurones. Each impulse is a separate signal. They travel along the neurones one after another.

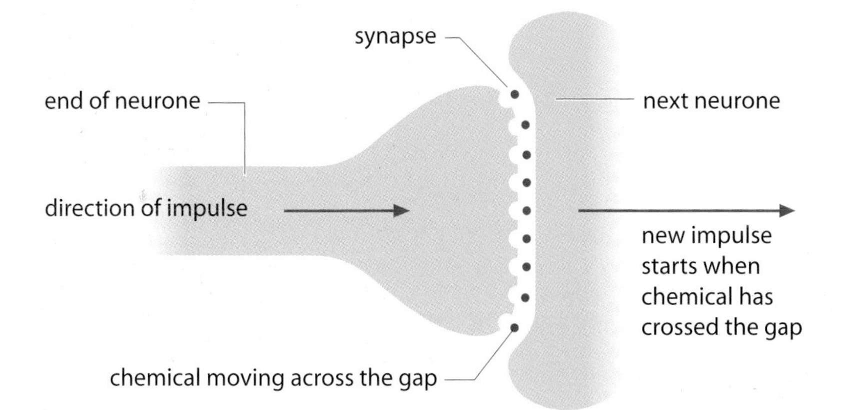

A chemical is produced when an impulse reaches the end of a neurone. The chemical diffuses across the gap to the next neurone. When it arrives it starts a new impulse in the next neurone.

C How a synapse works.

A neurone is actually a cell. It has a cell body containing a nucleus. The end of a neurone does not actually make contact with the next neurone. There is a tiny gap between them. This gap is called a **synapse**.

When an impulse gets to the end of a neurone, a chemical is released into the synapse. When it reaches the next neurone, the chemical triggers a new impulse. Only one end of a neurone can make the chemical, so impulses only travel one way along the neurones.

4 Why do impulses only travel one way along neurones?

5 Look at the photo of the athlete. What are the stimulus, receptor, response and effector here?

D Stimulus, receptor, response and effector in action.

P

Summary Exercise

Higher Questions

The human brain

By the end of these two pages you should be able to:

- describe the structure of the brain
- explain what the functions of the brain are
- H understand some of the things that can go wrong with the brain.

A Iwan's brain coordinates everything that is going on in his body.

At the start of a race Iwan's body is tensed and ready to go. Everything is coordinated and in perfect balance. He hears the starting gun and he's off. His actions are complicated and are coordinated by his brain.

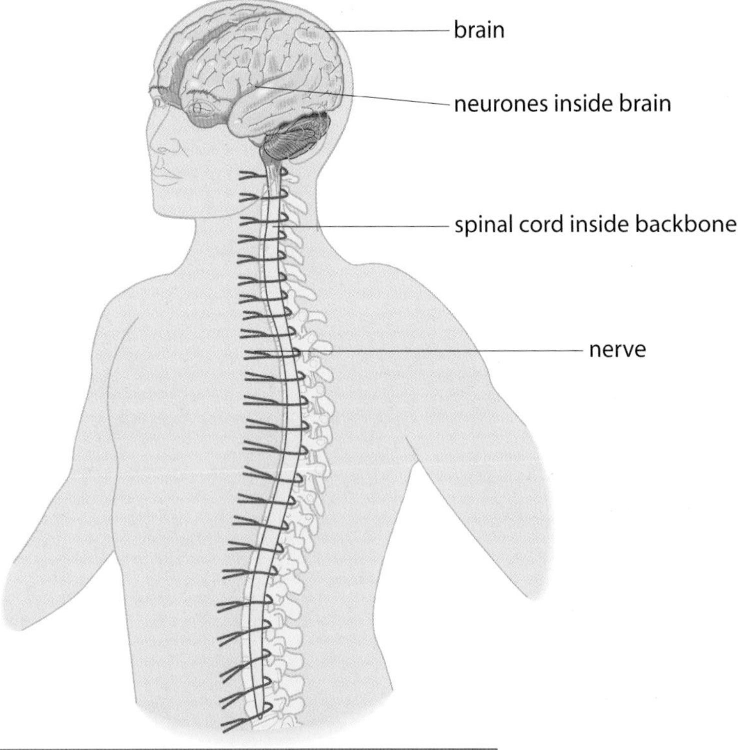

B The brain is connected to the rest of the body.

Iwan's brain receives impulses from his sense organs. It contains more than 10 million neurones. The neurones link up to coordinate the incoming and outgoing impulses. The outgoing impulses go to effectors like muscles so that his body carries out the actions he wants. Some of these actions are automatic, like his heartbeat. He has no control over them. The brain is connected to the spinal cord. This runs down inside the spine. The spine protects the delicate nervous tissue.

1 How are the brain and spinal cord connected to the rest of the body?

2 Why do you think that some actions are automatic?

The brain is so complex that we still do not know how much of it works. Different parts of the brain are responsible for different tasks. For example, the front of the brain controls thinking and memory and the back of the brain controls vision.

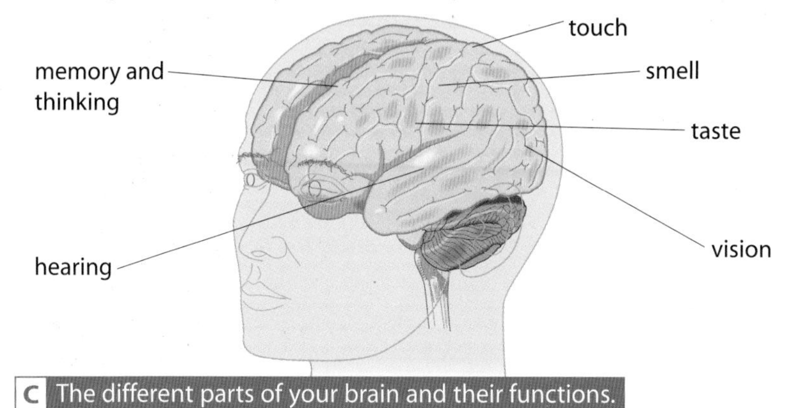

C The different parts of your brain and their functions.

3 A knock to the back of the head can cause blurred vision. Explain why.

4 What do neurones within the brain do?

H The way the brain functions can be disrupted. This usually means the transmission of impulses is affected in some way.

A **stroke** is when a blood clot or bleeding occurs in an artery that leads to or is inside the brain. This stops oxygen getting to the brain cells. Sometimes this lasts for only a short while and the patient can make a full recovery. If the oxygen supply is cut off for very long, cells will die. Unlike other body cells, dead brain cells cannot regrow. If a blood vessel in the brain bursts then blood builds up, which presses on the delicate cells of the brain, damaging them.

Brain tumours occur when extra cells grow unnecessarily inside the skull. They grow slowly and do not spread around the body. Pressure builds up inside the skull, causing headaches, damage to cells and unconsciousness. Brain **tumours** can also interfere with the way impulses travel between brain cells, and can cause seizures.

Parkinson's disease is caused by a problem with the impulses transmitted between one neurone and another, and between neurones and muscles. Symptoms include muscle tremors, stiffness, muscle rigidity and slow movement. There is no cure, but drugs and therapy can improve the patient's quality of life.

Some people have sudden bursts of electrical activity in their brain. This disrupts the normal transmission of impulses in the brain and can lead to a seizure – the brain stops working properly for a few moments. This is known as **epilepsy**. In some people the seizure leads to unconsciousness. They fall down, their body stiffens and they start to jerk uncontrollably. This kind of seizure is called a **grand mal**.

Higher Questions

Sense organs

By the end of these two pages you should be able to:

- explain that sense organs detect changes inside and around you
- explain how your body can respond to these changes.

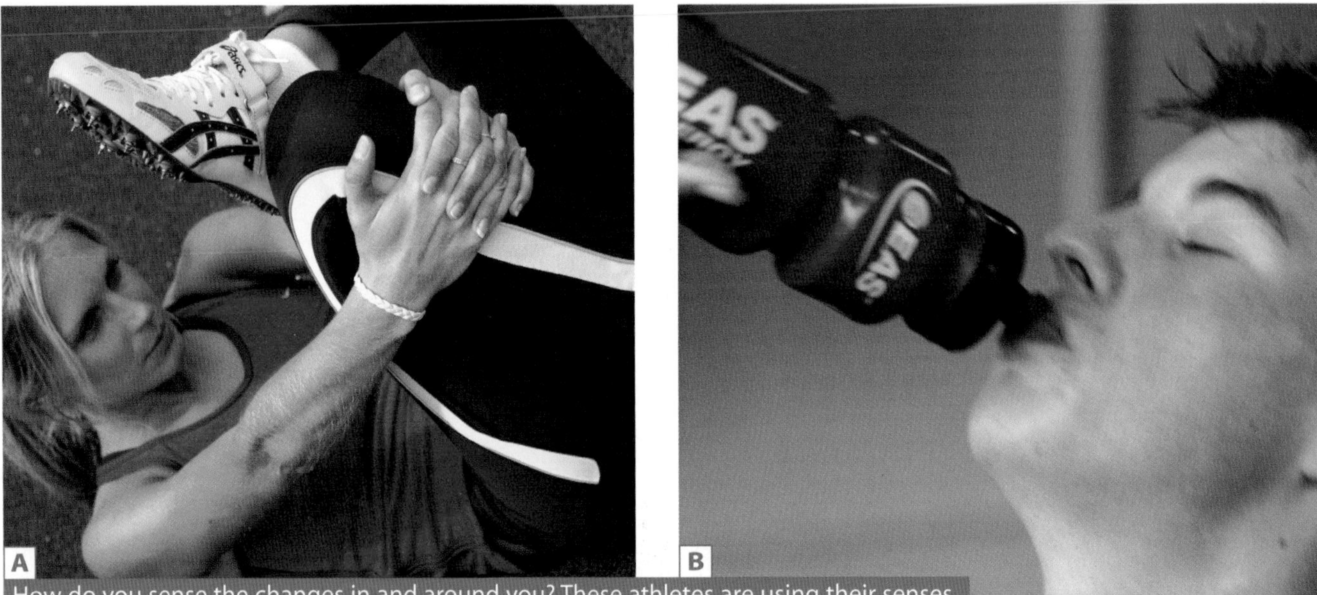

A B How do you sense the changes in and around you? These athletes are using their senses.

The five senses – sight, hearing, taste, touch and smell – are very important to an athlete. They detect stimuli and you respond. Finely tuned senses can make the difference between winning and coming last. You have sense organs to detect different types of stimulus.

Stimuli are changes that can be outside or inside your body. Sense organs contain receptors that detect these changes. These receptors send impulses to the central nervous system (CNS). The CNS can then respond to the change. Your main sense organs are your skin, eyes, ears, nose and tongue.

Your skin responds to pain, heat and cold, and pressure and touch. It contains lots of receptors, especially in your fingertips and in your lips.

1 Why do you have so many receptors in your fingertips and lips?

2 Why do you need receptors inside your body?

Your eyes respond to light and give you sight. They need to control the amount of light entering the eye to prevent damage. They also focus the light so what you see is clear for both near and far objects. Special receptor cells in the eye detect the light and send impulses to the brain.

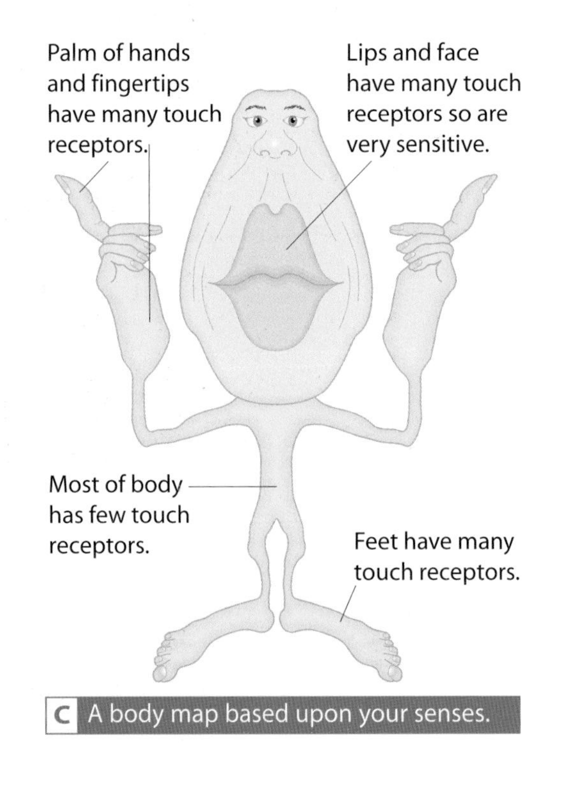

C A body map based upon your senses.

Your ears respond to sound vibrations and movement of your body. Sound waves are funnelled into the ear canal and cause the ear drum to vibrate, passing waves to the inner ear. Here tiny hairs detect the sound waves and convert them into impulses that are sent to the brain. These impulses are interpreted as sounds by the brain.

Your ears also help you keep your balance. You have three canals filled with fluid in the inner ear. The fluid moves if you move. Your movements are detected by tiny hairs in the fluid. The hairs send impulses to the brain. The impulses are interpreted as movements by the brain.

Your nose responds to chemicals in the air you breathe. These chemicals are smells. They dissolve in the moist lining of your nose. Receptors then detect them and send impulses to the brain. The impulses are interpreted as smells by your brain.

There are sensors on the tongue called taste buds. These are sensitive to some types of taste more than others. For example, the tip of your tongue detects sweet and salt, the centre detects sweet, the edges sour and the back bitter. Chemicals in your food must be dissolved before you can taste them. The taste buds send impulses to the brain.

D Dolphins use different senses to you.

Some animals have different senses to us. Bats use ultrasound to catch their prey. Dolphins use ultrasound to communicate. Some snakes detect their prey using infrared radiation, heat seeking in the dark.

3 How can an ear infection affect your balance?

4 How can you tell that you have touch sensors attached to the base of the hairs on your skin?

5 You see a banana. What would your senses tell you about the banana between seeing it and eating it?

P

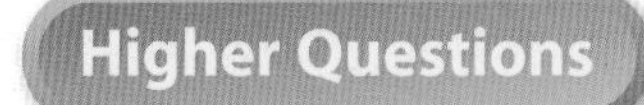

The eye

By the end of these two pages you should be able to:

- explain how your eye focuses light on the retina
- explain how your iris controls the amount of light entering your eye.

Imagine being an athlete with no sight. How would you cope with a 100 m race? You have a sighted person to run beside you. They are your eyes, and your ears take over. You trust your guide completely. Your sight is one of your most important senses. It is also one of the most complex.

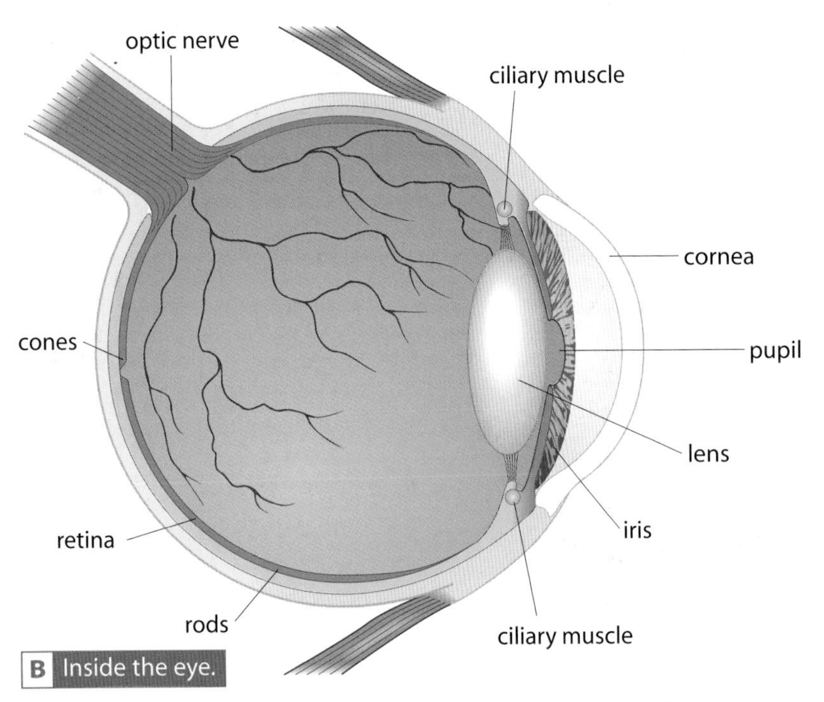

B Inside the eye.

The inner lining of the eye is the **retina**. It is made of light-sensitive cells. There are two types. **Rods** are receptors that work in dim light. **Cones** detect colour. Cones are concentrated at the back of the eye. This is the most sensitive part and can detect fine detail. When light hits these cells impulses are sent to the brain. The brain assembles all of these impulses and interprets them as a picture, telling you what you see.

When light enters the eye it must be focused on your retina. If it is not focused then what you see will be blurred. To focus the light, the light rays must be bent inwards so that they meet at a point. Most of this bending is done by the **cornea**. The cornea is curved into a lens shape.

Inside the eye there is another lens. This can change its shape slightly. It changes its shape so that you can see far and near objects. This is called **accommodation**.

A Eyes are not needed to be a sprinter. Blind sprinter Neville Bonfield ran 100 m in 12.49 s.

Accommodation – how you see distant objects

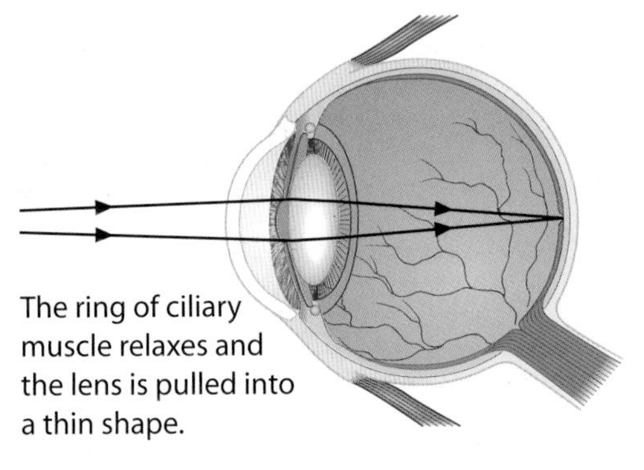

Accommodation – how you see near objects

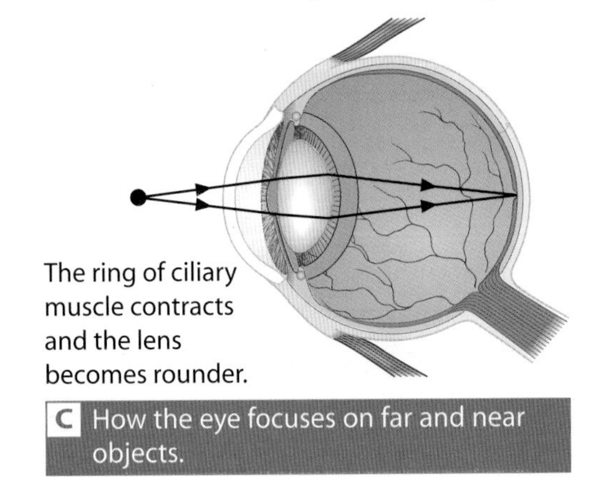

C How the eye focuses on far and near objects.

It is important that the amount of light entering your eye is controlled. Too much light can damage the delicate rods and cones in the retina. Not enough light makes it very difficult to see properly. Light enters through a hole in the centre of your **iris**. This hole is called the **pupil**.

1 How do your eyes see in colour?

2 What changes take place in your eyes so that you can read a book clearly?

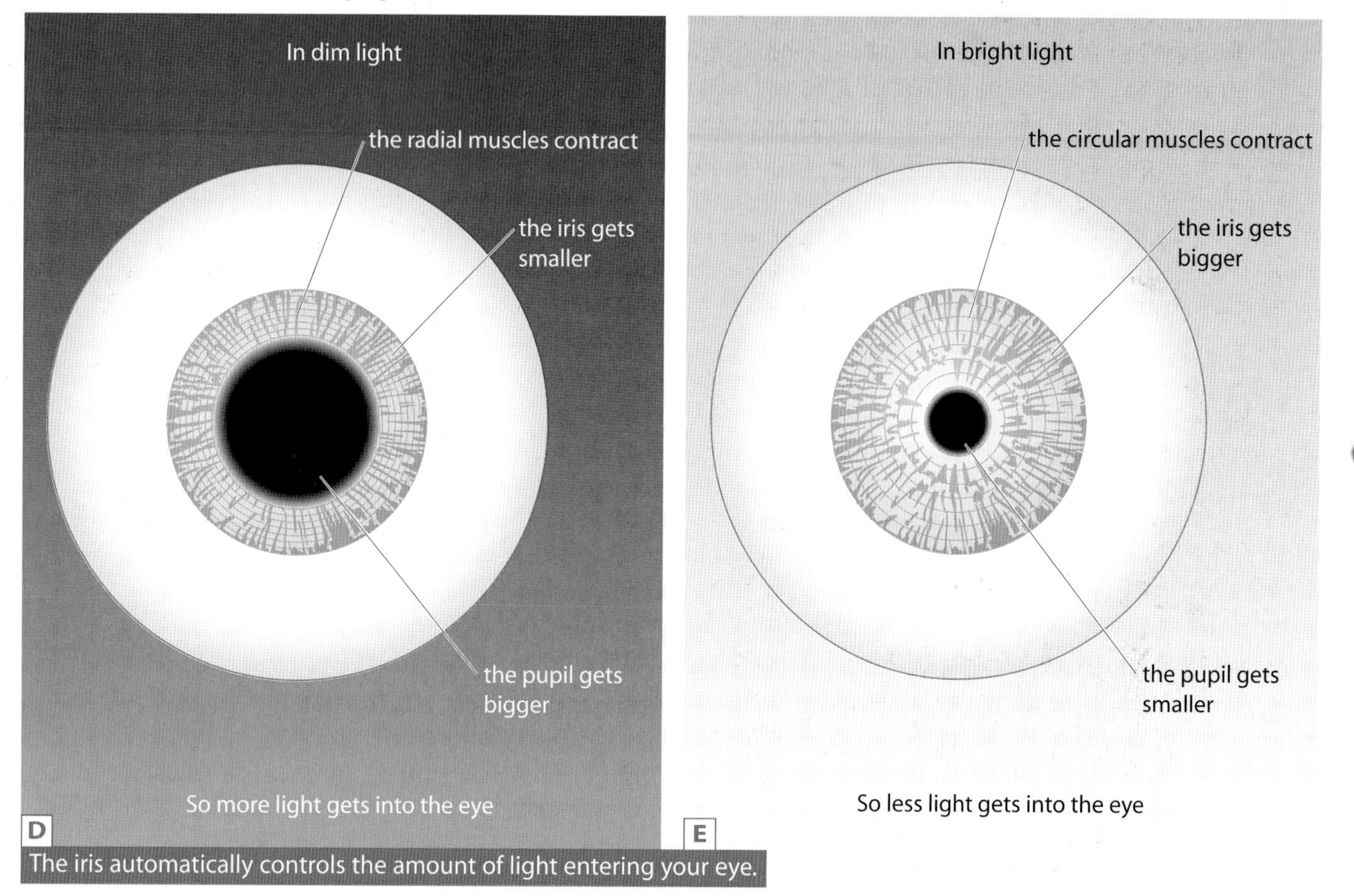

The iris automatically controls the amount of light entering your eye.

Your iris is made from two different types of muscle. There are circular muscles and radial muscles. There are two muscle types because one type makes the pupil smaller and the other type makes the pupil bigger.

In bright light

- the circular muscles contract
- the iris gets bigger
- the pupil gets smaller
- less light gets into the eye.

In dim light

- the radial muscles contract
- the iris gets smaller
- the pupil gets bigger
- more light gets into the eye.

This all happens automatically. It is called the **iris reflex**.

3 Why is it important that the amount of light entering your eye is controlled?

4 Where does most bending of the light entering your eye take place?

Summary Exercise

Higher Questions

Reflexes

By the end of these two pages you should be able to:

- explain the difference between voluntary and reflex responses
- explain the advantages of reflex responses in helping to safeguard your body.

A Your body keeps you safe with an automatic response to a stimulus – a reflex.

Have you ever wondered:

When travelling in a car, why do I duck down when a bird flies low over me?

Sometimes people have a choice about responding to a stimulus. Sometimes their bodies take over and respond automatically before they have a chance to do anything about it. The nervous system responds very quickly to a stimulus because it uses electrical impulses.

A **voluntary response** to a stimulus is one that involves the person. They use their brain to make a conscious decision. For example, they can decide whether they want to have a drink of water or stay in an armchair.

An **involuntary response** to a stimulus is an automatic response. A cough or a sneeze is an involuntary response. You have no conscious control over what happens. Involuntary responses are often called **reflex** responses.

A reflex response is very fast because it does not involve the brain. You do not think about it. So it involves only a few neurones. Reflex responses often protect the body. Here are some examples:

- If some food goes down the wrong way you cough to shift it out again.
- If you touch a hot object your hand will move away very fast to stop it being burnt.
- If you put your finger on a sharp object you pull away quickly.

1 What is the purpose of a reflex action?

2 Why are reflex actions automatic?

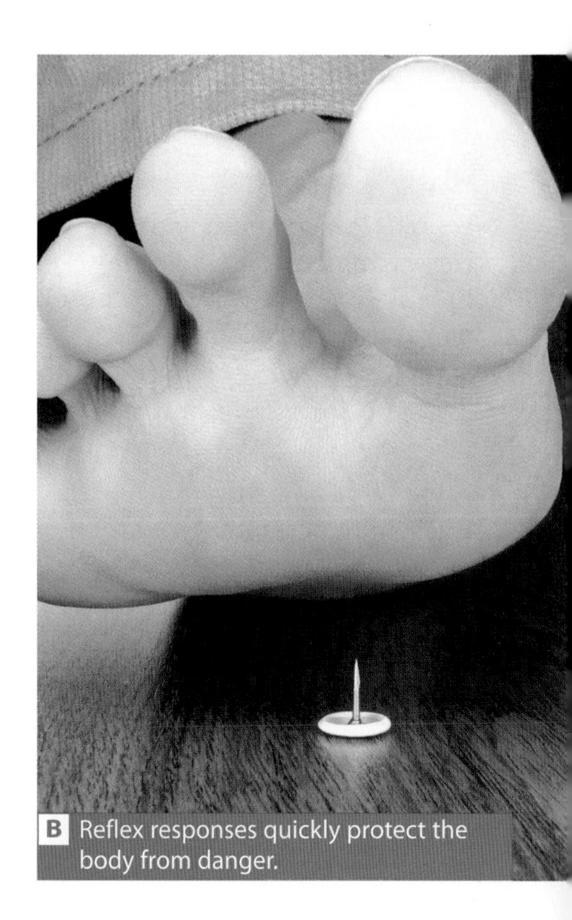

B Reflex responses quickly protect the body from danger.

The iris reflex is an involuntary response to changes in the amount of light entering your eyes. The amount of light is detected by the retina, and impulses are sent to the muscles of the iris. These open or close the pupil. This response is to protect the delicate cells in the retina from damage by too much light.

The accommodation reflex means that the eyes automatically refocus when they look at near or distant objects. It happens very quickly. This may help to protect the body from injury. It is very important that it works properly if the person is driving a car.

The ducking reflex is triggered when something moves very fast near someone's head. The movement is detected by their eyes and causes muscles to contract, making them duck. This helps to protect the most delicate part of the body – the head and the brain inside it.

3 Put a V or an I to show the type of response

Response	Voluntary (V) or involuntary response (I)
Drinking a glass of water	
Sweating in hot weather	
Making saliva when you smell food	
Blinking when dust gets in your eye	
Pulling your hand from a hot object	
Watching television	
Coughing if food goes down your windpipe	
Kicking when your knee is tapped	
Ducking when a ball goes near your head	
Sneezing in a dusty room	
Reading a book	
Shivering in the cold	

4 How can you tell that blinking is a reflex action?

P

S

Summary Exercise

Higher Questions

The reflex arc

By the end of these two pages you should be able to:

- explain the path of an impulse in a reflex arc
- explain how the iris reflex works.

One of the tests a physiotherapist will carry out on an athlete is the knee-jerk test. A sharp tap on the tendon just below the knee cap will make the leg kick. This is used as a test to see if the nervous system is working correctly.

This test is a good example of an involuntary response – you have no control over it. Remember that involuntary responses are often called reflexes. Most reflex actions are there to protect your body. For example, if you touch a hot object your hand moves away – fast!

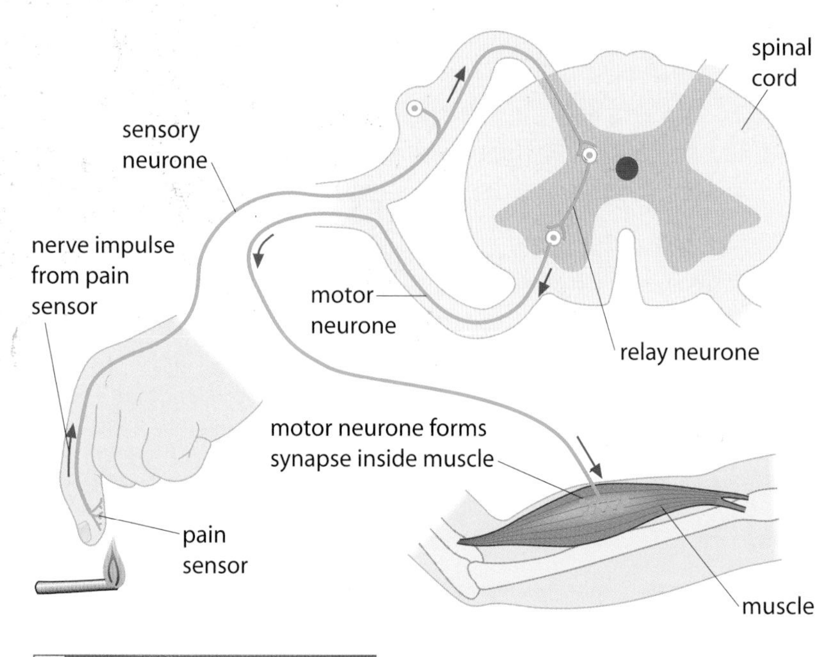

B The pathway of a reflex arc.

In your finger there are **pain sensors**. Heat is the stimulus that causes them to send impulses. The impulses travel along the sensory neurone. In the spinal cord, the sensory neurone connects with a **relay neurone**. The impulses travel across the synapse and along the relay neurone. At the end of the relay neurone there is a synapse with a motor neurone. The impulses cross to the motor neurone.

Where the motor neurone meets the muscle there is another synapse. The impulse crosses to the muscle, causing the muscle to contract. This action is called the response. The muscle is called an effector. The whole pathway is called a reflex arc.

A A physiotherapist carrying out a knee-jerk test.

So the reflex arc uses three neurones and three synapses. It starts with a stimulus and ends with a response. The reflex arc is completely automatic. It does not involve the brain or thinking. It is very fast.

1 Why is it important that reflexes are very fast?

2 What would happen to the speed of a reflex if you had to think about it?

P

C Blinking is a reflex action. It happens without you having to think about it.

An important example of a reflex arc is the iris reflex. Receptors in the retina detect the light. Impulses travel along sensory neurones to the CNS. From here they travel to the muscles in the iris. The iris reflex protects the delicate cells in the retina from being damaged by too much light.

Blinking is another reflex action. Anything touching the delicate cornea at the front of your eye causes your eyelids to close very fast.

Accommodation is a reflex that lets you focus your eyes on distant objects and near objects.

3 What are the steps in a reflex arc?

4 Why is blinking an important involuntary action?

5 A frisbee flies close to your head and you duck. What are the different stages that result in your nervous system causing you to duck?

P

Summary Exercise

Higher Questions

Hormones

By the end of these two pages you should be able to:

- describe the composition and transport function of the blood
- explain how hormones are chemical messengers inside your body
- explain how insulin controls the concentration of glucose in your blood
- explain the advantage of using human insulin produced by genetically modified bacteria.

A Steve Redgrave with his fifth Olympic gold medal.

Steve Redgrave won gold at five Olympic Games for rowing. In 1997 he was diagnosed with **diabetes**. He went on to win his fifth gold medal in the 2000 Olympic Games in Sydney.

Diabetes is a problem caused by a failure in the **endocrine system**. The endocrine system works alongside the nervous system. It uses chemical messages to control and coordinate the body. The chemical messages are called **hormones** and are produced by special organs called endocrine **glands**. Some of these hormones affect other organs, called **target organs**. Other hormones work on cells. Hormones are carried from the gland to their target in the blood plasma.

The electrical impulses produced by the nervous system work very quickly. Hormones work rather more slowly because they are carried by blood. Responses to hormones may last a few minutes or may last years.

Pituitary gland in the brain makes growth hormone, reproductive hormones and a hormone that controls water balance.

Thyroid gland makes **thyroxine**. Thyroxine controls the speed of our metabolism (chemical reactions).

Adrenal glands make **adrenaline**. This hormone helps your body cope with an emergency situation.

Pancreas makes **insulin** and **glucagon**. These control blood sugar levels.

Testes makes **testosterone**, a male sex hormone.

Ovaries make the female sex hormones **oestrogen** and **progesterone**.

B Where your glands are in your endocrine system.

The body contains about 5 litres of blood. Blood is a liquid with cells in it. There are two types of cell. Red blood cells carry oxygen around the body and white blood cells are part of the defence system against disease. The liquid part of blood is called plasma. Plasma is mostly water with substances dissolved in it. These substances include food molecules, chemical waste, hormones and blood proteins. The heart pumps blood around the body.

Have you ever wondered:

How do my hormones 'know' where to go?

Your body cells use a sugar called **glucose** to make energy. They need glucose in controlled amounts. Too much or too little glucose can be very dangerous.

Your **pancreas** makes a hormone called **insulin** if the level of blood glucose is too high. Insulin tells the liver to take the excess glucose out of the blood. The glucose is stored as glycogen, a type of sugar, in the liver. The glucose in the blood falls to its correct level.

When you exercise, your muscles use up a lot of glucose. If blood glucose falls, the pancreas makes another hormone, called glucagon. This tells the liver to convert some glycogen into glucose and put it back into the blood. Glucose in the blood rises to its correct level.

Sometimes the pancreas goes wrong and fails to produce enough insulin. The blood glucose level becomes dangerously high. This is called diabetes. It is treated by injecting insulin into the bloodstream.

Have you ever wondered:

Why people with diabetes inject themselves with products from bacteria?

Insulin needed by diabetics can be made from the pancreases of pigs or cattle. This kind of insulin does not suit some people and there are limited supplies. **Bacteria** have been **genetically modified** to make insulin. The human gene for insulin is put into the bacteria. The bacteria make human insulin, which is collected.

There are advantages in using genetically engineered insulin. The insulin is made in large quantities. It is unlikely to cause a reaction, unlike animal insulin, and no animals are involved so vegetarians and others can ethically use the insulin.

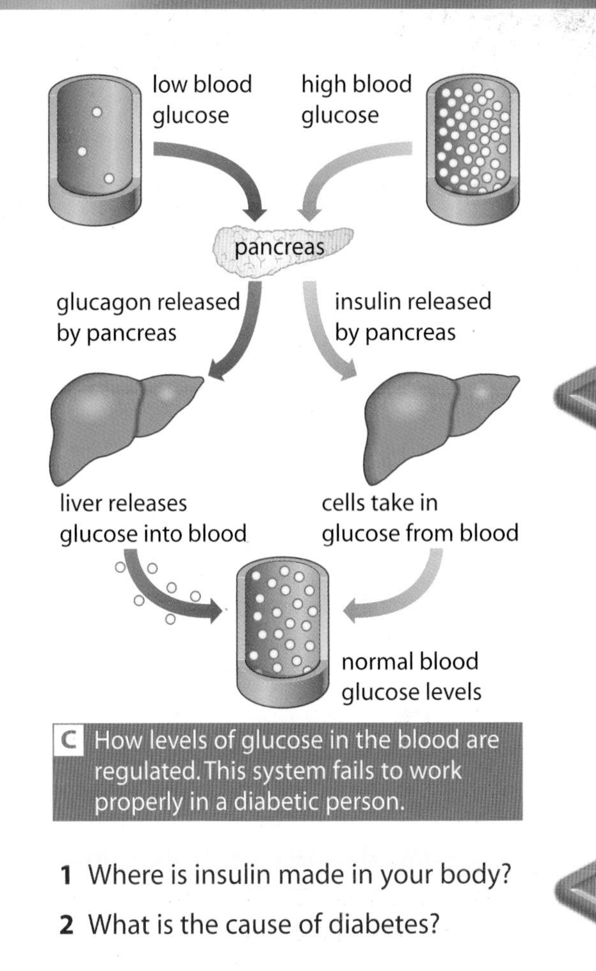

C How levels of glucose in the blood are regulated. This system fails to work properly in a diabetic person.

1 Where is insulin made in your body?

2 What is the cause of diabetes?

3 Why are two hormones needed to control glucose levels in the blood?

4 How is insulin transported around the body?

Higher Questions

B1b.3.9

Hormones and contraception

By the end of these two pages you should be able to:

- explain how manufactured sex hormones can be used for contraception
- H describe how reproduction is controlled by hormones
- explain how the menstrual cycle is controlled by hormones.

During the Olympic Games in 1988, Olympic sprinter Ben Johnson won gold in the 100 m. He was sent home a few days later in disgrace and stripped of his gold medal after testing positive for drugs.

He was using a drug which is a synthetic version of the male sex hormone **testosterone.** It works by increasing muscle growth. This is the same as happens when boys go through puberty. Some women athletes have been known to take the drug. There can be pretty unpleasant side effects. Some athletes must be desperate to win if they are prepared to risk this. They also risk being banned from their sport if they are caught by one of the random drug tests. An athlete never knows if they are going to be tested.

1 Why might increased muscle growth be useful to an athlete?

2 Why do most people think that using such drugs is cheating?

A Ben Johnson used banned substances to help him win.

Have you ever wondered:

How contraceptive pills work?

Sex hormones can be used to stop **pregnancy**. The pill contains sex hormones. It has to be taken by a woman over the course of a month and works by stopping the ovaries from releasing a new egg every month. This is called **contraception**.

3 How does the contraceptive pill work?

H

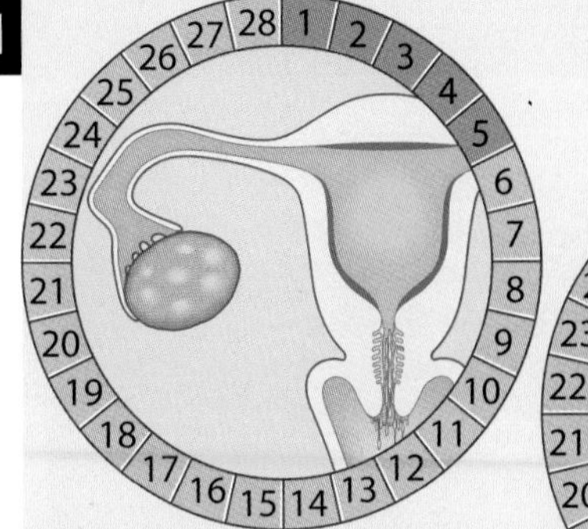

Menstruation takes place in the first 5 days of the menstrual cycle. The lining of the uterus breaks down.

The lining of the uterus starts to thicken and eggs develop in the ovaries.

Around the 14th day of the cycle an egg is released. It can be fertilised if sperm are present.

The egg will implant into the uterus lining if it is fertilised. If not it dies and the cycle starts again.

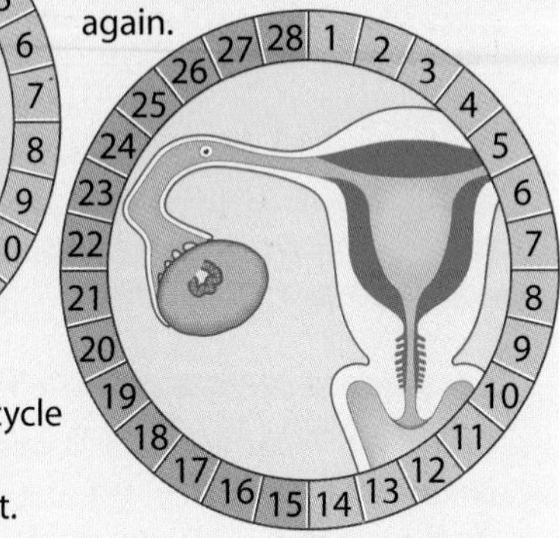

B The menstrual cycle.

There are two important female hormones. They are **oestrogen** and **progesterone**. Together they control the **menstrual cycle**.

The menstrual cycle starts with the breakdown of the lining of the uterus. Blood and cells pass out through the vagina. This is menstruation, but is usually called a period. As soon as the period has finished a new egg starts to develop in the woman's ovary. It grows inside a special fluid-filled ball called a follicle. About 14 days into the cycle, the follicle bursts and releases the egg into the oviduct. This is called ovulation.

If the egg is fertilised by a male's sperm, it implants into the uterus wall. It will develop into a baby. If the egg is not fertilised it dies and passes out through the vagina. The next period will then start, and the whole process begins again.

The process is controlled by hormones. As the egg develops in the follicle the ovary starts to make oestrogen. Oestrogen causes the lining of the uterus to start thickening and stops any more eggs from developing.

After ovulation, the follicle forms a yellow body. This starts to produce progesterone, which makes the uterus lining thicken even more.

So oestrogen and progesterone together make sure that the uterus is ready for a fertilised egg to implant. If a fertilised egg does implant, both hormones continue to be produced. This stops the next period from happening. If the egg is not fertilised, hormone production stops. The next period starts and the whole cycle starts again.

4 Why is it important that periods stop if a woman becomes pregnant?

5 What two things would happen if a woman's ovaries did not produce enough oestrogen?

P

Higher Questions

Hormones and fertility

By the end of these two pages you should be able to:

- explain how manufactured sex hormones can be used to treat infertility in women
- discuss the social and ethical implications of IVF treatment.

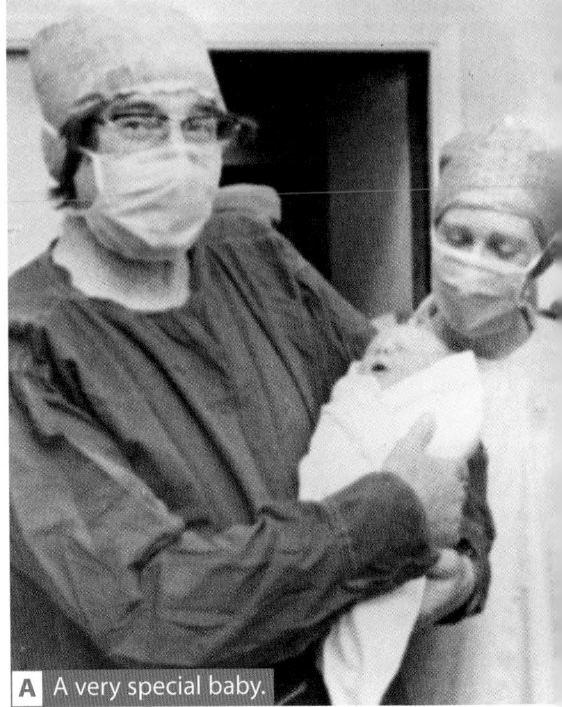

A A very special baby.

Louise Brown was born in July 1978. Louise was very special. She was the world's very first **IVF** baby. Most people call IVF babies test-tube babies, but in fact flat glass dishes are used. IVF stands for **in-vitro fertilisation**. In-vitro means in glass, which is where the test tube idea came from.

Some couples cannot have children. The woman might have blocked oviducts. The man might not make enough sperm. This is called **infertility**. IVF is used to treat infertility.

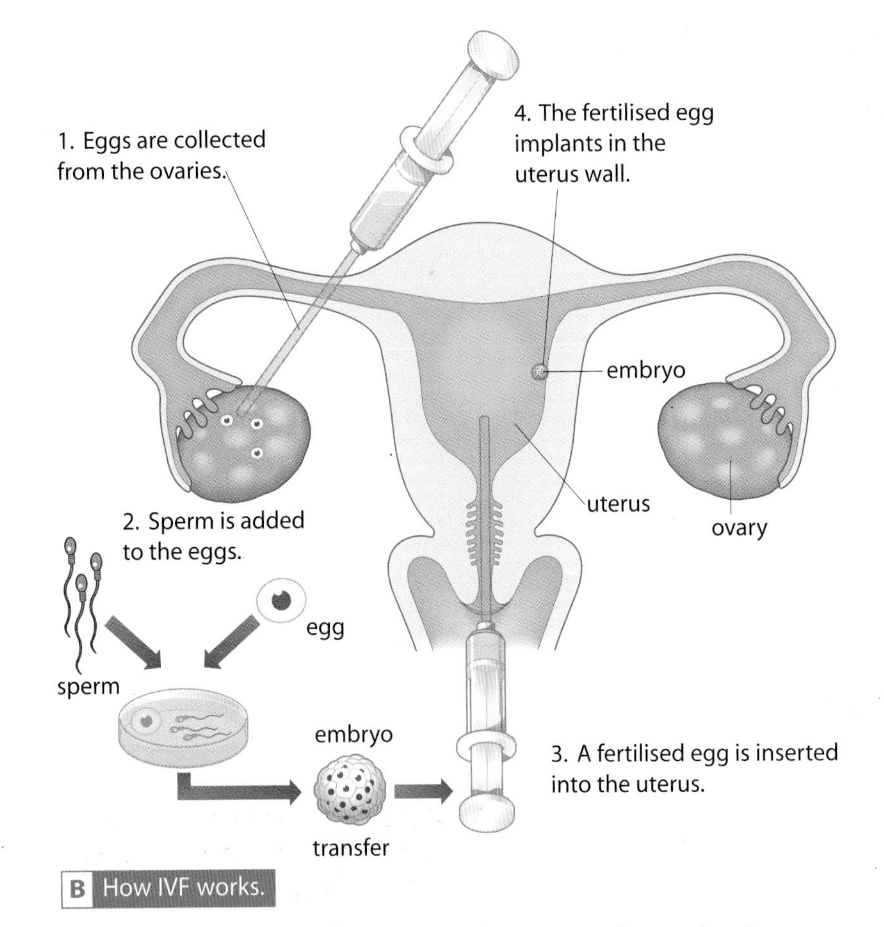

B How IVF works.

First the woman is injected with a hormone. This makes her produce lots of eggs at once. A tiny cut is made in her body wall. The doctor collects these eggs by sucking them out through a fine glass tube inserted into the cut. The eggs are kept alive in a special liquid. The liquid contains oxygen and food. Spare eggs may be frozen to use later if the first try fails.

Some of the eggs are mixed with semen from the father. They are kept in the solution for a few days. The doctor watches them using a microscope to see if any of the eggs are fertilised by the sperm and develop into embryos. One of the embryos is then placed into the woman's uterus. The embryo should develop into a baby just as if the egg had been fertilised inside the woman.

C The world's oldest mum.

In January 2005, Adriana Iliescu gave birth to a baby girl. Adriana was 66 years old. She had been having fertility treatment for months without success. Women who do not produce eggs can be given a fertility drug that contains a hormone that gets the ovaries to produce eggs. The hormones are manufactured sex hormones that mimic the way the woman's own hormones should be working. Often, more than one egg is produced. So twins, triplets, quadruplets or even more can be born!

This did not work for Adriana. She was producing eggs but they were not getting fertilised. Doctors used a technique called **artificial insemination**. Sperm from an anonymous donor was put inside her after **ovulation**. She became pregnant as a result. When her daughter is 14, Adriana will be 80.

IVF treatment has social and ethical implications. Many people think that 66 is far too old to become a mother. Some are concerned that it is not natural, others think that the mother will be unable to cope when the child is a teenager. Some people are concerned about the rights of a woman to be a mother against those of unused embryos. Other people ask whether enough is known about the long-term use of these techniques. Many women ask why it is acceptable for older men to become fathers, when for an older woman to become a mother is seen as unacceptable.

1 Why are the eggs kept in a solution containing oxygen and food during IVF treatment?

2 Why is the woman injected with a hormone during IVF treatment?

3 Why is fertility treatment a risky business for a woman who is not producing eggs?

4 What do you think about using fertility treatment for much older women?

Higher Questions

Questions

Multiple choice questions

1 A stimulus is
- **A** something that you detect changes with.
- **B** a change that can be detected.
- **C** your response to a change.
- **D** detected by a muscle.

2 Receptors send messages to the central nervous system along a
- **A** muscle fibre.
- **B** sensory neurone.
- **C** motor neurone.
- **D** reflex arc.

3 Your brain and spinal cord are parts of
- **A** a reflex arc.
- **B** your senses.
- **C** the central nervous system.
- **D** a synapse.

4 Messages carried by the nervous system are
- **A** hormones.
- **B** stimuli.
- **C** effectors.
- **D** electrical impulses.

5 Which of the following is missing from a reflex?
- **A** sensory neurone
- **B** stimulus
- **C** brain
- **D** effector

6 The brain
- **A** controls sense organs.
- **B** controls reflexes.
- **C** controls accommodation in the eye.
- **D** coordinates the actions of the body.

7 Effectors receive messages from the central nervous system along a
- **A** muscle fibre.
- **B** sensory neurone.
- **C** motor neurone.
- **D** reflex arc.

8 The amount of light entering your eye is controlled by your
- **A** iris.
- **B** retina.
- **C** cornea.
- **D** lens.

9 Focusing on near and far objects is a process called
- **A** ovulation.
- **B** fertilisation.
- **C** accommodation.
- **D** stimulation.

10 The iris reflex
- **A** protects the retina from too much light.
- **B** lets the retina detect coloured light.
- **C** changes the shape of the lens.
- **D** stops the eyeball from drying out.

11 Parkinson's disease is caused by
- **A** a sudden burst of excess electrical activity in the brain.
- **B** the brain being unable to coordinate muscle actions properly.
- **C** a shortage of insulin in the bloodstream.
- **D** a person losing consciousness and making jerky movements.

12 The level of sugar in your blood is controlled by
- **A** adrenaline.
- **B** insulin.
- **C** oestrogen.
- **D** progesterone.

13 Which hormone can be used as a contraceptive pill?
- **A** oestrogen.
- **B** adrenaline.
- **C** insulin.
- **D** testosterone.

14 Artificial insemination is
- **A** using oestrogen to make more eggs develop.
- **B** using sperm from a donor to fertilise an egg.
- **C** stopping the uterus wall from thickening.
- **D** a method of preventing pregnancy.

15 Fertility drugs work by
- **A** making menstruation start.
- **B** stopping the menstrual cycle.
- **C** making the ovaries develop eggs.
- **D** stopping the uterus wall from thickening.

Short-answer questions

Alex and Karim have found a website that has a reaction timer. After trying it out they look at the collected scores of all the people who have tried it. Alex says that you can improve reaction time by practising, but Karim says no, because it is a reflex action. You can help them by answering the questions.

What was your reaction time?

Reaction time/s	Number of people
0.05 – 0.09	419
0.10 – 0.14	610
0.15 – 0.19	1532
0.20 – 0.24	3420
0.25 – 0.29	3043
0.30 – 0.34	1721
0.35 – 0.39	738
0.40 – 0.44	334
0.45 – 0.49	150
0.50 – 0.54	112
0.55 – 0.59	71
≥0.60	188

A Reaction times for a lot of people.

D

P

1 Alex and Karim disagree about what the results show.
   - **a** What is the fastest range of reaction times?
   - **b** What is the most frequent range of reaction times?
   - **c** How many people have the most frequent reaction time?

2 Alex says that the reaction time involves impulses going to the brain and that we need to think about our response.
   - **a** What are the steps in the pathway of the impulses from stimulus to response?
     Stimulus → ______ → ______ → ______
   - **b** If reaction time were a reflex, which steps would be missing from the pathway?

3 Karim says he thinks that there are gaps between the neurones in the pathway.
   - **a** How do the impulses get across these gaps?
   - **b** Where do impulses go when they get to the end of a motor neurone?

4 Alex has discovered that scientists have measured the time between light being detected by your eye (seeing something) and your brain processing that information. The time is 0.12 s. Reflex actions are much faster than that. Use the table to explain whether reaction time is a reflex action or not.

Glossary

D

accommodation Changes in the thickness of the lens in the eye to focus on near and far objects.
artificial insemination Insertion of sperm from a donor into a woman to fertilise her egg.
bacteria A type of microorganism. Most are useful, some cause disease.
brain An organ that coordinates the actions of your body.
brain tumour A mass of unnecessary cells growing inside the skull.
central nervous system (CNS) The brain and spinal cord.
cones Light-receptor cells in the retina of the eye that detect colour.
contraception A method of preventing pregnancy.
contract The shortening of muscle cells to make the muscle shorter.
cornea The curved outer surface of the eye that focuses most of the light on the retina.
diabetes A disease where your body is unable to control the level of sugar in the blood.
effectors Muscles that contract in response to impulses from the nervous system.
electrical impulse A signal carried by the nerves in your body.
endocrine system The system of glands in the body that make hormones.
epilepsy A sudden burst of excess electrical activity in the brain causing a seizure.
genetically modified A cell that has had specific sections of DNA added to change what it produces.
gland An organ that produces a hormone.
glucose A simple sugar, a type of carbohydrate.
grand mal A form of epilepsy where the person loses consciousness and makes jerky movements.
hormone A chemical message produced by a gland to coordinate the body.
impulse An electrical signal carrying information through the nervous system.
infertility Where a couple are unable to have children. May be caused by blocked oviducts or not making enough sperms.
insulin A hormone made in the pancreas to control the level of sugar in the blood.
involuntary response *see* reflex
in-vitro fertilisation (IVF) The fertilisation of a human egg outside the body.
iris The coloured ring of muscle in the eye that controls the size of the pupil.
iris reflex A reflex action that controls the amount of light entering the eye.
menstrual cycle The cycle of preparing the uterus to receive a fertilised egg after ovulation.
motor neurone A neurone that carries impulses to an effector.
muscle Tissue made from cells which can contract, allowing movements to take place.
nervous system The nerves, brain and spinal cord inside the body.
neurone A single cell that carries impulses.
oestrogen A hormone that makes the lining of the uterus thicken and stops eggs from developing.
ovulation The release of an egg from the ovary.
pain sensor A receptor cell in the skin that responds to pain.
pancreas An organ in your abdomen that produces insulin.
Parkinson's disease A disease where the brain is unable to coordinate muscle actions properly.
pathway The route of an impulse between a receptor and an effector.
pregnancy The development of an embryo from fertilisation until birth.
progesterone A hormone that makes the uterus lining thicken after ovulation.
pupil The hole in the centre of the iris that lets light through to the retina.
reaction time The time between a sense organ detecting a stimulus and the muscles reacting.
receptor A special cell that detects a stimulus, like light, sound and heat.
relay neurone A neurone found in the central nervous system connecting a sensory neurone and a motor neurone as part of a reflex.
reflex An automatic response to a stimulus.
retina The inner lining of an eyeball containing light-detecting receptor cells called rods and cones.
rods Light-receptor cells in the retina of the eye that detect light intensity.
sense organ An organ that contains receptors that detect stimuli.
sensory neurone A neurone carrying impulses between a receptor cell and the central nervous system.
spinal cord The bundle of neurones inside the backbone.
stimulus Something you react to, like a sound, a hot object or something you see (plural: stimuli).
stroke A blood clot or bleeding in the brain, which causes brain cells to die.
synapse The gap between two neurones.
target organ The organ a particular hormone works on.
testosterone The male sex hormone.
tumour Cells growing to form an abnormal tissue.
voluntary response A response to a stimulus that you have to think about and can control.

Use, misuse and abuse

A Paul Sommerfeld is chairperson of TB Alert, Britain's national tuberculosis charity.

B TB is a growing problem in parts of London.

Paul Sommerfeld is chairperson of TB Alert. This is a charity that raises awareness of tuberculosis (TB) as a growing problem in Britain and across the world. Tuberculosis is not under control in many parts of the world. More than 5000 people a day die from the disease.

TB is becoming a major problem in London. It is a disease of poverty. People who are poor, live in overcrowded conditions and do not eat the right food are most at risk from TB.

To understand why, you will need to know more about microorganisms. You will also need to know more about drugs, and how some people misuse them.

In this topic you will learn that:

- the human body has three lines of defence against invading microorganisms
- immunisation and antibiotics are used against diseases caused by microorganisms
- the use and misuse of substances can affect the normal functioning of the body systems, affecting mental and physical health
- there are socio-economic reasons that contribute to ill health, and ethical considerations for the development of treatments.

Alex has made a grid. Add four more things you think you know, and four you want to know.

Five things *I think I know* about microorganisms	Five things *I want to know* about microorganisms
Bacteria and viruses are microorganisms	

TB in London

By the end of these two pages you should be able to:

- explain what causes tuberculosis
- explain how tuberculosis is spread
- H interpret data on the number of cases of tuberculosis over a period of time.

A People who live in overcrowded conditions may be at risk from TB.

Have you ever wondered:

Why is TB in the news again?

In 2004, newspapers reported that a young girl had been diagnosed with **tuberculosis** (TB). Her infant school in west London tested all of its pupils for TB.

Why is TB a problem again? London is a top destination for tourists. It is also one of the most important centres for trade. Millions of people head for London every year. If they are from or have visited a country where TB is a problem, they may bring it with them.

TB is caused by a bacterium. The **disease** used to be called consumption. The bacterium is called *Mycobacterium tuberculosis*. Mycobacteria are **bacteria** that are spore-like. This means they can survive in the body for many years in an inactive state, hidden away from your body's defences.

TB is spread through the air. When an infected person coughs or sneezes, tiny droplets spread into the air. These will contain TB bacteria. Other people may breathe them in. You need to breathe in only about 10 bacterial cells to be infected. Direct sunlight will quickly kill the bacteria.

B As chairperson of TB Alert, Paul spreads awareness that TB is caused by a bacterium that is breathed in.

1 What causes tuberculosis?

2 How does someone catch tuberculosis?

Most people who get TB do so because they spend a lot of time with an infected person. You will not catch TB on a bus or train.

TB bacteria like lots of oxygen. They prefer to grow in places like the lungs, but can be carried in the blood to other organs. TB often spreads to the lymph nodes in the neck and under the arms where it causes swollen glands. It can also affect bones, joints and kidneys.

The symptoms of the disease are

- coughing
- shortness of breath
- loss of appetite and weight loss
- extreme tiredness
- fever and sweating.

If left untreated the disease may kill you.

H

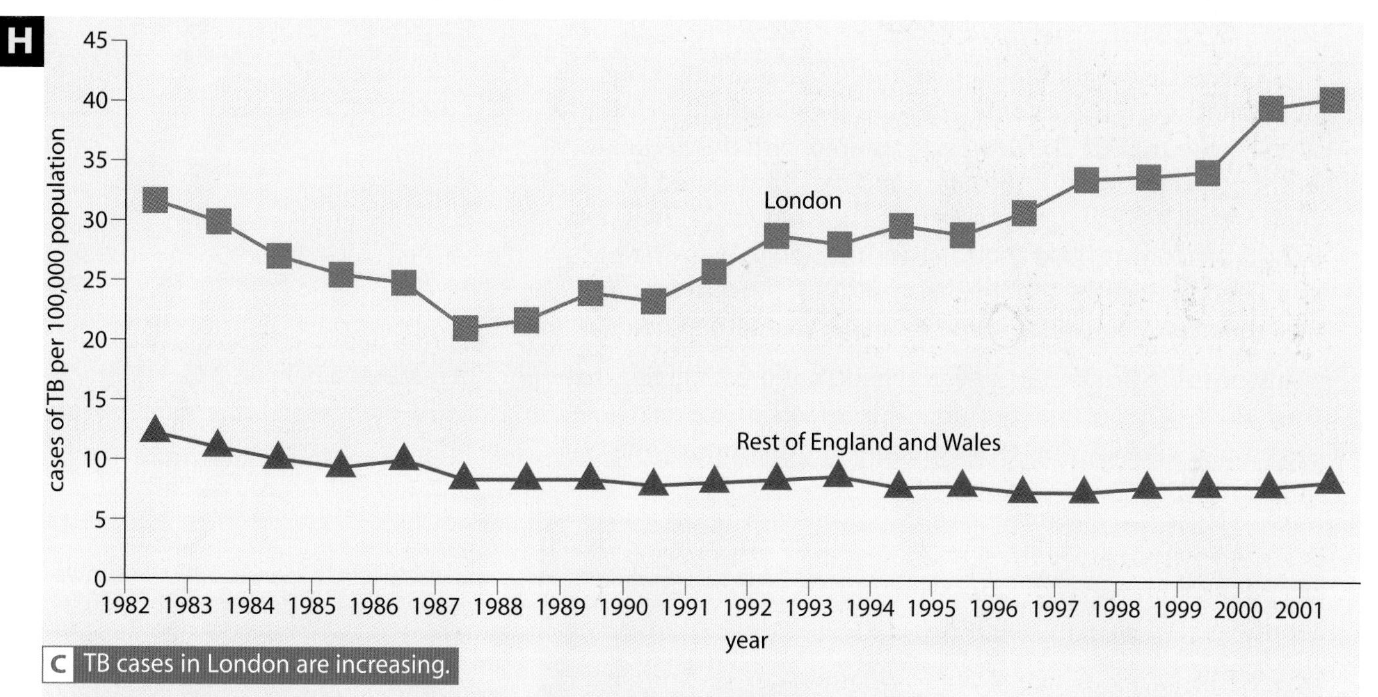

C TB cases in London are increasing.

TB had been in decline in London for years but from 1988 to 2005 the number of cases doubled. Forty per cent of all cases in the UK were in London.

AIDS is caused by a virus called HIV. HIV gradually destroys the body's defence system against **microorganisms**. For an AIDS sufferer, TB can be deadly. HIV and TB work together, so that the body has no defence against the TB bacterium. But with the right treatment the TB can be cured.

Worldwide, up to 50% of AIDS sufferers will also get TB, and nearly half of them will die from TB. Fourteen million people worldwide have both diseases and 5000 people a day die from TB.

3 What is AIDS?

4 Why is TB such a problem for someone who has AIDS?

5 Why is TB in the news again?

6 How can TB bacteria survive in the body for years undetected?

P

Summary Exercise

Higher Questions

Controlling tuberculosis

By the end of these two pages you should:

- know about how much money, time and effort goes into producing a new drug
- H be able to describe the prevention and control of TB
- know about the emergence of drug-resistant TB.

P

?

Have you ever wondered:

Why is it so expensive to produce a new drug?

Drugs are substances which alter the way the body works. They can be used to treat diseases like TB. A new drug can take 10–15 years and cost over £500,000,000 to develop.

First, chemicals are identified that might make useful drugs. It then takes several years of testing and development before the drug can be trialled. The drug is registered with the Regulatory Authority, who checks it for safety and may then agree to clinical trials. Human volunteers take the drug to test its safety, to find the correct dose and to check for side effects. Trials are also needed to prove that the drug works. If everything goes well the new drug will be made available to doctors.

P

Many possible drugs start development, but most fail because they are toxic, have unacceptable side effects or are too expensive. It is much better for people to use drugs properly and complete the treatment.

1 Why does it cost so much to develop a new drug?

2 Why are clinical trials important when testing a new drug?

H

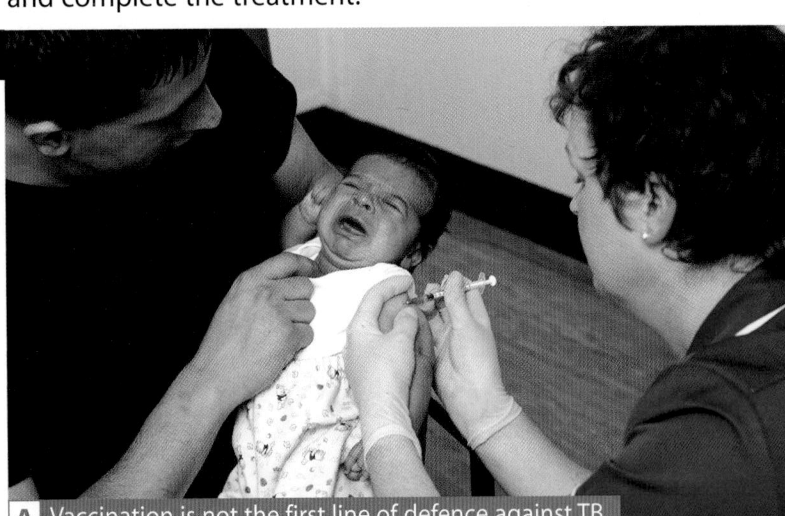

A Vaccination is not the first line of defence against TB.

V

You can be protected against TB. There is a **vaccination** called the BCG. This only works for about 75% of people in the UK. In tropical countries it is not very effective because many people are already infected. The best way to fight TB worldwide is to try and treat and cure those with the disease.

Antibiotics are drugs that kill bacteria. There are two problems with antibiotics:

- they kill useful bacteria as well as the harmful ones
- people stop taking antibiotics because they feel better.

By not completing the treatment, patients risk not killing all the harmful bacteria. Some bacteria are harder to kill than others. If the only bacteria to survive are those that are hard to kill then they will reproduce and make bacteria which are even harder to kill. Eventually they may develop **resistance** to the drug, which then no longer works. Then a new drug is needed to kill the resistant bacteria.

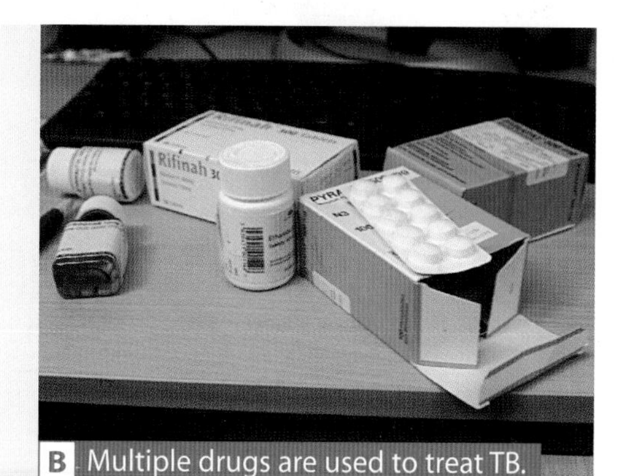

B Multiple drugs are used to treat TB.

Have you ever wondered:

Why won't your doctor give you antibiotics for a cold?

Multi-antibiotics are used to treat TB patients. This means that several different antibiotics are used at the same time. Treatment lasts for 6 months. This makes sure that all the TB bacteria are killed.

To make sure that patients complete their treatment a new system is used in many countries. This is called DOTS. This stands for Directly Observed Treatment, Short Course. A nurse watches the patient swallow every tablet to make sure that they complete their treatment.

C Paul knows that drug resistance is a real problem in the treatment of TB.

Drug-resistant TB has appeared in some countries. This is far more difficult to treat and needs new, more powerful antibiotics.

3 What are resistant bacteria?
4 Why are resistant bacteria becoming a serious problem?
5 What is an antibiotic?
6 What advice would you give to a friend who is on antibiotics and what reasons would you give?
7 Explain one of the problems of using antibiotics.

Higher Questions

Microorganisms and disease

By the end of these two pages you should:

- know that a pathogen is a disease-causing organism
- know about the different ways that microbes can pass from one person to another.

Have you ever wondered:

What is the difference between an infection and a disease?

A Maria and her family.

Maria is 12 years old. She lives in Peru. She shares her bedroom with six brothers and sisters and her grandma. Maria has TB.

The World Bank estimates that there are about 1.2 billion people like Maria living on less than $1 per day. That's less than 60 p. This is called poverty. You do not get enough to eat, and the food is of poor quality.

Poor nutrition and an inadequate diet weaken your **immune system**. This means you have a greater chance of **infection** and developing active TB. Overcrowded conditions make it easy for TB to be passed between people.

TB is caused by bacteria. Bacteria are microscopic **organisms**, sometimes called **microbes** or microorganisms. Microorganisms like the TB bacterium are called **pathogens**. This means disease-causing. Not all microorganisms are pathogens. The vast majority do not cause disease, and some are even helpful.

Have you ever wondered:

Are there more 'good' microorganisms than disease-causing ones?

1 What are the reasons that the rest of Maria's family are at risk from TB?

2 What is a pathogen?

Microorganisms are transmitted from one person to another in lots of ways. These methods of **transmission** fall into two groups: direct contact and indirect contact.

Direct contact means that microorganisms are passed by actual body contact. Direct contact can be by two different methods: horizontal and vertical transmission.

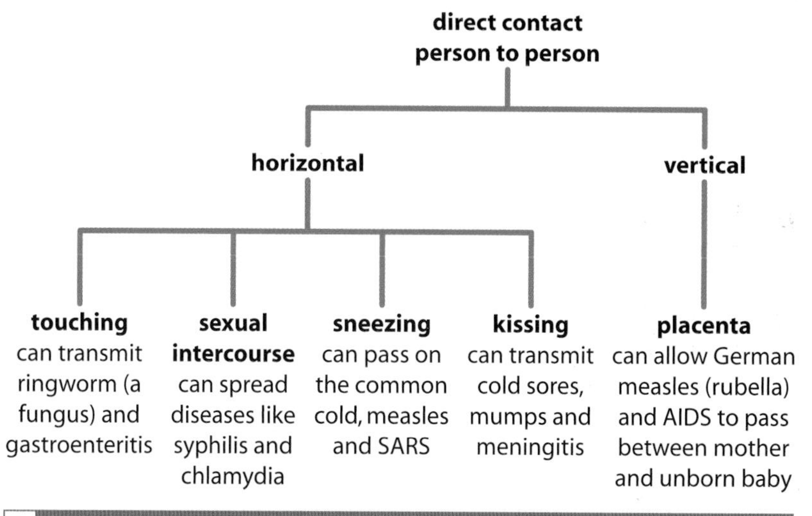

B How microorganisms can be transmitted from one person to another.

Horizontal transmission means microbes passed on by touching, sexual intercourse, kissing and sneezing. Vertical transmission is when microorganisms pass through the **placenta** from mother to unborn baby.

Indirect contact means that other humans are not involved. There are two main ways indirect contact can spread disease: vehicle-borne and vector-borne. Borne means 'carried by' so water-borne means carried by water.

Vehicle-borne means that the microorganisms are carried in a physical way. For example:

- TB and influenza are both carried in the air: you breathe them in.
- Your food can carry microorganisms. Gastroenteritis and *Salmonella* can be transmitted in food. This is why proper storage and cooking are so important.
- Some microorganisms are carried in water. Cholera is spread when human faeces get into drinking water.
- Objects like needles, thorns and splinters can penetrate the skin and then any disease-causing microorganisms can enter your body. Tetanus and HIV can enter the body in this way.

Vector-borne means that the microorganisms are carried by another animal, often an insect. This might be microorganisms transferred to your food on the feet of flies. Microorganisms can be injected into your blood in the saliva of mosquitoes.

3 Why is it important that food is cooked for the correct amount of time?

4 How does vertical transmission of disease work?

5 Why is it important to wash your hands with soap and hot water after going to the toilet?

6 Make a table with two columns. Label them *Direct contact* and *Indirect contact*. Complete the table by putting all the ways that microorganisms can be transmitted into the correct column.

Higher Questions

P

The first line of defence

By the end of these two pages you should be able to:

- explain that the body has physical barriers as the first line of defence against microbes
- describe how each of the physical barriers works
- describe how chemical barriers are also used in this first line of defence.

A Iwan has about 2 m^2 of skin.

An average adult has 2 m^2 of skin: about the same area as one side of your bedroom door. This amount of skin contains about 300 million cells. One of the most important jobs of the skin is to keep microorganisms out.

Your body has two kinds of **barrier** against microorganisms. There are physical barriers like your skin and tiny hairs in your breathing system. You also have **chemical barriers** like the acid in your stomach and an antiseptic in your tears.

Your skin is an important physical barrier. The cells of the outer layer contain keratin, a protein that makes the cells really tough.

Your skin is covered in hairs. Each hair has an oil gland which produces an oily antiseptic liquid called sebum. Sebum kills many microorganisms and the oil keeps the skin supple and waterproof. It is a chemical barrier.

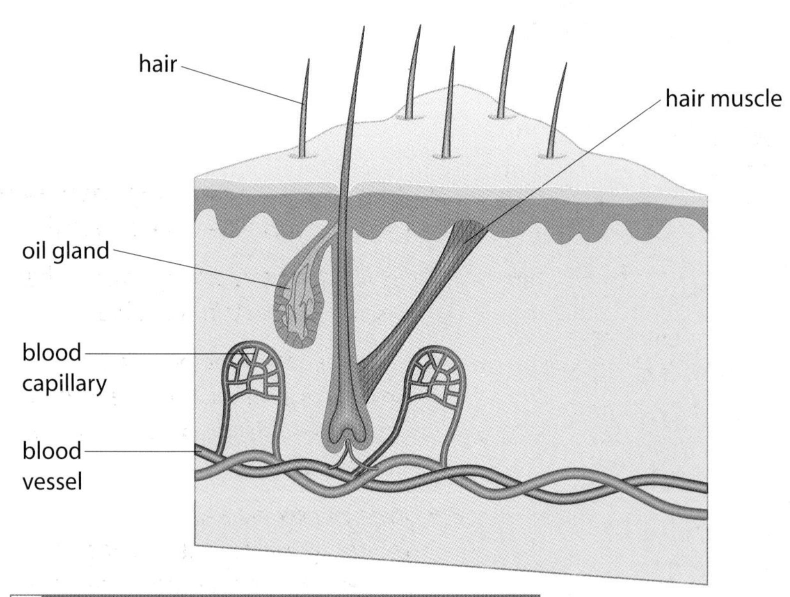

B Your skin is a barrier against microorganisms.

You have useful microorganisms that live on your skin. They are also found in your alimentary canal, where your food is digested. They compete with pathogens, using up nutrients and space. This helps prevent pathogens from multiplying.

1 Why is your skin the most important first line of defence against microorganisms?

2 How is your skin maintained so it continues to work as a barrier?

You also have membranes lining the inner surface of your body. These form another physical barrier inside your mouth, nose, alimentary canal and breathing system. The membranes contain special cells that make mucus. Mucus is a thick, sticky liquid that traps microorganisms.

In the nose, nasal hairs form a physical barrier to keep dust and larger microorganisms from entering your **gaseous exchange** tract. The lining of the breathing system has some other special cells. These cells have tiny hair-like projections called **cilia**. These beat all the time, moving the mucus, dirt and microorganisms up your windpipe to your throat. You can then either swallow the mucus or cough it out.

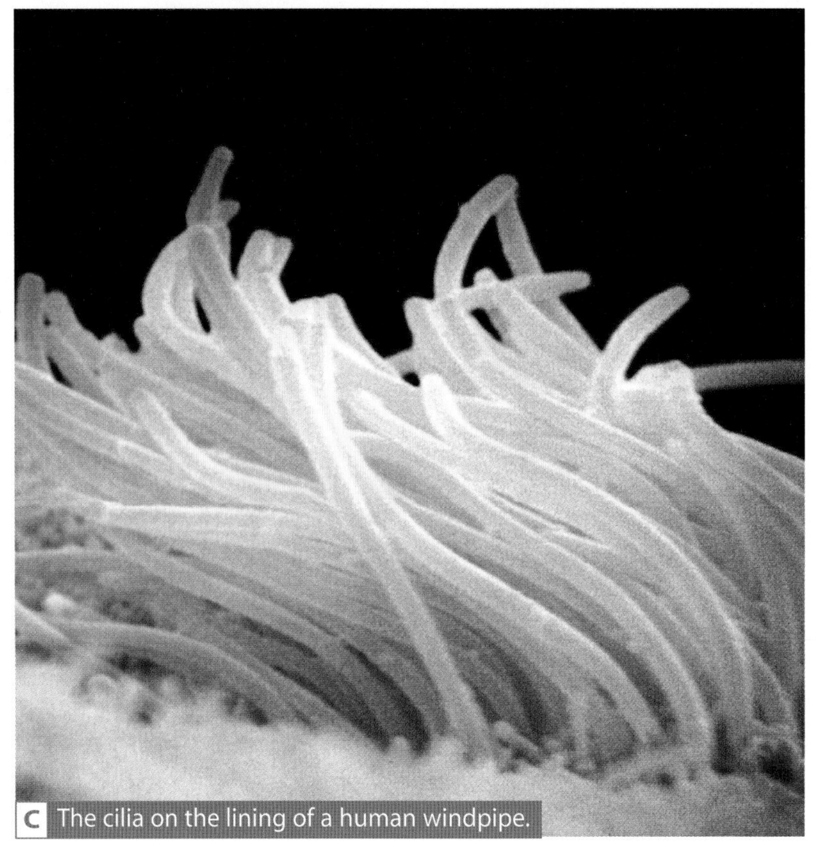

C The cilia on the lining of a human windpipe.

The lining of your stomach has different special cells. These make hydrochloric acid and are part of your body's chemical barrier. The acid kills any microorganisms in your food or in the mucus that you swallow.

When you cry, your tears contain an enzyme called **lysozyme**. This is another chemical barrier and it destroys any microorganisms on the surface of your eyes. When you blink it helps to spread the liquid all over the surface of your eye. This is a chemical barrier against microorganisms. Lysozyme is also found in your saliva and mucus.

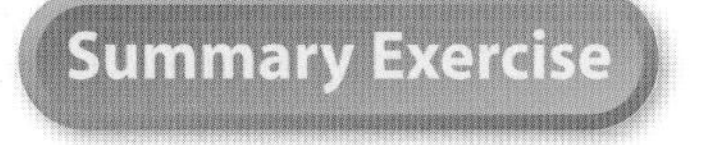

3 Why is crying a very important part of your body's defence system?

4 How are any microorganisms that get into your stomach killed?

5 How do coughing and sneezing protect against microorganisms? How does this work as a defence mechanism?

6 Imagine you are are a microorganism trying to infect someone. Describe how the skin, nasal hairs, mucus, cilia and tears can stop you.

B1b.4.5

The second line of defence

By the end of these two pages you should be able to:

- explain why infection can lead to inflammation
- explain how white cells defend against microbes
- explain how the TB bacterium overcomes this line of defence.

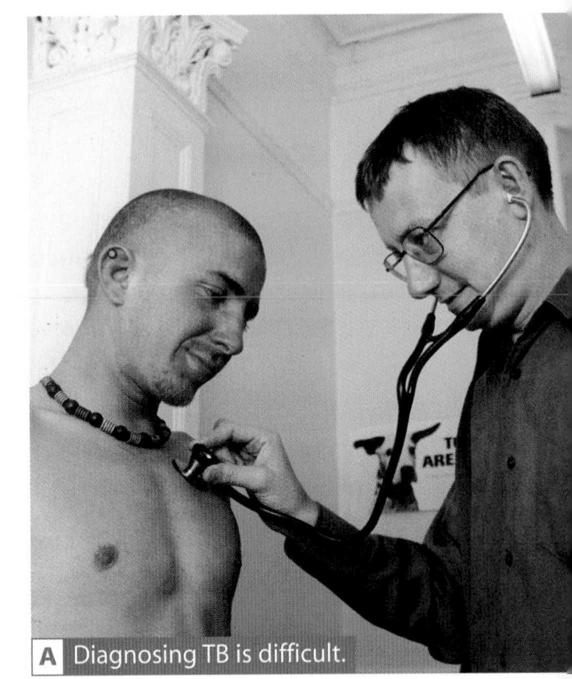

A Diagnosing TB is difficult.

Some people can have TB bacteria inside them but not be ill. They can't spread it to others and have no symptoms but they can develop the disease in later life, so they need treatment. TB bacteria can hide inside the body's defence systems.

If microorganisms get through your physical and chemical barriers then your body has a second line of defence. This is found in the blood **circulatory system** where there are **white blood cells**, carried around the bloodstream.

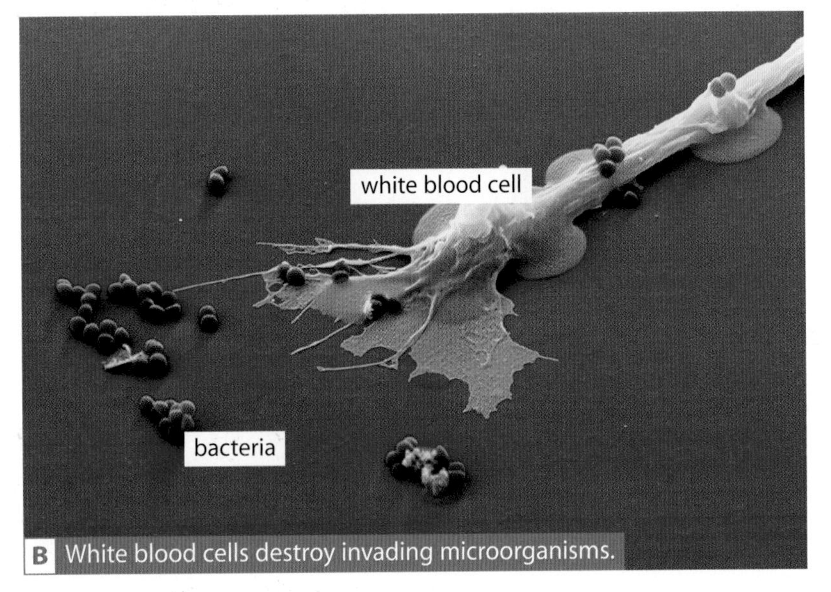

B White blood cells destroy invading microorganisms.

When microorganisms invade the body the white blood cells move towards them. The white cells then swallow up (ingest) the invading microorganisms and digest them using enzymes.

This is called the **non-specific immune system** because the response is the same for all pathogens.

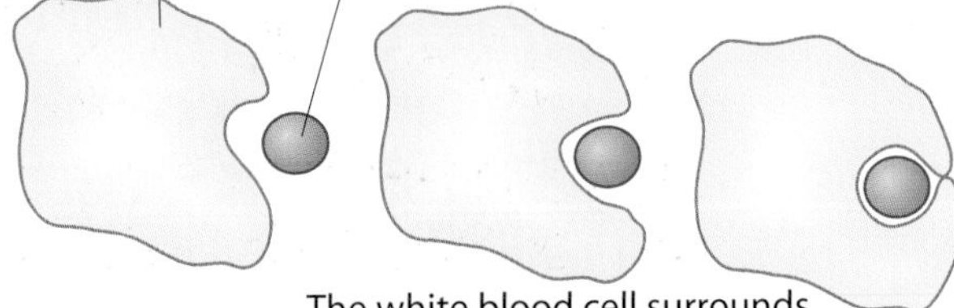

C White blood cells ingest bacteria.

TB bacteria can be breathed in and can enter the body through the lungs. They are ingested by the white blood cells but are resistant to the enzymes that should destroy them. Once inside the white blood cells the bacteria become dormant, or inactive. They can stay like this for years. This is known as latent TB infection. The microorganisms are inside the body but there are no symptoms and the person is not infectious.

The bacteria can be activated years later either by an infection with HIV, poor diet, poor living conditions or a weakened immune system. The infection will then develop into full TB.

1 Why is TB infection potentially dangerous?

2 Why is the second line of defence called the non-specific immune system?

As well as being breathed in like TB, pathogens can also enter the body if the skin is damaged. The damaged skin cells release chemicals. This makes blood vessels around the area swell. They let more blood flow through. This extra blood makes the skin in the area feel warm, and look red. This is all part of the healing process.

The tissue around the damage swells because fluid leaks from the blood capillaries. The swelling and the effect of the chemicals released causes pain in the damaged area. White blood cells move into the area. They ingest bacteria and dead cells.

The redness, warmth, swelling and pain together are called **inflammation**. The response to an infection caused by damage to the skin is called the inflammatory response. The damage is quickly cleared up so that new tissue can grow to repair the break in the skin.

Sometimes a yellowish fluid forms, called pus. It includes white blood cells, living and dead bacteria, dead tissue and blood plasma. If this happens under the skin the pus becomes trapped and builds up to form an abscess.

3 What happens to bacteria when they are ingested by white blood cells?

4 Why do white blood cells move to a cut in your skin?

5 How does inflammation help the healing process?

6 How can inactive TB bacteria inside the body be activated?

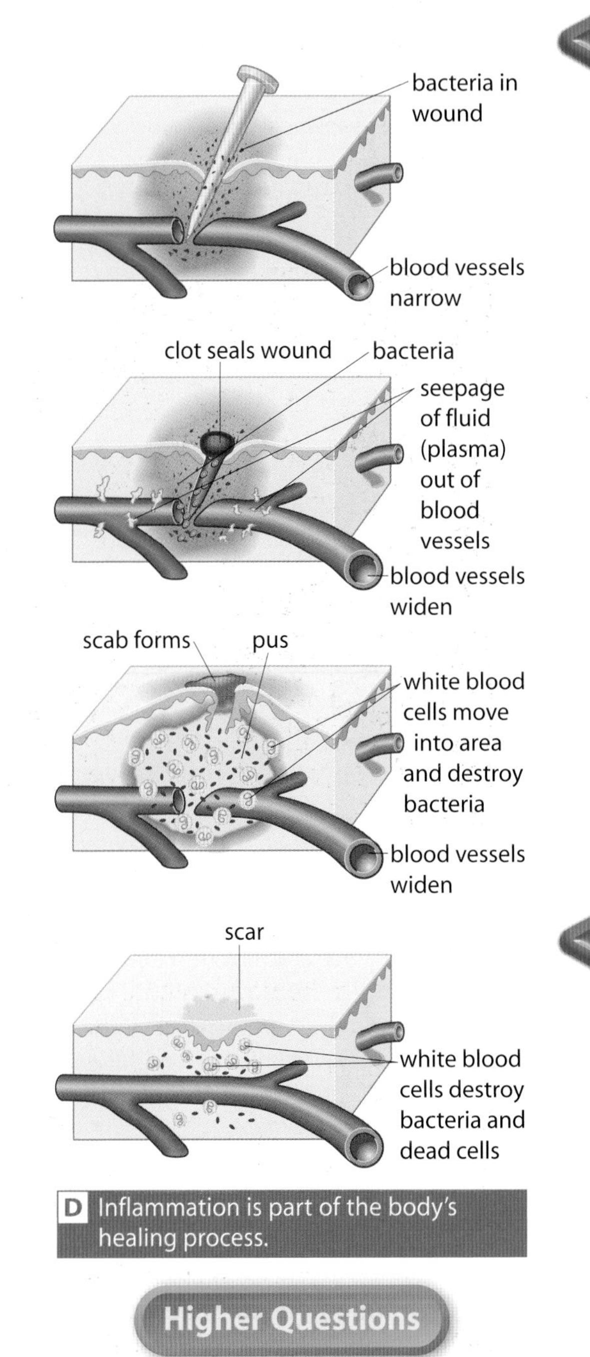

D Inflammation is part of the body's healing process.

Summary Exercise

Higher Questions

The third line of defence

By the end of these two pages you should be able to:

- explain how the immune system works
- understand the difference between antibodies and antigens
- explain how vaccination gives you immunity.

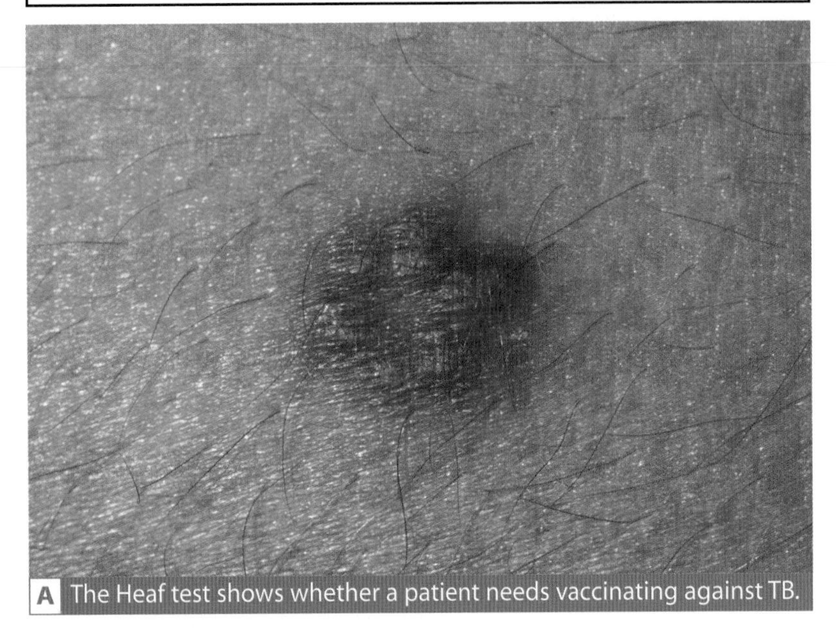

A The Heaf test shows whether a patient needs vaccinating against TB.

A BCG vaccination can be used against TB. A small amount of TB protein is injected onto the forearm. A few days later a nurse examines the arm. A raised red area shows the test is 'positive'. The red marks show that there is already immunity to TB and the BCG vaccination is not given. A 'negative' response with no raised red area means that a vaccination is needed.

Until 2005 the BCG vaccination was given to all school pupils 10–14. It was really only effective on children, lasted about 10 years and didn't work on about 25% of children.

Then the Government announced that the vaccination would no longer be used for all children. In future only children whose parents were born abroad, in high-risk countries, would have the BCG.

To understand how vaccinations work you need to know about the third line of defence against infection. This is your immune system. White blood cells in the bloodstream are the main part of the immune system. White blood cells produce molecules called **antibodies**. Antibodies help you to fight pathogens and are produced in millions of different forms.

Your immune system is activated when a pathogen gets through your first two lines of defence.

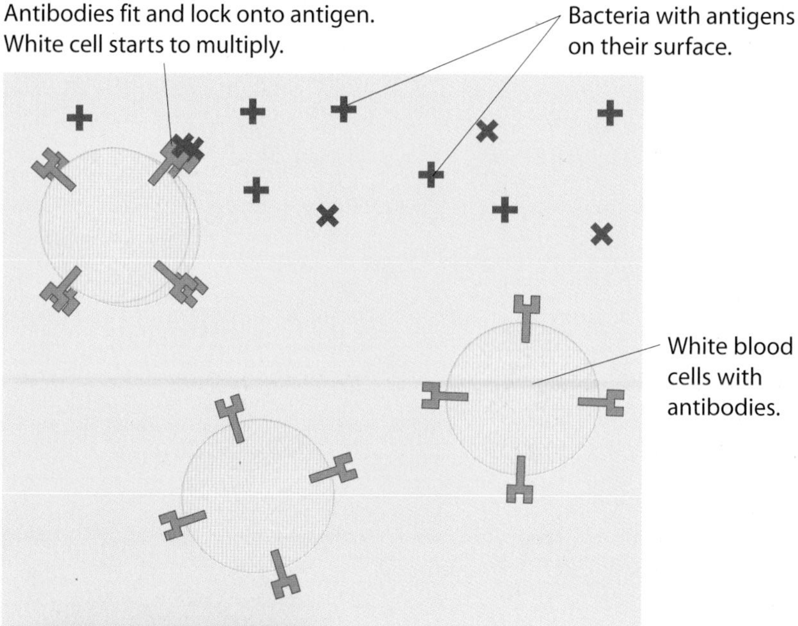

B Antigens and antibodies.

When they find **antigens** of **foreign bodies**, the white blood cells try different antibodies until they find one that exactly matches the antigen. The specific antibody locks onto the antigen. Then other white blood cells destroy the pathogen. Every antigen needs its own special antibody. There will be a specific antibody for every antigen – millions of different ones. So the third line of defence is called the **specific immune system**.

1 Why does a positive test for TB mean that you do not need a vaccination?

2 What is the body's third line of defence called?

Vaccination involves injecting a small, harmless amount of infection into you. This causes your immune system to create antibodies, to be ready in case the real pathogen invades. If enough people in a community are immunised against certain diseases, then it is more difficult for the disease to get passed between those who aren't immunised. This is known as herd immunity. This only works for diseases that pass from one person to another. If nearly all people are immunised the disease cannot spread and may disappear.

In 1998 there was a scare about the safety of the MMR vaccination, and lots of parents refused to let their children have it. Since then cases of mumps, measles and rubella have increased enormously. It will take a long time to restore people's trust in the vaccination.

Summary Exercise

3 What is meant by a foreign body?

4 Why is it important that as many people as possible are immunised against diseases like measles?

5 What is the difference between an antigen and an antibody?

6 How would you explain what herd immunity was to your younger brother or sister?

Higher Questions

Types of drugs

By the end of these two pages you should know:

- how drugs affect the body
- the differences between stimulants, sedatives and painkillers
- how drugs can affect reaction time.

Have you ever wondered:

How do different drugs affect people differently?

A survey of sportsmen and women showed that 36% of athletes use **caffeine** supplements. This rises to 50% in rugby union players.

Caffeine is a drug found in tea, coffee, chocolate and some cola drinks. A drug is a substance which alters the way that the body works. Caffeine improves alertness and **reaction time**. Some energy drinks contain caffeine.

A Energy and caffeine-based drinks are useful for an athlete.

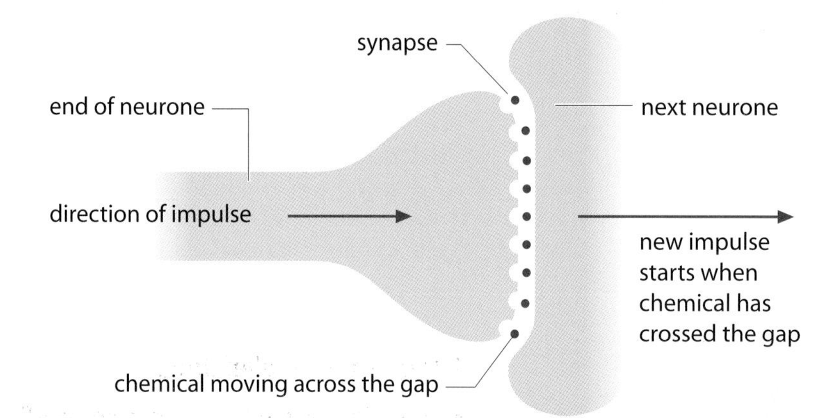

A chemical is produced when an impulse reaches the end of a neurone. The chemical diffuses across the gap to the next neurone. When it arrives it starts a new impulse in the next neurone.

B Impulses are carried from one neurone to the next across a gap called a synapse.

In the nervous system, information is carried by cells called **neurones**. There is a tiny gap between neurones called a synapse. Drugs like caffeine are called **stimulants**. They work by increasing the rate at which impulses cross synapses. So the speed of the nervous system is increased. This may lead to increased heart rate, faster breathing and muscles contracting more easily.

C Drugs can give some athletes an advantage.

Cocaine and amphetamines are powerful stimulants. Their use can lead to **addiction**. Another problem is that when the effect wears off you may feel 'down' and tired. Some people can get addicted to coffee and get headaches when they don't drink it.

Drugs may be used to help control epilepsy. One group of these drugs are called **barbiturates**. They work by slowing down the nervous system, helping to stop the bursts of electrical activity that cause epilepsy.

Alcohol and barbiturates are examples of **sedatives**. They slow down the nervous system and may increase reaction time. This is why people under the effect of alcohol may stagger, or slur their speech. Sedatives may be used to reduce stress or to help people sleep.

Use of sedatives may lead to dependence – you can become addicted to them. Doctors are very careful when treating epilepsy to make sure that this does not happen. Sedatives can also cause tolerance. Tolerance means that your body gets used to the drug and you need more to have the same effect.

Pain-relief drugs include aspirin, **paracetamol**, ibuprofen, heroin and morphine. Many pain-relief drugs block impulses travelling to the part of the brain that deals with pain. Some, including aspirin, block impulses mainly where the pain occurs.

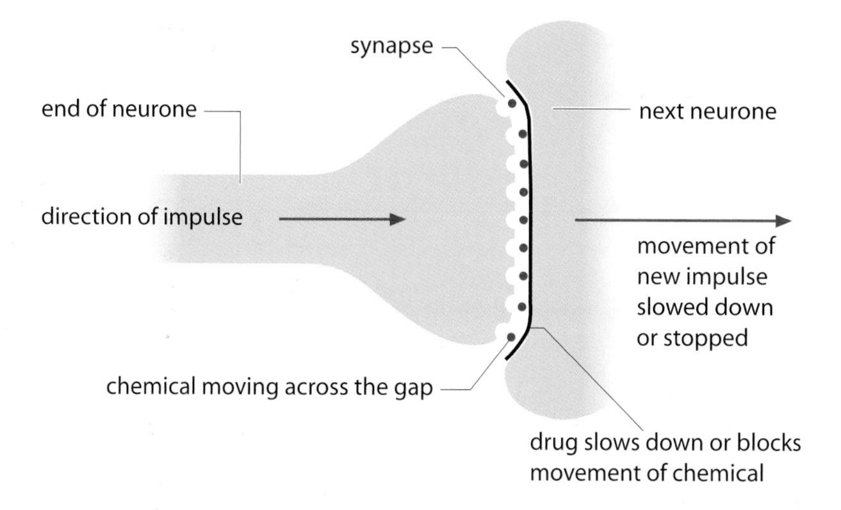

D Impulses can be blocked by some drugs to reduce pain.

Stimulants, sedatives and painkillers all work on parts of the nervous system. They affect the speed of impulses travelling along neurones. They can change how the body and mind work. Some people abuse these drugs and this will have serious effects on their physical or mental health.

1 A well-known cola drink originally contained cocaine. Why do you think this was removed from the recipe in 1906?

2 What useful effects does caffeine have on your body if you are an athlete?

3 What does a sedative do to the body?

4 Explain two disadvantages of sedatives.

5 Name a pain-relief drug.

6 How does aspirin provide pain-relief?

7 What do sedatives do to the nervous system?

8 Why are doctors very careful when using barbiturates to treat epilepsy?

Summary Exercise

Higher Questions

Pain-relief

By the end of these two pages you should be able to:

- discuss how drugs can be used to relieve pain
- discuss the dangers of overdose when using paracetamol
- discuss the dangers of addiction to pain-relief drugs.

Have you ever wondered:

Why are some drugs considered good for your body and others bad?

Paracetamol is the most widely used medicine for pain and fever relief. It is safe at the correct dose and has almost no side effects. Paracetamol blocks some of the impulses that carry information in the brain. These impulses are the ones to do with pain. But taking more than the stated dose can severely damage the liver and stop it working. This can cause death unless a liver transplant is available.

Taking more than the stated amount is called an **overdose**. An overdose of paracetamol must be treated within 12 hours. Restricting sales of paracetamol and clear labelling of medicines containing paracetamol has reduced the number of overdoses.

There are many types and strengths of painkiller. Paracetamol is good for headaches and fever. For some people a much stronger painkiller is needed.

1 Why is paracetamol such a useful painkiller?

2 Why might some people need a more powerful painkiller?

3 What happens to the body if too much paracetamol is taken?

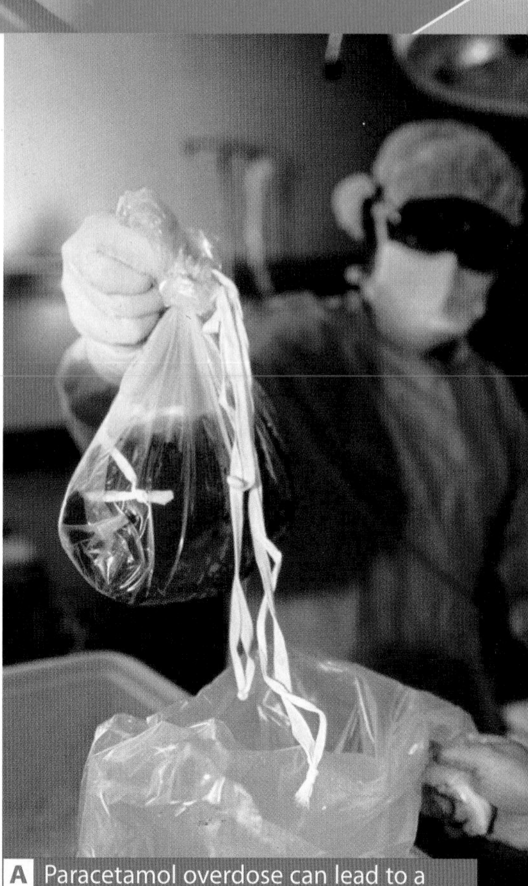

A Paracetamol overdose can lead to a liver transplant.

B The opium poppy produces a powerful painkiller.

C Heroin addiction causes social problems.

Morphine is a very strong painkiller. People who use it for long periods of time can become addicted to it. It is often used for people in severe pain, like cancer patients.

Heroin was manufactured from morphine, but it is even more dangerous. Morphine and heroin are called **opiates** because they originally came from the opium poppy. Most opiates are now manufactured. Heroin is very dangerous because it is addictive. The effect wears off and the user needs more. They may get withdrawal symptoms like severe cramps and excess sweating. It also produces tolerance. You have to keep taking more just to get the same effect. This can lead to an overdose.

Heroin is usually injected. Users often share needles for injection and risk transferring diseases to each other. HIV and hepatitis B are spread in this way. HIV is a **viral infection** that is spread by transferring blood or body fluids from one person to another.

4 What is meant by withdrawal symptoms?

5 Why is it dangerous for drug addicts to share needles?

Cannabis is a drug that can be smoked. It can make people feel more relaxed. People used to think it was safer than tobacco. Now we know that it can cause cancer and bronchitis just like **tobacco**. Research is being done to see if some of the chemicals in cannabis might be effective painkillers. It may give some relief for sufferers of diseases like multiple sclerosis, but many doctors remain unsure. The use of cannabis as a painkiller is still illegal.

Have you ever wondered:

Why are the uses of some substances controlled by law?

Summary Exercise

D This criminal justice team leader regularly sees the problems associated with cannabis use.

6 Why are doctors very careful when prescribing morphine to people in severe pain?

7 Why are doctors still arguing about the use of cannabis as a painkiller?

Higher Questions

Drug misuse and abuse

By the end of these two pages you should be able to:

- describe the main effects of solvent and alcohol abuse
- describe how abuse of drugs may produce abnormal behaviour.

When someone talks about **solvent** abuse, you probably think they mean sniffing glue. But people sniff all kinds of things. This could be gas refills and lighters, air fresheners, aerosols, spray paint, thinners or correcting fluid. All these things contain chemicals. Some of these chemicals dissolve other substances. This is why they are called solvents.

When solvent is breathed in, it enters the lungs. It quickly gets into the bloodstream and then the brain. Most solvents work a bit like alcohol. They slow down breathing and heart rate. They are sedatives. Other effects include nausea, vomiting, severe headaches and blackouts.

Some people die instantly because the solvent causes heart failure. Others put the solvent into a plastic bag and then put their head into the bag to sniff the solvent. They can suffocate inside the plastic bag.

Inhaling gases from an aerosol may freeze the throat and lung tissue because the gases are very cold. Longer-term use may result in lung irritation and diseases like bronchitis.

A Some people work full-time to help those who abuse drugs.

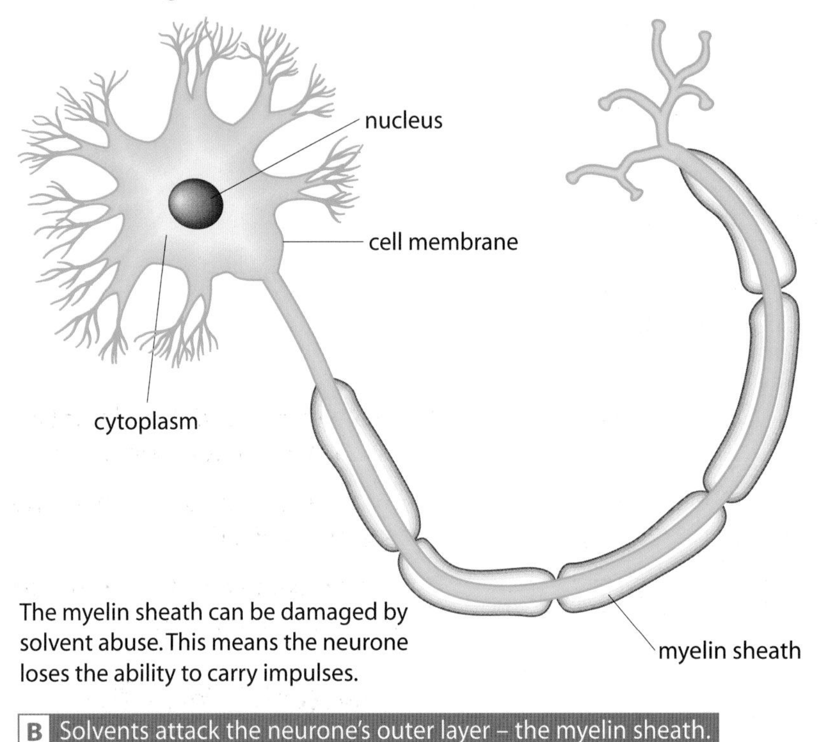

B Solvents attack the neurone's outer layer – the myelin sheath.

Solvents attack the myelin sheath around neurones. This leads to problems with the nervous system. Memory loss, brain damage, and kidney or liver failure may follow.

1 Why are some drugs called solvents?

2 Why do some people die instantly when sniffing solvents?

3 What effect can solvent abuse have on the lungs?

4 What effect does solvent abuse have on neurones?

C Alcohol is a drug that can be abused.

Alcohol is a part of many people's lives. They enjoy a drink. But alcohol is a drug and, like all drugs, it can be abused. Alcohol is a sedative. It slows down the reactions of your body. It also affects the parts of the brain devoted to physical coordination and making judgements.

Small amounts of alcohol can have a relaxing effect. With more drink the drinker has increased reaction time, slurred speech and loss of coordination. They may become angry or violent. More alcohol results in vomiting. Alcohol can even make people unable to walk, become unconscious and slip into a coma. If people try to drink and drive their reaction time and judgement will be affected. They are more likely to have an accident.

Some people become dependent on alcohol. They need more to get the same effect. This is called tolerance. Dependence on drink is called alcoholism.

Too much drinking will eventually lead to brain damage and liver disease. The liver tissue becomes scarred and healthy cells are replaced by fat and scar tissue. The liver is less able to do its job of removing toxins from the blood. A liver transplant may be the only way to save the person.

5 What effect do sedative drugs like alcohol have on reaction time?

6 What effect does too much alcohol have on the liver?

7 Why is it dangerous to use any kind of drug and then drive?

Higher Questions

Tobacco

By the end of these two pages you should be able to:

- describe the effects of tobacco on your breathing and circulation
- use data on drug misuse to make presentations to different audiences.

A Tobacco has serious effects on breathing and circulation.

A national survey showed that one in six smokers have a cigarette within five minutes of waking up in the morning. Half of all smokers smoke within the first half hour.

This kind of information can come from secondary sources like websites. Data can be used in presentations to show the main effects of drug misuse. By changing the way the data are presented, information can be understandable to people of different ages and backgrounds. There are many other ways of using ICT to present data.

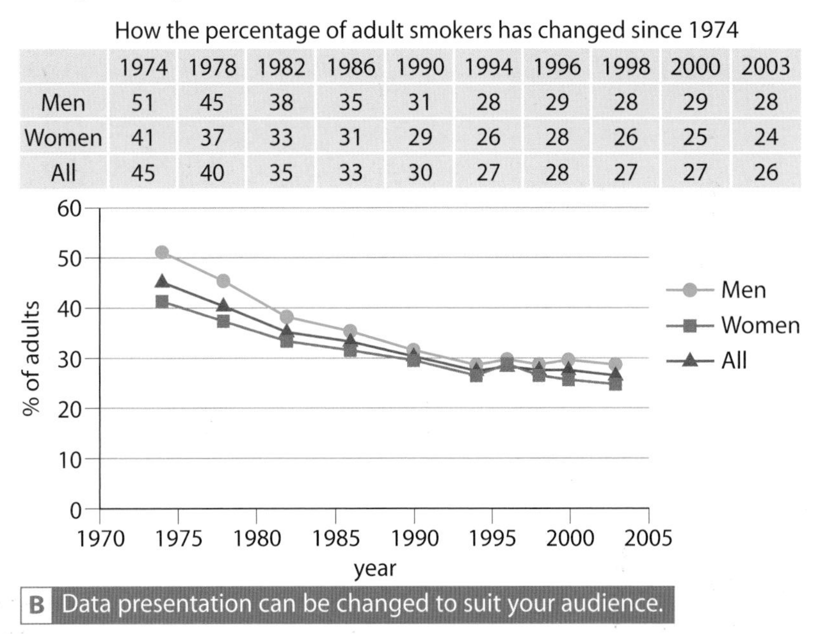

How the percentage of adult smokers has changed since 1974

	1974	1978	1982	1986	1990	1994	1996	1998	2000	2003
Men	51	45	38	35	31	28	29	28	29	28
Women	41	37	33	31	29	26	28	26	25	24
All	45	40	35	33	30	27	28	27	27	26

B Data presentation can be changed to suit your audience.

Cigarettes contain a drug called nicotine. When people smoke, the nicotine enters their lungs and gets into their bloodstream. Within 8 seconds it arrives in their brain. There it works on the neurones that control the feeling of pleasure. The effect soon wears off, so people take another puff to get the effect back. The smoke has a serious effect on their gaseous-exchange and circulatory systems.

Smokers must keep smoking tobacco to get the pleasurable effects. If they don't, they may become tense and irritable. They just have to have another one. This is called craving. They are now addicted. Nicotine has very strong craving effects. It is incredibly difficult to give up.

1 Why do smokers have to light up as soon as they wake up in the morning?

2 How does nicotine get to the brain so quickly?

The nicotine in tobacco makes the heart beat faster and narrows blood vessels. This can increase blood pressure and cause heart disease.

Tobacco contains other substances as well. The smoke contains carbon monoxide, a poisonous gas, which is taken up by the blood instead of oxygen. So the blood carries less oxygen than it should.

Nicotine and carbon monoxide together make the blood clot more easily. This can block the coronary arteries in the heart muscles. This damages the heart muscles and can lead to a heart attack.

Tar from burning tobacco contains hundreds of chemicals. It irritates the air passages and makes them narrower. This causes 'smoker's cough'. Several of the chemicals are known to cause cancer in the lungs.

3 How does smoking tobacco increase blood pressure?

4 What causes 'smoker's cough'?

5 How can smoking tobacco give a person cancer?

A smoker may suffer from bronchitis. Their air passages become inflamed. The cilia lining the air passages stop beating. The mucus, tar, dirt and bacteria stay in the air passages and the lungs.

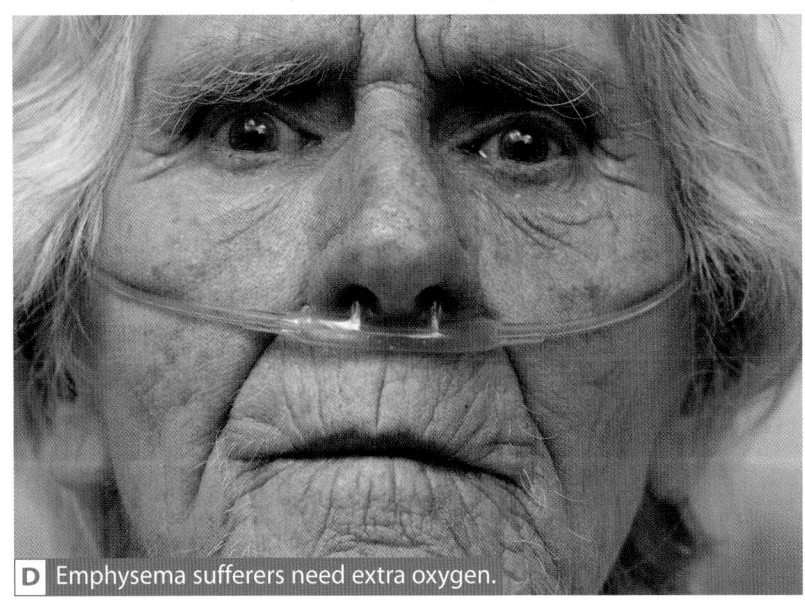

D Emphysema sufferers need extra oxygen.

Chemicals in smoke weaken the air sacs in the lungs, damaging the lung tissue. There is less surface area for gaseous exchange, causing the disease emphysema. People with emphysema cannot get enough oxygen and get breathless. In severe cases they may need to have pipes leading into their nose from an oxygen cylinder.

nicotine causes blood vessels to narrow

carbon monoxide reduces the amount of oxygen carried by red blood cells

carbon monoxide and nicotine together make blood clot more easily

C Carbon monoxide in tobacco smoke reduces oxygen in the blood.

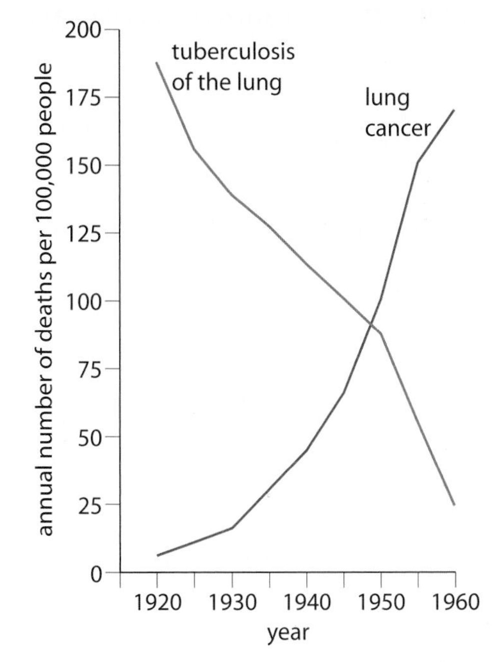

6 Look at the graph. What can you say about the number of deaths due to lung cancer and tuberculosis between 1920 and 1960?

7 What reasons can you give for the trends shown in the graph?

8 One in ten school pupils between ages 11 and 15 smoke regularly. What could you say to them about their habit?

Higher Questions

Questions

Multiple choice questions

1 Tuberculosis is caused by a
- **A** bacterium.
- **B** fungus.
- **C** virus.
- **D** vector.

2 Tuberculosis is particularly deadly when you also suffer from
- **A** measles.
- **B** AIDS.
- **C** cholera.
- **D** influenza.

3 Tuberculosis is spread through
- **A** air.
- **B** touching.
- **C** water.
- **D** food.

4 Lysozyme is
- **A** found in the stomach and kills microorganisms.
- **B** found in pathogens and kills microorganisms.
- **C** found in cilia and kills microorganisms.
- **D** found in tears and kills microorganisms.

5 The third line of defence against pathogens is called
- **A** the non-specific response.
- **B** the specific immune response.
- **C** the vector response.
- **D** the vertical transmission response.

6 Microorganisms which cause disease are called
- **A** bacteria.
- **B** microbes.
- **C** pathogens.
- **D** viruses.

7 Vertical transmission of disease means
- **A** disease passed on by contact with other people.
- **B** disease passed through the placenta.
- **C** disease passed on through water and food.
- **D** disease passed on by mosquitoes.

8 Horizontal transmission of disease means
- **A** disease passed on by contact with other people.
- **B** disease passed through the placenta.
- **C** disease passed on through water and food.
- **D** disease passed on by mosquitoes.

9 Vector-borne transmission of disease can mean
- **A** disease passed on by contact with other people.
- **B** disease passed through the placenta.
- **C** disease passed on through water and food.
- **D** disease passed on by mosquitoes.

10 Vehicle-borne transmission of disease means
- **A** disease passed on by contact with other people.
- **B** disease passed through the placenta.
- **C** disease passed on through water and food.
- **D** disease passed on by mosquitoes.

11 Antibodies
- **A** are found on the surface of pathogen cells.
- **B** are formed by red blood cells.
- **C** are found on the surface of cilia.
- **D** lock onto the antigens of pathogen cells and destroy them.

12 Drugs which increase your alertness and heart rate are called
- **A** sedatives.
- **B** stimulants.
- **C** barbiturates.
- **D** painkillers.

13 Breathing in smoke from tobacco may narrow the arteries in the body. This may lead to
- **A** bronchitis.
- **B** emphysema.
- **C** increased blood pressure.
- **D** lung cancer.

14 Painkillers work by
- **A** blocking some nervous impulses.
- **B** increasing reaction time.
- **C** speeding up impulses in the nervous system.
- **D** decreasing reaction time.

15 Tobacco contains a drug called nicotine which may
- **A** cause 'smoker's cough'.
- **B** give someone tuberculosis.
- **C** give someone bronchitis.
- **D** cause addiction in a person.

H Short-answer questions

Asif has found two graphs about tuberculosis. He is not sure what they mean. You can help him by answering the questions below.

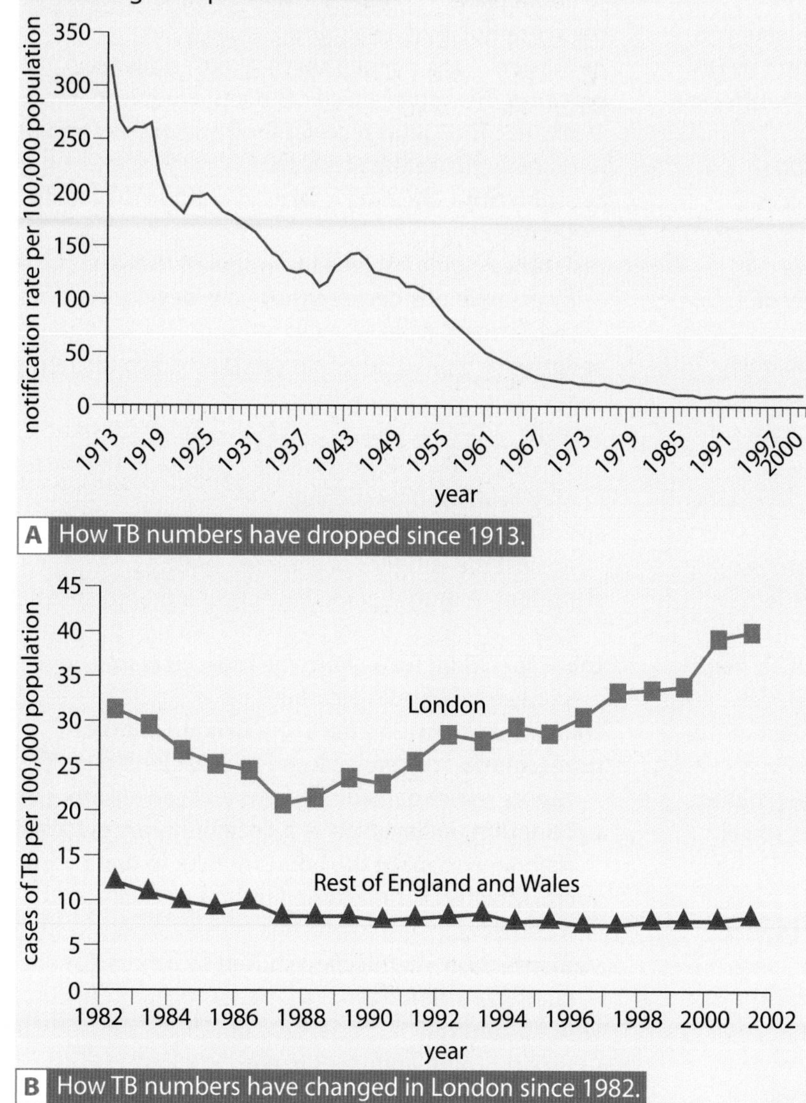

A How TB numbers have dropped since 1913.

B How TB numbers have changed in London since 1982.

1 Asif is looking at the first graph.

- **a** Describe the changes in the pattern of the numbers of cases of TB since 1913.
- **b** The BCG vaccination started in 1950. What effect did it have on the pattern of TB cases?
- **c** Why have the numbers of cases not gone down to zero?

2 Asif looks at the second graph, which shows cases of TB in London since 1982.

- **a** What has happened to the number of cases of TB in London since 1982?
- **b** How does this compare with the numbers of cases of TB in the rest of England and Wales?

3 **a** Asif is not sure of the reasons for this change in London. What do you think the reasons are?

- **b** Why is it very important to finish a course of antibiotics?

Glossary

addiction When someone is dependent on a drug and can't do without it.
AIDS Acquired immune deficiency syndrome – when the immune system in the body has been destroyed by HIV.
alcohol A drug produced by yeast. Has a sedative effect on the body.
antibiotic A drug used to kill bacteria in the body.
antibody A protein produced by white blood cells to destroy a pathogen.
antigen Protein markers on the surface of every cell which identify a cell as belonging to the body or a foreign body.
bacteria A tiny single-cell organism. Most are harmless, some useful, a few are pathogenic.
barbiturate A sedative that slows down the nervous system.
barrier Keeps things out; for example, the skin is a barrier against microorganisms.
caffeine A stimulant drug which increases alertness.
cannabis An illegal drug that may have pain-relief properties.
chemical barrier The body uses chemicals as a barrier against microorganisms.
cilia Tiny hair-like structures that move mucus out of the lungs and windpipe.
circulatory system Your heart, blood vessels and blood.
disease Caused by microorganisms and makes you ill.
drug A chemical designed to change the way your body works.
foreign body Something which is not part of the body.
gaseous exchange The process by which your lungs take in oxygen and expel carbon dioxide.
immune system The system inside the body for destroying invading microorganisms.
infection When microorganisms invade the body.
inflammation One of the body's responses to infection.
lysozyme An enzyme found in tears that destroys bacteria, one of the body's chemical barriers.
microbe Another word for microorganism.
microorganism Tiny organisms like bacteria, viruses and fungi.
neurone A single cell that carries impulses.
non-specific immune system The response of the white blood cells to a pathogen. The response is the same for all pathogens.
opiate A drug from the opium family, including opium and heroin.
organism A living thing.
overdose Taking more of a drug than is recommended. Can lead to death.
pain-relief Blocking the impulses that tell the body about pain.
paracetamol A general-purpose painkiller.
pathogen A microorganism that causes disease.
placenta The organ which connects a foetus to its mother. The foetus receives food and oxygen and gets rid of waste via the placenta.
reaction time The time it takes your body to react to a stimulus.
resistance Some bacteria in a population are not killed by an antibiotic because they have developed resistance to it.
sedative A type of drug that can slow down the nervous system.
solvent A chemical used to dissolve other substances. Solvents include aerosols, spray paint and thinners. May be abused by sniffing.
specific immune system The body makes antibodies against the antigens of a foreign body.
stimulant A drug that increases your alertness and heart rate.
tobacco A drug made from the tobacco plant and smoked. It contains nicotine.
transmission Passing from one thing to another.
tuberculosis A disease caused by a bacterium that causes severe damage to lungs and can lead to death.
vaccination An injection of a dead or harmless form of a pathogen to make the body immune to that pathogen.
vector-borne Disease carried by an insect like a mosquito.
viral infection An infection caused by a virus. Cannot be cured by antibiotics.
white blood cell A special type of cell in your blood that is part of your immune system.

Patterns in properties

A A forensic scientist at a crime scene.

Forensic scientists analyse samples to find out what substances they do or do not contain. They need to do this carefully because they may have to present their results in court. For example, they may need to discover if a car driver had been drinking alcohol. They do this by testing a blood or urine sample. Forensic scientists use complex equipment to obtain accurate results but often the chemistry behind the tests is simple.

To understand the chemistry involved, it helps to know how the different elements are arranged in the periodic table. We can use ideas about the periodic table to predict what will happen in chemical reactions.

In this topic you will learn that:

- all chemical elements are made up of atoms which consist of nuclei and electrons
- different elements have different properties related to their position in the periodic table
- atoms join together to form molecules and compounds
- the names of simple chemical compounds can be predicted from their formulae.

Look at these statements and sort them into the following categories:

I agree, I disagree, I want to find out more

- All metals are shiny and tough.
- All non-metals are unreactive.
- There are useful patterns in the way the elements are arranged in the periodic table.
- It is difficult to analyse a substance to see which metal it contains.
- During a chemical reaction, elements can change into other elements.
- Molecules are made from lots of different atoms joined together.

A map of the elements

By the end of these two pages you should be able to use the periodic table to find:

- the symbol of an element
- the metals and non-metals
- the alkali metals and the transition metals
- the halogens and the noble gases.

Forensic scientists must correctly identify the different substances found at the scene of a crime. To do this, they need to know about **elements** and **compounds** and their different chemical properties. There are about 116 different elements. An element is a substance that cannot be turned into anything simpler or another element using a **chemical reaction**. Elements can join together in different ways to form billions of different compounds. With so many different elements we need ways to keep track of them, and to understand what they are like and what they do. This is where **chemical symbols** and the periodic table help.

Have you ever wondered:

Is the periodic table really a map of what you're made of?

1 Approximately how many different elements are there?

2 What is a chemical element?

Key: relative atomic mass — 1 H hydrogen 1 — atomic number

Period \ Group	1	2											3	4	5	6	7	0
1								1 H hydrogen 1										4 He helium 2
2	7 Li lithium 3	9 Be beryllium 4											11 B boron 5	12 C carbon 6	14 N nitrogen 7	16 O oxygen 8	19 F fluorine 9	20 Ne neon 10
3	23 Na sodium 11	24 Mg magnesium 12											27 Al aluminium 13	28 Si silicon 14	31 P phosphorus 15	32 S sulphur 16	35.5 Cl chlorine 17	40 Ar argon 18
4	39 K potassium 19	40 Ca calcium 20	45 Sc scandium 21	48 Ti titanium 22	51 V vanadium 23	52 Cr chromium 24	55 Mn manganese 25	56 Fe iron 26	59 Co cobalt 27	59 Ni nickel 28	63.5 Cu copper 29	65 Zn zinc 30	70 Ga gallium 31	73 Ge germanium 32	75 As arsenic 33	79 Se selenium 34	80 Br bromine 35	84 Kr krypton 36
5	85 Rb rubidium 37	88 Sr strontium 38	89 Y yttrium 39	91 Zr zirconium 40	93 Nb niobium 41	96 Mo molybdenum 42	(98) Tc technetium 43	101 Ru ruthenium 44	103 Rh rhodium 45	106 Pd palladium 46	108 Ag silver 47	112 Cd cadmium 48	115 In indium 49	119 Sn tin 50	122 Sb antimony 51	128 Te tellurium 52	127 I iodine 53	131 Xe xenon 54
6	133 Cs caesium 55	137 Ba barium 56	139 La lanthanum 57	178 Hf hafnium 72	181 Ta tantalum 73	184 W tungsten 74	186 Re rhenium 75	190 Os osmium 76	192 Ir iridium 77	195 Pt platinum 78	197 Au gold 79	201 Hg mercury 80	204 Tl thallium 81	207 Pb lead 82	209 Bi bismuth 83	(209) Po polonium 84	(210) At astatine 85	(222) Rn radon 86
7	(223) Fr francium 87	(226) Ra radium 88	(227) Ac actinium 89	(261) Rf rutherfordium 104	(262) Db dubnium 105	(266) Sg seaborgium 106	(264) Bh bohrium 107	(277) Hs hassium 108	(268) Mt meitnerium 109	(271) Ds darmstadtium 110	(272) Rg roentgenium 111							

A The periodic table.

Each element has its own chemical symbol. A chemical symbol always starts with a capital letter. Some symbols are just one letter, such as O for oxygen and C for carbon. But most symbols are made of two letters, such as Ca for calcium and Al for aluminium. Notice that the second letter is always lower case, so Mg is the correct symbol for magnesium but MG is not. All the symbols for the elements are arranged in a chart called the **periodic table**.

3 What are the chemical symbols for sodium, iron, chlorine and neon?

4 What is the name of the element with the chemical symbol W?

Look at the periodic table. There is a zig-zag line that starts between B for boron and Al for aluminium. Anything to the left of this line is a metal. Anything to the right is a non-metal. You can see that elements like sodium and iron are metals but elements like chlorine and neon are non-metals.

5 Are most of the elements metals or non-metals?

Similar elements are found together in the periodic table. The left-hand column contains very reactive metals called the **alkali metals**. Lithium and sodium are alkali metals. They both float on water and react with it to make an alkali and hydrogen gas.

The large rectangular section in the middle contains most of the other metals, called the **transition metals**. Iron and gold are transition metals. They sink in water and are much less reactive than the alkali metals. Mercury, the only metal that is liquid at room temperature, is also a transition metal.

The right-hand column contains non-metals called the **noble gases**. Helium and neon are noble gases. Like all noble gases they are very unreactive. The column next to the noble gases contains very different non-metals, called the **halogens**. Chlorine and bromine are halogens. They are poisonous and very reactive.

6 Suggest one advantage and one disadvantage of using chemical symbols instead of chemical names.

7 Why must most chemical symbols have two letters rather than one?

8 Approximately what fraction of the elements are non-metals?

Higher Questions

C1a.5.2

Conducting heat

By the end of these two pages you should be able to:

- explain some uses of iron, copper, silver and gold.

Metals have certain **properties**. Metals are:

- shiny
- solid at room temperature (except for mercury)
- good conductors of heat and electricity
- **malleable**
- **ductile**.

Most of the metals that you use every day are transition metals, found in the middle of the periodic table. They include metals such as iron, copper, silver and gold. The transition metals are less reactive than the alkali metals, and most of them are hard and tough.

1 What are the chemical symbols for iron, copper, silver and gold?

1	2											3	4	5	6	7	0
							H										He
Li	Be											B	C	N	O	F	Ne
Na	Mg											Al	Si	P	S	Cl	Ar
K	Ca	Sc	Ti	V	Cr	Mn	Fe	Co	Ni	Cu	Zn	Ga	Ge	As	Se	Br	Kr
Rb	Sr	Y	Zr	Nb	Mo	Tc	Ru	Rh	Pd	Ag	Cd	In	Sn	Sb	Te	I	Xe
Cs	Ba	La	Hf	Ta	W	Re	Os	Ir	Pt	Au	Hg	Tl	Pb	Bi	Po	At	Rn
Fr	Ra	Ac	Rf	Db	Sg	Bh	Hs	Mt	Ds	Rg							

A The transition metals.

Iron is the cheapest and most widely used metal. But most iron that we use has other elements mixed with it, such as carbon and vanadium. A mixture of a metal with other elements is called an **alloy**. An alloy of iron containing less than 2 per cent carbon is called steel. Steel is hard and strong, so is useful for making girders for buildings and bridges. Thin sheets of steel can be pressed into shape, making them useful for car body panels.

B Steel is very useful for the construction industry.

Iron and most steels form **rust** when in contact with air and water. This is called **corrosion** and it damages the metal. Stainless steel contains chromium and other metals such as nickel. It resists corrosion, making it useful for kitchen sinks and cutlery.

2 Why would iron be a poor choice for making a water pipe?

3 Steel objects are often painted to stop them rusting. Suggest why razor blades are made of stainless steel, rather than painted steel.

Copper is a very good conductor of electricity. It is also malleable and ductile. These properties make it useful for electrical wiring. Copper does not corrode very much when it comes into contact with air and water. This makes it useful for water tanks and pipes.

4 Explain why it is important for electrical wiring that copper is malleable and ductile.

5 Copper is a very good conductor of heat. Why is this useful for making pans?

C Copper is widely used for plumbing.

Silver costs about 250 times more than copper per kilogram and 1000 times more than iron. However, it does not easily corrode in air or water, so it stays shiny. This makes it useful for jewellery. Silver is a better conductor of electricity than any other metal. Silver and silver alloys are used for printed circuit boards and electrical contacts.

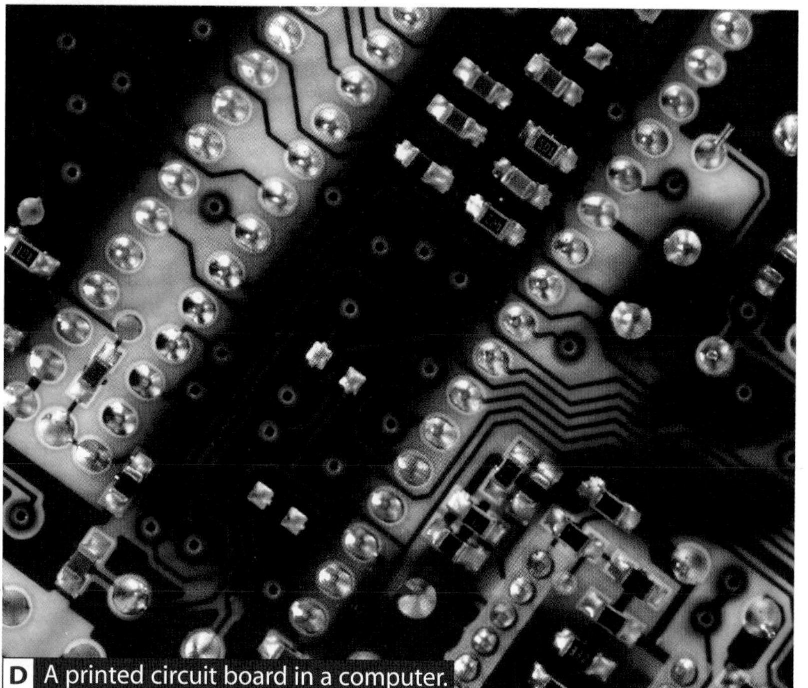

D A printed circuit board in a computer.

Gold has a rich yellow colour. It has a high **density**, so pieces of gold are heavy for their size. Gold costs about 60 times more than silver, but it does not react with air or water at all, so it is used for jewellery. Gold is almost as good as copper at conducting electricity. It is useful for electrical contacts and connecting wires for computer chips.

6 Why is copper used for electrical wiring instead of silver?

7 Suggest two disadvantages of using gold in space missions.

E This astronaut's visor has a thin gold coating to reduce the Sun's glare.

Summary Exercise

Higher Questions

Colourful chemistry

By the end of these two pages you should:

- know that solutions containing transition metals form coloured precipitates with sodium hydroxide solution
- be able to work out which metal is present in solution from the results of a reaction with sodium hydroxide solution

Forensic scientists perform chemical analysis on samples to find out which substances they contain. Using complex equipment, forensic scientists obtain accurate results from small samples. Many of these **analytical** tests are based on simple chemical reactions.

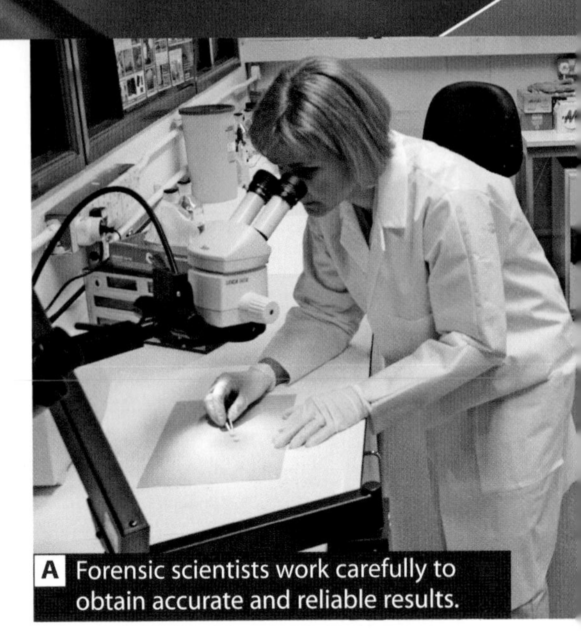

A Forensic scientists work carefully to obtain accurate and reliable results.

1 What is chemical analysis?

Sugar and many other substances are **soluble**. They dissolve in water to form a clear **solution**. Some solutions are colourless but others are coloured. **Insoluble** substances, such as sand and chalk, do not dissolve in water. When they are mixed with water they form a cloudy **suspension**, often with a layer of **sediment** at the bottom.

2 What is the difference between a soluble substance and an insoluble substance?

3 What is a precipitate?

Two solutions may react together to form an insoluble substance. This is a **precipitation** reaction, and the insoluble substance is called a **precipitate**.

B Salt is soluble (left) but sand is insoluble (right).

Transition metal compounds are coloured, and the soluble ones form coloured solutions. Some transition metals were named because of this property. For example, iridium is named after the Greek word *iris*, which means 'rainbow'.

Different transition metal compounds form different coloured solutions. A precipitation reaction happens when these are mixed with sodium hydroxide solution. Coloured precipitates form, called metal hydroxides. Different transition metals form different coloured precipitates. This is used in chemical analysis to see if a sample contains a particular transition metal.

C Copper sulphate solution is blue and nickel sulphate solution is green.

4 Which transition metal do you think is named after the Greek word *chroma*, which means 'colour'?

Copper sulphate solution forms a cloudy pale blue precipitate of copper hydroxide when it reacts with sodium hydroxide solution. This is the equation for the reaction:

copper sulphate + sodium hydroxide → copper hydroxide + sodium sulphate

$$CuSO_4 + 2NaOH \rightarrow Cu(OH)_2 + Na_2SO_4$$

5 You cannot see any sodium sulphate formed in the reaction above. Is sodium sulphate soluble, or insoluble? Explain your answer.

D Copper sulphate and sodium hydroxide solutions react to form the pale blue precipitate seen in the centre test tube.

The table shows the colours of the precipitates formed by some transition metals when they react with sodium hydroxide solution.

Transition metal	Colour of precipitate
cobalt	blue (turns grey when left standing)
copper	pale blue
iron(II)	dark green (turns orange-brown when left standing)
iron(III)	orange-brown
manganese	pale brown
nickel	green
zinc	white

Transition metals often form compounds that have different **formulae**, even though the same elements are present. The roman numbers next to iron help us to tell the compounds apart. For example, iron(II) chloride is $FeCl_2$ but iron(III) chloride is $FeCl_3$.

Have you ever wondered:

Why are 'chemical' names such as 'J_2O' and 'O_2' so good for advertising?

6 A solution of a transition metal compound forms an orange-brown precipitate with sodium hydroxide solution. Which transition metal is likely to be present?

7 Explain how you could test a solution to see if it contained nickel. Describe what you would see if it did contain nickel.

E Iron(II) compounds form green precipitates that gradually turn orange-brown.

Higher Questions

Atomic structure

By the end of these two pages you should know that:

- an atom consists of neutrons and positive protons in a nucleus surrounded by negative electrons
- all atoms of the same element have the same number of protons.

Have you ever wondered:

Can chemists turn cheap metal into gold?

Medieval alchemists attempted to change lead into gold, but we now know that an element cannot be turned into another element using a chemical reaction. To understand why, it helps to know about **atoms** and atomic structure.

1 In which part of the periodic table are lead and gold placed?

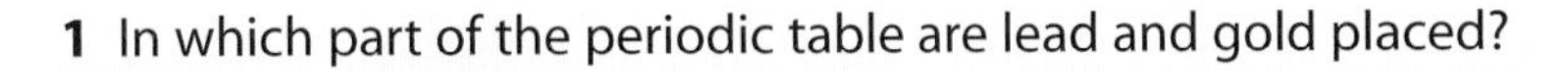

A A medieval alchemist at work.

You, and all the different substances around you, are made from tiny **particles** called atoms. Atoms are far too small for you to see. For example, an atom of gold is about six hundred million times smaller than a football. This makes it very difficult to study atoms. Until the beginning of the last century atoms were imagined to be tiny hard balls that could not be split apart. The word atom comes from the Greek word *atomos*, meaning 'indivisible'. However, experiments showed that atoms themselves are made from even smaller particles.

2 What are atoms?

B A cluster of gold atoms on a layer of carbon atoms, detected using a scanning tunnelling electron microscope.

There is a **nucleus** at the centre of each atom that contains most of its mass. The nucleus contains two types of particle: **protons** and **neutrons**. Protons are positively charged. Neutrons are electrically neutral and have no charge. Clouds of smaller particles called **electrons** surround the nucleus. Electrons are negatively charged. Every atom contains an equal number of protons and electrons. So an atom also has an equal number of positive and negative charges, so it is neutral overall.

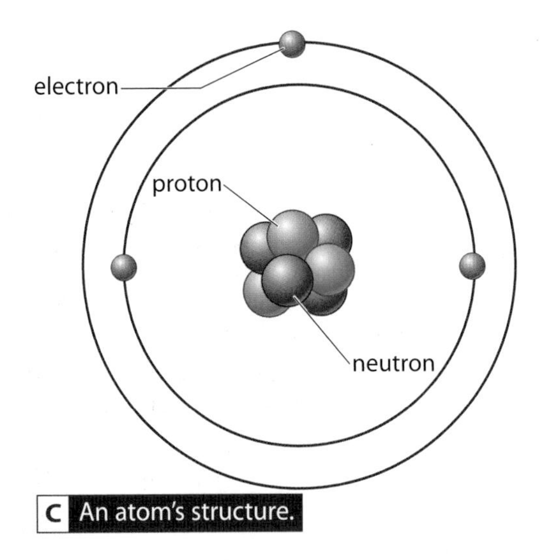

C An atom's structure.

3 The nucleus of a lead atom contains 82 protons. How many electrons will it have?

All the atoms of a particular element have the same number of protons. For example, every gold atom has 79 protons and every lead atom has 82 protons.

Each element has a number called the **atomic number**. This is equal to the number of protons in the nucleus, and also equal to the number of electrons surrounding the nucleus. The atomic number of gold is 79, and the atomic number of lead is 82. No two elements have the same atomic number. The periodic table shows the atomic number for each element (see page 106).

4 The atomic number of gold is 79. Write down the number of
a protons.
b electrons.

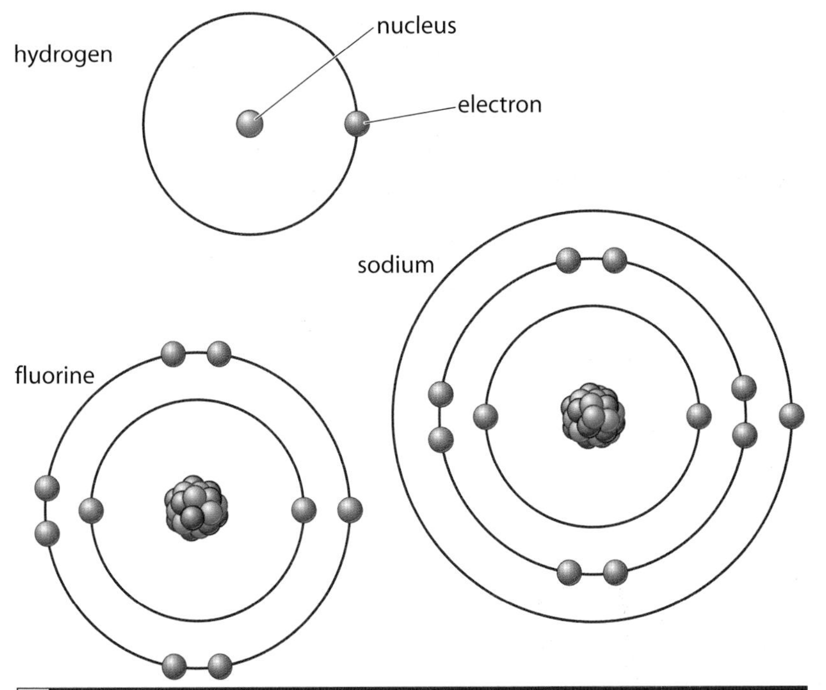

D The atoms of different elements have different numbers of electrons.

The elements in a row, or **period**, of the periodic table are arranged in order of increasing atomic number. For example, the period beginning with potassium, K, increases in atomic number from 19 to 36. This is how the elements are arranged in the modern periodic table.

5 Describe the structure of an atom. Include a labelled diagram to show where the protons, neutrons and electrons are found.

6 Compare the atoms of two different elements, such as aluminium and iron.
a What is the same about the atoms?
b What is different about the atoms?

3	4	5	6	7	0
11 B boron 5	12 C carbon 6	14 N nitrogen 7	16 O oxygen 8	19 F fluorine 9	20 Ne neon 10
27 Al aluminium 13	28 Si silicon 14	31 P phosphorus 15	32 S sulphur 16	35.5 Cl chlorine 17	40 Ar argon 18

atomic number

E Part of the periodic table. Atomic number increases from left to right.

Summary Exercise

Higher Questions

Group 1 – the alkali metals

By the end of these two pages you should know that:

- some chemical reactions give out heat
- some chemical reactions take in heat
- chemical reactions happen at different rates
- the alkali metals become more reactive as their atomic number increases.

A vertical column in the periodic table is called a **group**. Each group has a different number. The left-hand column is group 1, and it contains a family of very reactive elements called the alkali metals. It includes metals such as lithium, sodium and potassium. The alkali metals have the properties you would expect metals to have. They are:

- shiny when freshly cut
- solid at room temperature
- good conductors of heat and electricity.

However, they share some properties that are very different from other metals.

1	2											3	4	5	6	7	0
							H										He
Li	Be											B	C	N	O	F	Ne
Na	Mg											Al	Si	P	S	Cl	Ar
K	Ca	Sc	Ti	V	Cr	Mn	Fe	Co	Ni	Cu	Zn	Ga	Ge	As	Se	Br	Kr
Rb	Sr	Y	Zr	Nb	Mo	Tc	Ru	Rh	Pd	Ag	Cd	In	Sn	Sb	Te	I	Xe
Cs	Ba	La	Hf	Ta	W	Re	Os	Ir	Pt	Au	Hg	Tl	Pb	Bi	Po	At	Rn
Fr	Ra	Ac	Rf	Db	Sg	Bh	Hs	Mt	Ds	Rg							

A The alkali metals.

1 What is a group in the periodic table?

2 Write the chemical symbols for lithium, sodium and potassium.

Unlike most transition metals, alkali metals are very soft. They have low densities. Lithium, sodium and potassium are less dense than water, so they float on water, instead of sinking.

B The alkali metals are soft and easily cut with a knife.

The alkali metals are much more reactive than transition metals such as iron. They are silvery when freshly cut but very quickly tarnish and turn dull. This is because they react rapidly with oxygen in the air, forming metal oxides. They also react violently with water. To stop these reactions happening, the alkali metals are stored in oil, which stops air and water from reaching them.

C The alkali metals have to be stored in oil.

3 Describe three differences between the properties of the alkali metals and the transition metals.

All the alkali metals react with water to form an **alkaline** metal hydroxide and hydrogen gas, but the reaction becomes increasingly rapid and violent as you go down the group. A gradual change in a property like this is called a **trend**.

D Lithium reacts steadily with water.

Lithium releases a steady stream of hydrogen bubbles when it reacts with water. Sodium melts to form a shiny ball that whizzes around on the surface of the water. If a drop of water is added to a piece of sodium, so much heat is produced by the reaction that the metal bursts into orange flames. Potassium bursts into lilac flames when it touches water. It darts around on the surface, crackling and giving off sparks. This is the equation for potassium reacting with water:

$$\text{potassium} + \text{water} \rightarrow \text{potassium hydroxide} + \text{hydrogen}$$
$$2K + 2H_2O \rightarrow 2KOH + H_2$$

Reactions that give out heat energy to the surroundings are called **exothermic** reactions. The reactions of the alkali metals with water are exothermic reactions. Other reactions take in heat energy from the surroundings, such as the test tube, and their temperature falls. These are called **endothermic** reactions.

E Sodium reacts vigorously with water.

4 What is a trend in a property?

5 What is the difference between an exothermic and an endothermic reaction?

6 Lithium reacts with water to form lithium hydroxide and hydrogen. Write the word equation for this reaction.

7 Describe a simple test you could do to see if a reaction is exothermic or endothermic. Name the piece of equipment you would use, and what you would observe when using it.

F Potassium reacts violently with water.

Higher Questions

C1a.5.6

Group 0 – the noble gases

> ***By the end of these two pages you should be able to:***
> - describe the noble gases as chemically inert compared with other elements
> - interpret data describing the properties of helium, neon and argon
> - explain the uses of the noble gases.

The elements in group 0 are very unreactive non-metals. They are all gases at room temperature and include helium, neon and argon.

The group 0 elements used to be called the **inert** gases. Inert chemicals do not react with anything, and the group 0 elements are very unreactive. However, they do form compounds under extreme conditions, so they are now called the noble gases. This is to show that, usually, they are too 'posh' to react with other elements.

2 What does inert mean?

3 Look back at the last two pages, then give two differences between the noble gases and the alkali metals.

The alkali metals become more reactive as you go down the group. It is difficult to see any trend in the reactivity of the noble gases, but they clearly become denser as you go down the group.

Helium is less dense than air, so party balloons filled with helium will rise. Helium is also used as a lifting gas for weather balloons. These carry scientific instruments into the air to let weather forecasters collect data about conditions.

4 Look at the graph. Which noble gases could not be used as lifting gases?

C Helium is used as a lifting gas.

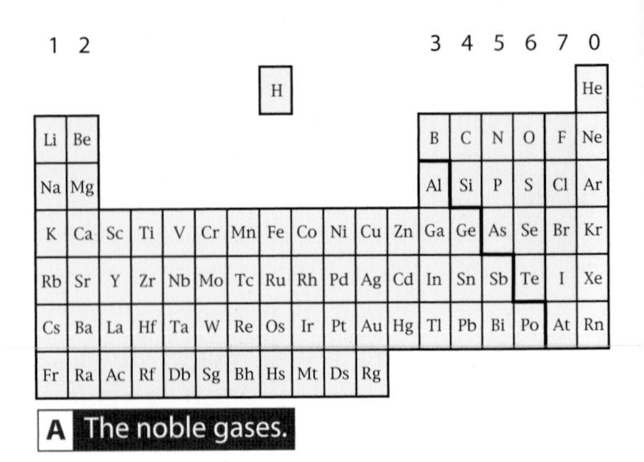

A The noble gases.

1 Write the chemical symbols for helium, neon and argon.

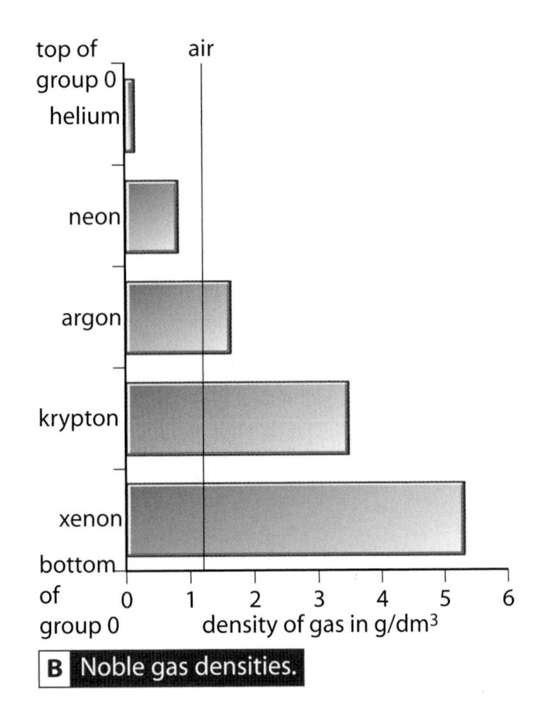

B Noble gas densities.

Airships are also filled with helium. Early in the last century, airships were filled with dangerously explosive hydrogen instead.

D The Hindenburg was a huge airship filled with hydrogen. It exploded as it arrived at Lakehurst in the USA in 1937.

Neon lights are widely used for advertising signs. They are glass tubes containing neon, which glows bright red when electricity is passed through it. Other colours are produced using mixtures of different gases or coloured glass.

Argon is the third most common gas in the air after nitrogen and oxygen. It forms just under 1 per cent of the air, so it is the cheapest noble gas. Welders use argon as an inert atmosphere to stop the hot steel from reacting with oxygen. Light bulbs contain argon, instead of air, to stop the hot metal filament inside burning.

Gas lasers produce light by passing electricity through gases. Different mixtures of noble gases produce different colours, with different uses. Red helium–neon lasers are used in supermarket barcode readers, while green argon lasers are used by eye surgeons.

Have you ever wondered:

What are the chemicals they use for laser light shows?

6 a Suggest two reasons why argon is used in welding.
b Where could you find argon at home? Why is it there?

7 What colour is the light from helium–neon lasers?

Summary Exercise

5 Why is helium a safer lifting gas than hydrogen?

P

E Neon lights give a spectacular display at night.

F Argon stops hot metals from reacting with oxygen.

?

P

G The lasers in laser light shows contain noble gases.

Higher Questions

Group 7 – the halogens

By the end of these two pages you should be able to:

- recall some properties of the halogens
- interpret data describing the properties of chlorine and iodine, and explain their uses.

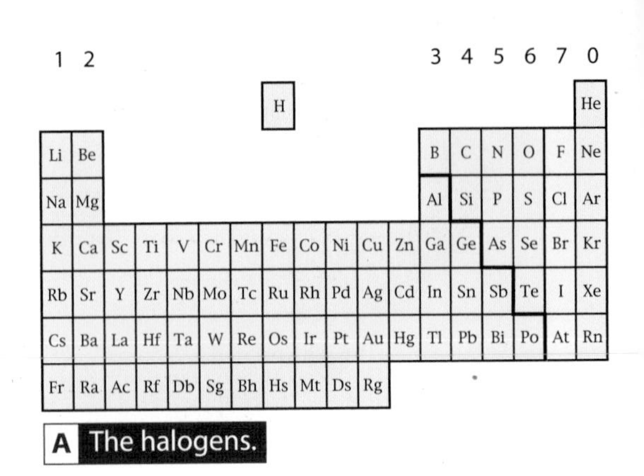

A The halogens.

The group to the left of the noble gases is group 7, and it contains a family of very reactive non-metals, called the halogens. It includes elements such as fluorine, chlorine, bromine and iodine. They have the typical properties of non-metals:

- they have low melting and boiling points
- they are poor conductors of heat and electricity.

1 Write the chemical symbols for fluorine, chlorine, bromine and iodine.

The halogens exist as **molecules** in which two atoms are joined together by a chemical **bond**. Molecules like this are called **diatomic molecules**. The chemical formulae for halogen molecules contain a small 2 to show that they contain two atoms. The formula for chlorine gas is Cl_2.

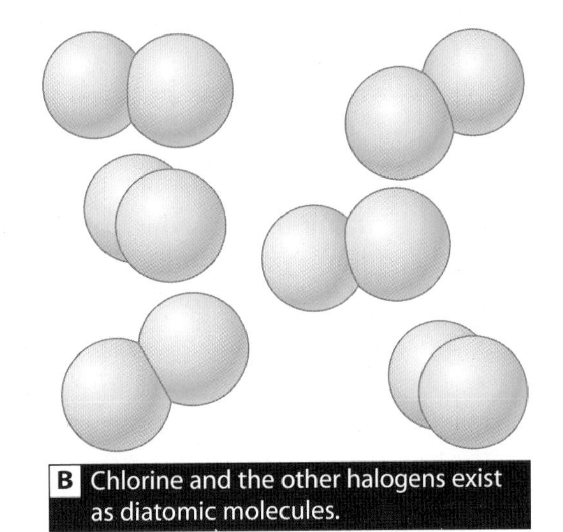

B Chlorine and the other halogens exist as diatomic molecules.

The halogens are coloured gases or form coloured vapours. Fluorine is very pale yellow, chlorine is yellow-green, bromine is red-brown and iodine vapour is dark purple. The halogens become darker as you go down the group.

There is also a trend in their state at room temperature. Fluorine and chlorine, at the top of the group, are gases. Bromine is the only liquid non-metal at room temperature. Iodine, near the bottom, is solid at room temperature.

C The halogens are coloured.

2 Astatine is the element at the bottom of group 7. Suggest its colour and give reasons for your answer.

The states of the halogens change like this because the melting and boiling points of the halogens increase as you go down the group.

Room temperature is usually about 20°C. The melting and boiling points of fluorine and chlorine are below this, so fluorine and chlorine are gases. The melting point of bromine is below room temperature but its boiling point is above room temperature, so it is a liquid. The melting and boiling points of iodine are above room temperature, so it is a solid.

3 a What is the trend in the boiling points of the halogens?
b What state do you expect astatine to have at room temperature? Explain your answer.

Halogen	Melting point in °C	Boiling point in °C
fluorine	−220	−188
chlorine	−101	−34
bromine	−7	59
iodine	114	184

D Melting and boiling points of some halogens.

The halogens are **toxic** and must be handled with care. However, this property makes them useful for killing harmful bacteria. Chlorine is used to **sterilise** drinking water and water in swimming pools, making them safe for you to use. Iodine is used in antiseptics to treat wounds.

Have you ever wondered:

Why is chlorine so good at protecting you from other people's bugs in a swimming pool?

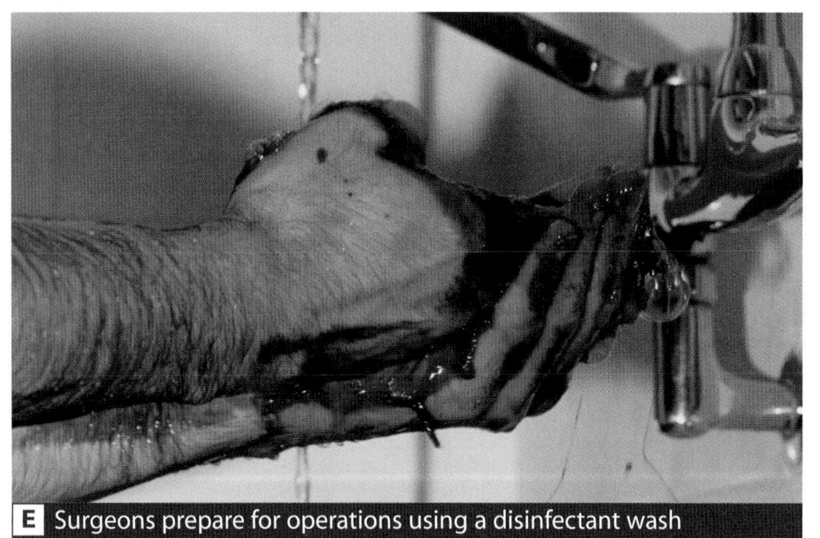

E Surgeons prepare for operations using a disinfectant wash containing iodine.

4 What does toxic mean?

5 What is a diatomic molecule?

6 What state would each halogen be in at 120°C?

7 Why would it be dangerous to put too much chlorine into a swimming pool?

Summary Exercise

Higher Questions

The salt formers

By the end of these two pages you should be able to:

- describe how the reactivity of the halogens changes as you descend the group
- explain the displacement reactions of the halogens with solutions of other halides.
- H explain the use of -ide and -ate endings in chemical names.

Have you ever wondered:

If potassium is like sodium, can you put potassium chloride on your chips?

'Halogen' comes from Greek words meaning 'salt former'. A salt is formed when a metal reacts with a halogen. For example, sodium and chlorine react together to form sodium chloride, which is common salt. Potassium and chlorine react together to form potassium chloride. This tastes like common salt, so it is added to common salt to make 'low-sodium salt', which is safer for people with high blood pressure.

The halogens decrease in reactivity as you go down group 7. This is the opposite of the trend in group 1. Fluorine, at the top of the group, is the most reactive halogen. In fact it is the most reactive non-metal of all. It even attacks glassware, so you will not see it used at school.

Fluorine reacts explosively with hydrogen when they are mixed, even in the cold and dark. However, the other halogens need light or heat to start their reaction with hydrogen. Chlorine reacts with hydrogen in sunlight or if it is warmed. Bromine reacts when heated to around 200°C. Iodine and hydrogen do not completely react together, even when heated.

The difference in reactivity between the halogens is used in bromine manufacture. Over half a million tonnes of bromine are produced worldwide each year. It is used by the oil industry and in fire-retardant plastics. Sea water contains bromine compounds such as potassium bromide. When chlorine is bubbled through sea water, it pushes out bromine. This is the word equation for the reaction, with the formula equation underneath:

chlorine + potassium bromide → potassium chloride + bromine

$$Cl_2 + 2KBr \rightarrow 2KCl + Br_2$$

This is a **displacement** reaction. The chlorine pushes out or **displaces** the bromine from the potassium bromide and takes its place. It works because chlorine is more reactive than bromine.

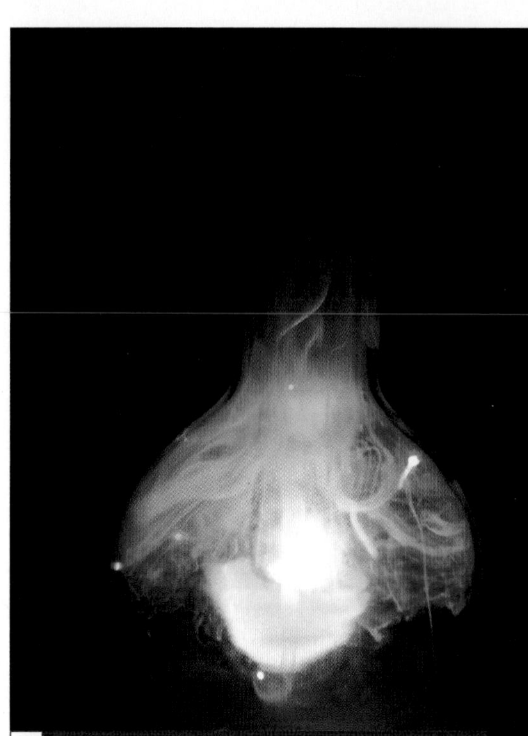

A Sodium and chlorine react very vigorously.

1 Why is fluorine difficult to store safely?

2 Write the names of the halogens in order of reactivity, starting with the most reactive.

3 Astatine is at the bottom of group 7. Would you expect it to react with hydrogen? Give reasons.

4 Why does chlorine displace bromine from sea water?

Displacement reactions are used in the laboratory to show differences in reactivity of the halogens. Weak solutions of chlorine, bromine or iodine are added to solutions of potassium chloride, bromide or iodide. There is a colour change if displacement happens, because displaced bromine or iodine turn the mixture light brown.

A more reactive halogen will displace a less reactive halogen from its compounds. So chlorine can displace bromine and iodine. Bromine can displace iodine but not chlorine, and iodine cannot displace chlorine or bromine.

B Bromine is extracted from sea water using displacement by chlorine.

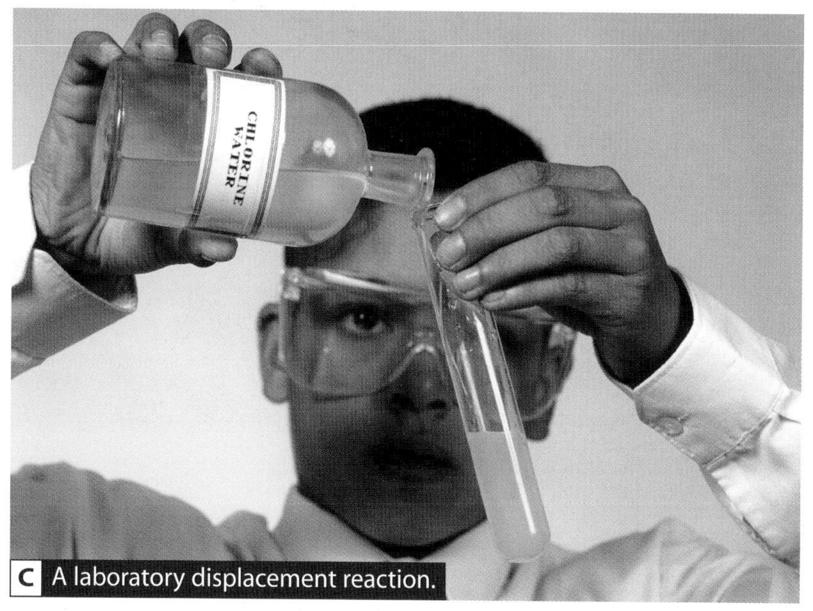

C A laboratory displacement reaction.

H Sodium chloride (NaCl) is table salt but sodium chlorate ($NaClO_3$) is a weedkiller. Both contain sodium and chlorine but only sodium chlorate contains oxygen. The ending '-ate' shows that oxygen is present. Here are some more examples.

'-ide' compounds		'-ate' compounds	
CaC_2	calcium carbide	$CaCO_3$	calcium carbonate
NH_4I	ammonium iodide	NH_4IO_3	ammonium iodate
FeS	iron(II) sulphide	$FeSO_4$	iron(II) sulphate

5 Explain in your own words the rule for deciding if one halogen will displace another halogen from its compounds.

6 **a** Which is more reactive, fluorine or chlorine?
b Will fluorine displace chlorine from sodium chloride solution? Explain your answer.

Summary Exercise

Higher Questions

Using the periodic table

By the end of these two pages you should be able to:

- recall that elements in the same group have similar properties
- understand that the periodic table can be used to predict the discovery of new elements
- H use data to show how elements are added to the periodic table.

The elements in a period, or row, are arranged in order of increasing atomic number. There is usually little similarity between the elements in a period. The elements in a group, or column, have similar properties. For example, the group 1 elements are very reactive metals. The group 7 elements are very reactive non-metals and the group 0 elements are very unreactive non-metals.

Have you ever wondered:

Which combination of chemicals makes the most violent explosion?

Key: relative atomic mass (top), symbol, atomic number (bottom) — e.g. 1 H 1

Period	Group 1	2											3	4	5	6	7	0
1									1 H 1									4 He 2
2	7 Li 3	9 Be 4											11 B 5	12 C 6	14 N 7	16 O 8	19 F 9	20 Ne 10
3	23 Na 11	24 Mg 12											27 Al 13	28 Si 14	31 P 15	32 S 16	35.5 Cl 17	40 Ar 18
4	39 K 19	40 Ca 20	45 Sc 21	48 Ti 22	51 V 23	52 Cr 24	55 Mn 25	56 Fe 26	59 Co 27	59 Ni 28	63.5 Cu 29	65 Zn 30	70 Ga 31	73 Ge 32	75 As 33	79 Se 34	80 Br 35	84 Kr 36
5	85 Rb 37	88 Sr 38	89 Y 39	91 Zr 40	93 Nb 41	96 Mo 42	(98) Tc 43	101 Ru 44	103 Rh 45	106 Pd 46	108 Ag 47	112 Cd 48	115 In 49	119 Sn 50	122 Sb 51	128 Te 52	127 I 53	131 Xe 54
6	133 Cs 55	137 Ba 56	139 La 57	178 Hf 72	181 Ta 73	184 W 74	186 Re 75	190 Os 76	192 Ir 77	195 Pt 78	197 Au 79	201 Hg 80	204 Tl 81	207 Pb 82	209 Bi 83	(209) Po 84	(210) At 85	(222) Rn 86
7	(223) Fr 87	(226) Ra 88	(227) Ac 89	(261) Rf 104	(262) Db 105	(266) Sg 106	(264) Bh 107	(277) Hs 108	(268) Mt 109	(271) Ds 110	(272) Rg 111							

A The modern periodic table.

1 What is the rectangular section between groups 2 and 3 called?

The atomic number of the elements increases as you go down a group. In a period the atomic number increases by one each time, but in a group it increases in jumps.

Several attempts at organising the elements into a periodic table were made during the nineteenth century. The most successful attempt was made in 1869 by a Russian scientist called Dmitri Mendeleev. At the time nobody knew about atomic number, so Mendeleev arranged the elements in order of atomic mass (a measure of how heavy the atoms are). He also

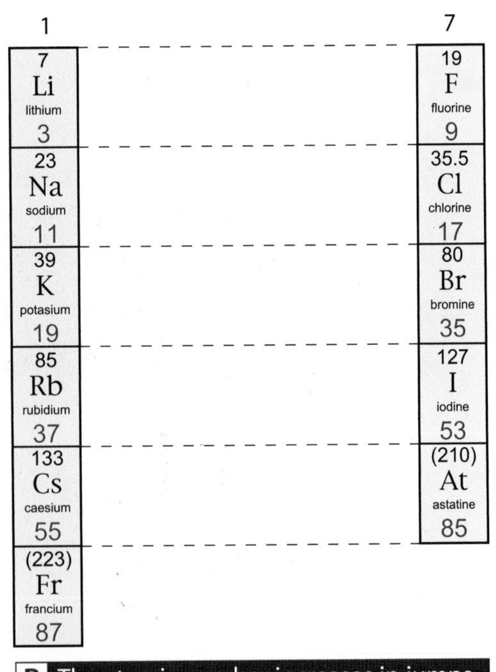

B The atomic number increases in jumps going down a group.

arranged similar elements in groups, as in the modern periodic table. Fewer elements had been discovered then, including none of the noble gases. Mendeleev had to leave gaps in his table to get everything to line up neatly. These were spaces for elements that had not yet been discovered.

2 In what order did Mendeleev put the elements?

3 What were the gaps in Mendeleev's table for?

Mendeleev predicted the properties of some undiscovered elements. One of these was in the gap between silicon and tin, which he called eka-silicon. Later, in 1886, germanium was discovered. It fitted the gap and its properties were very close to the ones that Mendeleev had predicted. This was very powerful evidence to support his periodic table.

H

	eka-silicon	germanium
atomic mass	72	72.6
density in g/cm³	5.5	5.32
melting point in °C	high	938
colour	grey	grey

D The properties of germanium are very close to the ones predicted for eka-silicon.

4 Eka-aluminium was another of Mendeleev's predicted elements. Suggest where its gap was.

New elements are still being discovered. The modern periodic table is arranged in order of increasing atomic number, not increasing atomic mass. But it has gaps in it, just like Mendeleev's table, ready to be filled when the elements that fit there have been found. We can predict their properties by studying the elements around the gap and in the same group.

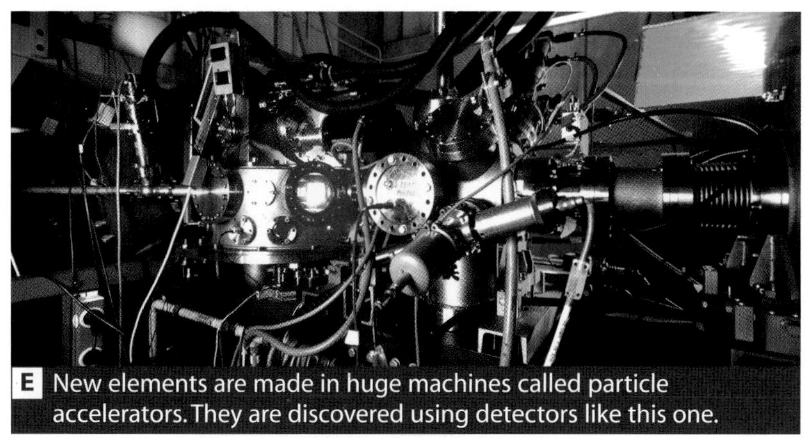

E New elements are made in huge machines called particle accelerators. They are discovered using detectors like this one.

C Dmitri Mendeleev (1834–1907) made the first periodic table.

5 Which group of elements was missing from Mendeleev's table?

6 What is the difference between a group and a period in the periodic table?

7 **a** Describe two similarities between the elements in group 1.

b Describe two similarities between the elements in group 7.

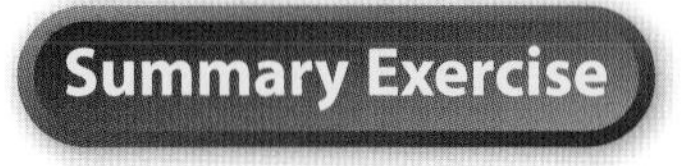

Higher Questions

Forensic science

By the end of these two pages you should be able to:

- interpret data such as the colours formed by copper, iron and zinc with sodium hydroxide
- explain how to use flame tests to identify the presence of a particular metal
- use analytical data to identify substances found at a crime scene.

So far, the police had just one suspect for the break-in at the chemicals factory. When escaping, the thief had broken a container of copper sulphate used to kill algae in fish tanks. Could the forensic scientist link the suspect to the crime?

Solutions of transition metals form coloured precipitates when sodium hydroxide solution is added to them. These precipitation reactions are a way to **analyse** samples to see if they contain a particular transition metal. The table shows the colours of the precipitates formed by copper, iron and zinc.

Transition metal	Colour of precipitate
copper	pale blue
iron(II)	dark green (turns brown when left standing)
iron(III)	orange-brown
zinc	white

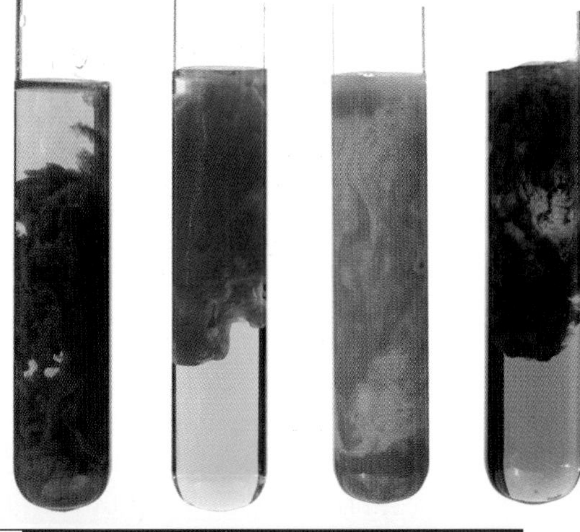

A Transition metals form coloured precipitates.

1 A sample produces an orange-brown precipitate. Which metal is likely to be present in the sample?

2 Samples of dust from the suspect's clothing produced a white precipitate when mixed with sodium hydroxide solution. What does this tell you?

Alkali metals do not form a precipitate with sodium hydroxide solution. A **flame test** is used to analyse a sample to see if it contains an alkali metal.

When a solution containing a metal compound is heated in a flame, the metal causes the flame to change colour. Different metals produce different coloured flames. Copper always gives a blue flame tinged with green, while sodium always gives a bright orange flame. This is the same orange colour that is given off by street lamps at night, as these contain sodium vapour.

3 Why are flame tests needed to identify sodium in a sample?

B Street lamps contain sodium.

Flame tests are easy to do but you must work carefully to get reliable results. The diagram shows you how to do a flame test.

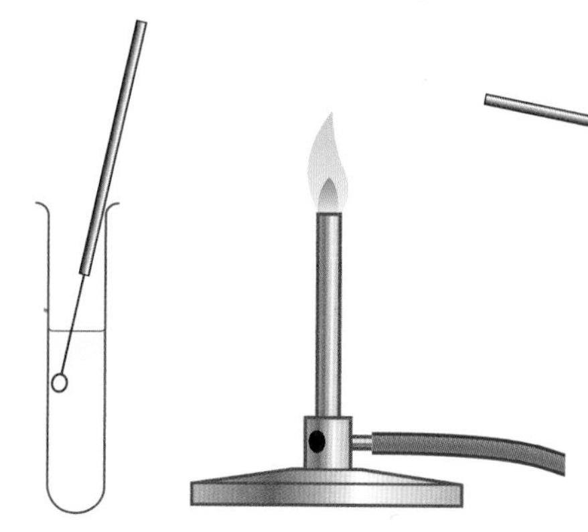

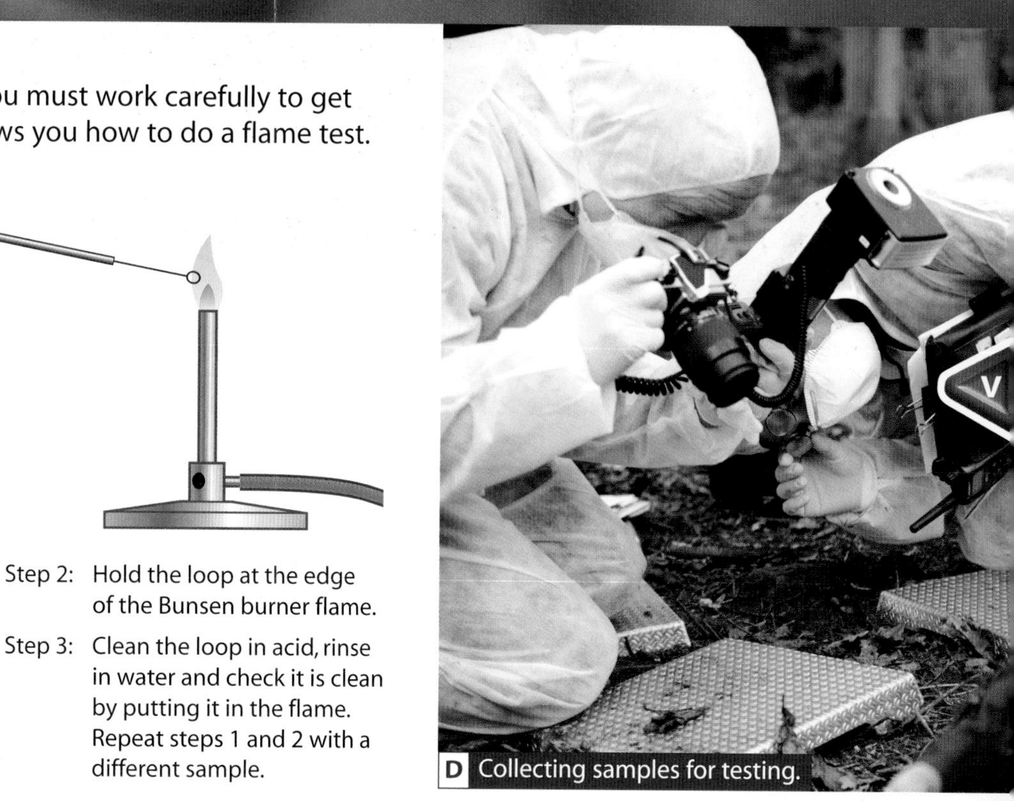

Step 1: Dip the clean flame test loop in the sample solution.

Step 2: Hold the loop at the edge of the Bunsen burner flame.

Step 3: Clean the loop in acid, rinse in water and check it is clean by putting it in the flame. Repeat steps 1 and 2 with a different sample.

C How to perform a flame test.

D Collecting samples for testing.

4 Why should the loop be rinsed with water before checking that it is clean?

The table shows the flame test colours produced by several different metals.

Metal	Flame test colour
lithium	red
sodium	intense yellow
potassium	lilac
calcium	brick red
barium	pale green
copper	green blue

5 A sample produces a pale green flame in a flame test. Which metal is likely to be present?

The colours from some metals can hide the colours from other metals, so it is important to prepare a sample and perform its flame test carefully. For example, the red colour from lithium can be hidden by the orange from sodium.

Have you ever wondered:

How can forensic scientists identify traces of substances at a crime scene?

6 A sample from a crime scene produced a lilac flame in a flame test. Why is the sample unlikely to contain sodium?

7 When some sodium hydroxide solution was added to a sample from a crime scene, a white precipitate formed. Why is the sample unlikely to contain iron?

8 Explain how you could use a flame test to see if a sample from a crime scene contained copper, rather than calcium.

Summary Exercise

Higher Questions

Questions

D

D

Multiple choice questions

1 In the modern periodic table the elements are arranged in order of
- **A** atomic number.
- **B** atomic mass.
- **C** their name.
- **D** their density.

2 In the periodic table
- **A** columns are called families and rows are called lines.
- **B** columns are called periods and rows are called groups.
- **C** columns are called families and rows are called periods.
- **D** columns are called groups and rows are called periods.

3 In the periodic table
- **A** most elements are non-metals and are found on the right.
- **B** most elements are metals and are found on the right.
- **C** most elements are metals and are found on the left.
- **D** most elements are non-metals and are found on the left.

4 Compared to the alkali metals, the transition metals are
- **A** more reactive and less dense.
- **B** less reactive and less dense.
- **C** less reactive and denser.
- **D** more reactive and denser.

5 Why is copper used in electricity cables?
- **A** It is shiny.
- **B** It is a good conductor of electricity.
- **C** It is a good conductor of heat.
- **D** It does not rust.

6 Why is steel used for building bridges?
- **A** It is strong.
- **B** It rusts easily.
- **C** It is a good conductor of heat.
- **D** It is expensive compared to most metals.

7 In an atom
- **A** the nucleus is negatively charged and surrounds the electrons.
- **B** the nucleus is positively charged and surrounds the electrons.
- **C** the electrons are negatively charged and surround the nucleus.
- **D** the electrons are positively charged and surround the nucleus.

8 Carbon has an atomic number of 6. What does this mean?
- **A** Carbon atoms have six nuclei at their centre.
- **B** Carbon atoms have six electrons at their centre.
- **C** Carbon atoms have six protons.
- **D** Carbon atoms have three protons and three electrons.

9 Which is the least reactive alkali metal?
- **A** lithium
- **B** sodium
- **C** potassium
- **D** francium

10 Which gas is produced when sodium reacts with water?
- **A** hydrogen
- **B** helium
- **C** oxygen
- **D** carbon dioxide

11 Which of the following is a correct trend for the noble gases?
- **A** They become less reactive going down the group.
- **B** They become darker going down the group.
- **C** They become denser going down the group.
- **D** They become more expensive to produce going down the group.

12 Neon is used in neon lights because
- **A** it burns with a red flame.
- **B** it gives off light when electricity passes through it.
- **C** it is coloured.
- **D** it is less dense than air.

13 How many atoms are there in a bromine molecule?
A one
B two
C three
D four

14 Which is the most reactive halogen?
A bromine
B chlorine
C fluorine
D iodine

15 Which of the following mixtures will not react?
A Chlorine and potassium fluoride solution.
B Chlorine and potassium bromide solution.
C Chlorine and potassium iodide solution.
D Bromine and potassium iodide solution.

Short-answer questions

1 Paul thought that all the alkali metals should float on water but Alison thought that this was wrong. They collected some data about the densities of the alkali metals and put it into a table.

Element	Atomic number	Density in g/cm^3
lithium	3	0.54
sodium	11	0.97
potassium	19	0.86
rubidium	37	1.53
caesium	55	1.88
francium	87	could not find

A Sodium floats on water.

a Using graph paper, or a graph-plotting program, plot a graph to show their data. Plot atomic number on the horizontal axis and density on the vertical axis. Leave space for francium even though they could not find any data for it.
b Water has a density of 1 g/cm^3. Objects sink if their density is more than this. Who was correct, Paul or Alison? Explain your answer fully.
c Draw a curve of best fit on the graph.
(i) What is the general trend in density down the group?
(ii) Use your graph to estimate the density of francium.

2 Nathif needed to analyse a sample of crystals to find out which metal they contained. He dissolved some of the crystals and then did two different analytical tests.
a No precipitate formed when he added some sodium hydroxide solution. Is the metal likely to be a transition metal or an alkali metal? Explain your answer.
b Nathif then did a flame test on some of the sample and saw a red flame. Which metal could be in the sample?

D

P

3 Heidi needed to find out which metal was present in a different sample of crystals. She dissolved some of the crystals so that she could analyse them.
a When she added some sodium hydroxide solution a green precipitate formed. Which two metals might have been in the sample?
b After a while, Heidi noticed that the precipitate had started to turn brown. Which metal was actually in the sample?

Glossary

alkali metal An element found in group 1 of the periodic table. Alkali metals include lithium, sodium and potassium.
alkaline A substance that turns universal indicator paper blue and has a pH greater than 7.
alloy A mixture of metals or metals and non-metals.
analyse To study a substance carefully to find out which chemicals it contains.
analytical Analytical tests are experiments to find out which chemicals a substance contains. The results of these tests are called analytical data.
atomic number The number of protons (positively charged particles) in a nucleus.
atoms Particles that are the smallest part of an element that can exist.
bond A chemical join between two atoms.
chemical reaction A chemical change in which new substances are formed but there is no change in the number of atoms of each element.
chemical symbol A code for an element consisting of one, two or sometimes three letters.
compound A substance containing two or more different elements chemically joined together.
corrosion The reaction between a metal and air or water.
density The mass of a substance compared to its volume. Substances with a high density feel very heavy for their size.
diatomic molecule A molecule made from two atoms chemically joined together.
displacement A type of chemical reaction in which a more reactive element pushes out or displaces a less reactive element from its compound.
displaces Pushes out.
ductile Can be stretched into wires.
electron A negatively charged particle that surrounds the nucleus in an atom.
element A substance made of only one type of atom. An element cannot be turned into another element or anything simpler using a chemical reaction.
endothermic A reaction that takes in energy from its surroundings.
exothermic A reaction that gives out energy, usually in the form of heat energy.
flame test An analytical test to find out which metal is present in a substance. Different metals produce different colours when held in a Bunsen burner flame.
formula The chemical code for a substance. A formula contains the symbols for the different elements in a substance. Plural: formulae.
group A vertical column of the periodic table.
halogen An element found in group 7 of the periodic table. Halogens include chlorine, bromine and iodine.
inert Unreactive.
insoluble Will not dissolve in water.
malleable Can be bent or hammered into shape without breaking.
molecule A particle made from atoms joined together by chemical bonds.
neutron Electrically neutral particle found in the nucleus of most atoms.
noble gas An element found in group 0 of the periodic table. Noble gases include helium, neon and argon.
nucleus Positively charged centre of the atom, containing protons and neutrons.
particles Tiny pieces that are often too small to see.
period A horizontal row of the periodic table.
periodic table A chart in which all the elements are arranged.
precipitate The insoluble solid formed in a precipitation reaction.
precipitation A reaction in which a solid is formed when two solutions are mixed together.
properties The characteristics of a substance. Chemical properties describe how the substance reacts with other substances. Physical properties include information such as colour, melting point and state.
proton Positively charged particle found in the nucleus of all atoms.
rust A type of iron oxide formed when iron reacts with oxygen and water.
sediment A solid that sinks to the bottom of a liquid without dissolving.
soluble Will dissolve in water.
solution The mixture formed when a substance dissolves in water.
sterilise To kill microorganisms such as bacteria.
suspension A solid mixed evenly in a liquid without dissolving.
symbol *see* chemical symbol
toxic Toxic substances are poisonous. They may cause death if they are swallowed, breathed in or come into contact with the skin.
transition metal An element found between groups 2 and 3 of the periodic table. Transition metals include copper, iron and gold.
trend A gradual change.

Making changes

A Chemicals are used to make fireworks.

Firework manufacturers use many different substances in their fireworks. Some substances produce exciting colours that light up the night sky. Others react together explosively to launch the firework into the air, to provide sheets of flame or loud bangs. Similar chemical reactions are also needed to extract metals from their ores, and to make gases in the laboratory and the substances you use at home. The food you eat is made of chemicals and it too has been produced using chemical reactions.

To understand the chemistry involved, it helps to know some types of reaction and what they produce. It also helps to know which metals are more reactive than others.

In this topic you will learn that:

- similar elements or compounds react in similar ways
- predictions can be made about the products of reactions, based on knowledge of similar situations
- addition of oxygen to a substance is oxidation and loss of oxygen from a substance is reduction
- extraction of metals depends on their reactivity.

Look at these statements and sort them into the following categories:

I agree, I disagree, I want to find out more

- All metals can be found in small lumps in the ground.
- Carbon dioxide is only useful for putting out fires.
- Organic food does not contain any chemicals.
- Most common metals were discovered a very long time ago.
- It is difficult to make and collect gases safely in the laboratory.
- You can tell that a substance is dangerous just by looking at it.

V

D

C1a.6.1

Oxygen and oxidation

By the end of these two pages you should:

- know how to collect oxygen produced in reactions
- know a simple laboratory test for oxygen
- be able to explain that oxidation is the addition of oxygen to a substance.

Which element makes up about a fifth of the air, nearly half the Earth's **crust** and over 85 per cent of the water in the world's oceans? The answer is oxygen, and it forms around 61 per cent of you, too.

Oxygen gas is colourless and has no smell. It is only slightly soluble in water, but just enough dissolves so that fish and other living things in seas and rivers can survive. Without oxygen, you could not release energy from your food and fires would not burn. When a substance reacts with oxygen, it gains oxygen and becomes **oxidised**. Some **oxidation** reactions are fast, such as **combustion** or burning. Others are slow, such as rusting.

1 Give three properties of oxygen.

2 What happens when a substance becomes oxidised?

A The combustion of a sparkler is an exothermic reaction.

B Rusting is a slow oxidation reaction.

Oxygen is easily produced in the laboratory from a substance called hydrogen peroxide. This is a colourless soluble liquid that is also used to bleach hair. Hydrogen peroxide slowly breaks down to form water and oxygen. This is the word equation for the reaction:

hydrogen peroxide → water + oxygen

A reaction like this, where a substance breaks down to form two or more other substances, is called a **decomposition** reaction. A small amount of black manganese dioxide powder makes the hydrogen peroxide decompose rapidly to give off lots of oxygen bubbles.

3 Give one use of hydrogen peroxide.

4 a What is a decomposition reaction?
b Write the word equation for the decomposition of hydrogen peroxide.

Have you ever wondered:

How do you collect and test gases?

You can collect gases in a **gas syringe**, but you can also collect oxygen over water. Diagram C shows this happening. The bubbles push the water out as they go into the gas jar. You can collect oxygen like this because it is only slightly **soluble**, so very little of it dissolves. This method does not work for gases that are very soluble in water, such as ammonia, because the gas dissolves instead of being collected.

How can you be sure that you have collected oxygen and not a different gas? Remember that oxygen helps substances burn. This is the basis of a simple laboratory test for oxygen. A wooden splint is lit and then gently blown out so that it glows. The glowing splint relights when it is put into a test tube of oxygen.

5 Describe the laboratory test for oxygen in your own words.

Remember that oxidation is the addition of oxygen to a substance. Very **reactive** metals such as magnesium are easily oxidised. A piece of magnesium ribbon burns in air with a bright white flame, and with an even brighter flame in oxygen. The magnesium is oxidised to a white powder called magnesium oxide. This is the word equation for the oxidation of magnesium:

magnesium + oxygen → magnesium oxide

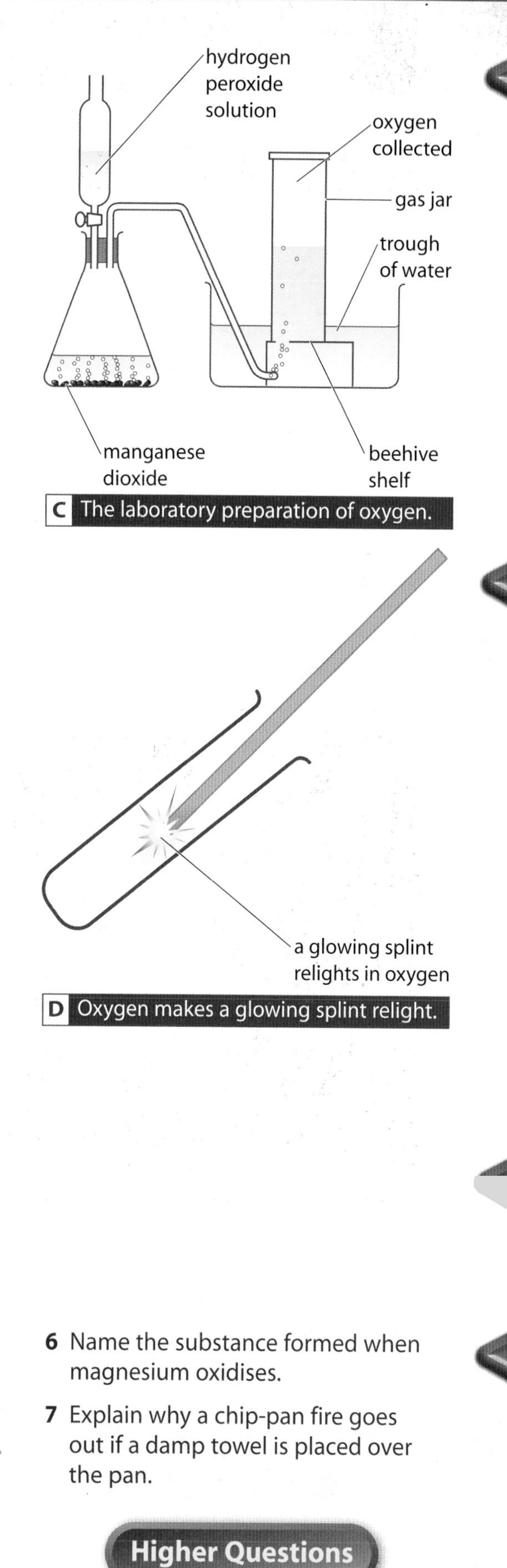

C The laboratory preparation of oxygen.

D Oxygen makes a glowing splint relight.

E Magnesium burns in oxygen with a white flame.

6 Name the substance formed when magnesium oxidises.

7 Explain why a chip-pan fire goes out if a damp towel is placed over the pan.

Summary Exercise

Higher Questions

The reactivity series

By the end of these two pages you should:

- know how to make and collect hydrogen
- know a simple laboratory test for hydrogen.

Hydrogen gas is given off when a metal reacts with an acid such as hydrochloric acid or sulphuric acid. Hydrogen gas is colourless and has no smell. It is nearly **insoluble** in water, so almost none of it dissolves. Hydrogen is the least dense gas and a balloon filled with it rises rapidly into the air.

1 Give three properties of hydrogen.

2 Which gas is given off when a metal reacts with an acid?

A Hydrogen bubbles are given off rapidly when magnesium reacts with acid.

Some metals react violently with acids, so it would be very dangerous to add lithium, sodium or potassium to hydrochloric acid or sulphuric acid. Other metals, such as aluminium and iron, react more slowly with acids. Some, including copper and gold, do not react with hydrochloric acid or sulphuric acid.

You can put the metals in order of their reactivity with an acid. This is called a **reactivity series**. Hydrogen is shown so that you can predict which metals will react with acids and which will not. Metals that are more reactive than hydrogen will react with acids, but metals that are less reactive than hydrogen will not.

3 Name two metals that will not react with acids.

4 Name two metals that will react more violently than aluminium with hydrochloric acid.

As well as hydrogen, a **salt** is formed when a metal reacts with an acid. The name of the salt depends upon the particular metal and acid used. The first part of its name comes from the metal, and the second part comes from the acid.

	Metal	Reaction with acids
most reactive	potassium sodium calcium	React violently with acids
	magnesium aluminium zinc iron tin lead	React, but the reaction becomes less violent as you go down the series
	hydrogen	
least reactive	copper silver gold platinum	Do not react with acids

B The reactivity series indicates how a metal will react with acids.

So, hydrochloric acid makes chloride salts. For example, magnesium reacts with hydrochloric acid to make magnesium chloride. Sulphuric acid makes sulphate salts. For example, sodium reacts with sulphuric acid to make sodium sulphate.

5 Which salt is formed when
  a magnesium reacts with sulphuric acid?
  b sodium reacts with hydrochloric acid?

Hydrogen is usually prepared in the laboratory by reacting zinc with sulphuric acid. The word and formula equations for the reaction are:

zinc + sulphuric acid → zinc sulphate + hydrogen

$$Zn + H_2SO_4 \rightarrow ZnSO_4 + H_2$$

Zinc is chosen because it is reactive enough to produce hydrogen quickly, but not so quickly that it would be dangerous.

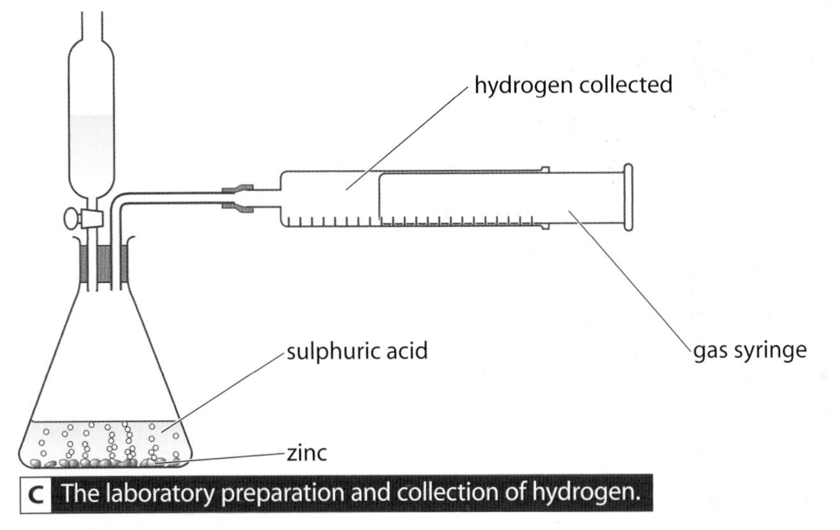

C The laboratory preparation and collection of hydrogen.

Like oxygen, hydrogen can be collected in a gas syringe or collected over water. As it is less dense than air, hydrogen can also be collected by **upward delivery**. Figure D shows how this works. The rising hydrogen pushes the air out of the test tube. This method does not work for gases that are denser than air, such as chlorine.

Hydrogen is highly **flammable**. This is the basis of a simple laboratory test for hydrogen. A test tube of hydrogen ignites with a 'pop' when a lighted splint is placed in it.

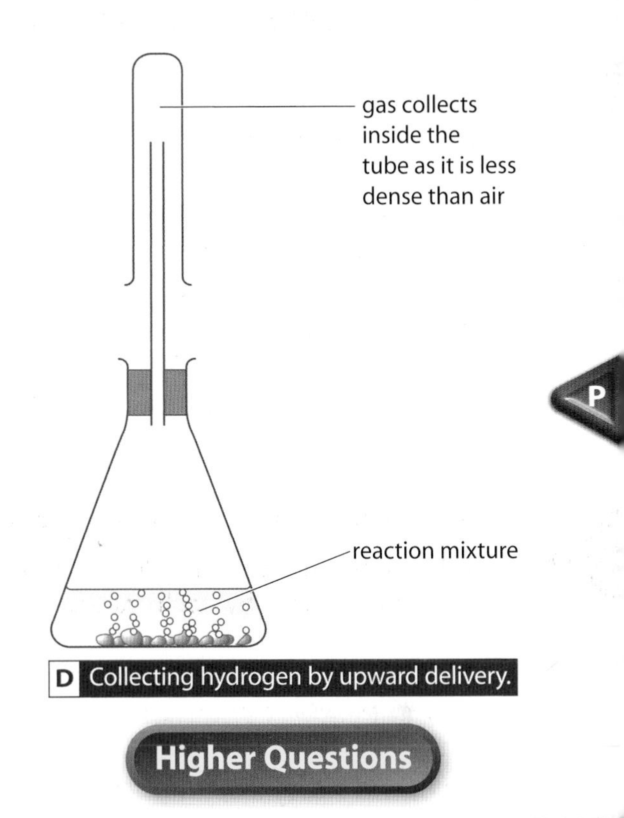

D Collecting hydrogen by upward delivery.

6 Describe the laboratory test for hydrogen in your own words.

7 Write the word equation for the reaction between iron and hydrochloric acid.

8 Why is hydrogen made in the laboratory by reacting zinc with sulphuric acid, rather than by reacting sodium with sulphuric acid?

9 Explain why hydrogen can be collected by upward delivery, and why it can be collected over water.

Higher Questions

C1a.6.3

Ores

By the end of these two pages you should be able to:

- recall that the least reactive metals are found uncombined in the Earth's crust
- explain that some metals occur as their oxides and can be extracted by reaction with carbon
- explain that reduction is the loss of oxygen from a substance
- explain that most metals have to be extracted from their ores, found in the Earth's crust
- relate the order of reactivity of metals to the stability of their ores, and to the method of extraction.

Have you ever wondered:

Did people always have metals?

Metals and their compounds are found in the Earth's crust. If metals in a rock are worth extracting, the rock is called an **ore**. Most ores contain only a small proportion of the desired metal. A lot of waste is produced during extraction.

A Copper ore is removed from the Earth's crust in huge mines like this.

1 What is an ore?

Metals with low reactivity do not react with oxygen. They do not tarnish or turn dull. Gold and platinum are like this. They are found in the Earth's crust **uncombined** with other elements. The least reactive metals were usually discovered first because people found shiny pieces of them on the ground or in streams. Gold was discovered thousands of years ago, even though it is rare.

2 Why are gold and platinum used to make jewellery?

Metals with high reactivity react easily with oxygen and become oxidised. For example, iron is oxidised to iron oxide. Metals like iron, lead and copper are found in the Earth's crust as their oxides. Chemical reactions are needed to extract them.

When a substance gains oxygen, we call it oxidation. When a substance loses oxygen, we say that **reduction** happens and the substance is **reduced**. For example, iron oxide can be reduced to iron.

3 What is copper oxide reduced to?

B Small pieces of gold can sometimes be found on river beds.

Iron, lead and copper can be extracted from their oxides using carbon. When the metal oxide is heated with carbon, the metal oxide is reduced and the carbon is oxidised. The equations for the reduction of lead oxide using carbon are:

$$\text{lead oxide} + \text{carbon} \rightarrow \text{lead} + \text{carbon dioxide}$$
$$2PbO + C \rightarrow 2Pb + CO_2$$

The carbon dioxide gas escapes, leaving the metal behind.

4 What is produced when iron oxide reacts with carbon?

Carbon can be used to extract a metal from its oxide if the metal is less reactive than carbon. This is why iron, lead and copper can be produced from their oxides using carbon. This does not work if the metal is more reactive than carbon.

5 Why can aluminium not be extracted from aluminium oxide using carbon?

The least reactive metals have the least **stable** oxides, so the oxides are easily reduced. The most reactive metals have the most stable oxides, and a lot of energy is needed to reduce them. Metals more reactive than carbon are usually extracted using electricity, so the most reactive metals were usually discovered last. For example, aluminium was only discovered around two hundred years ago. It is extracted using electricity.

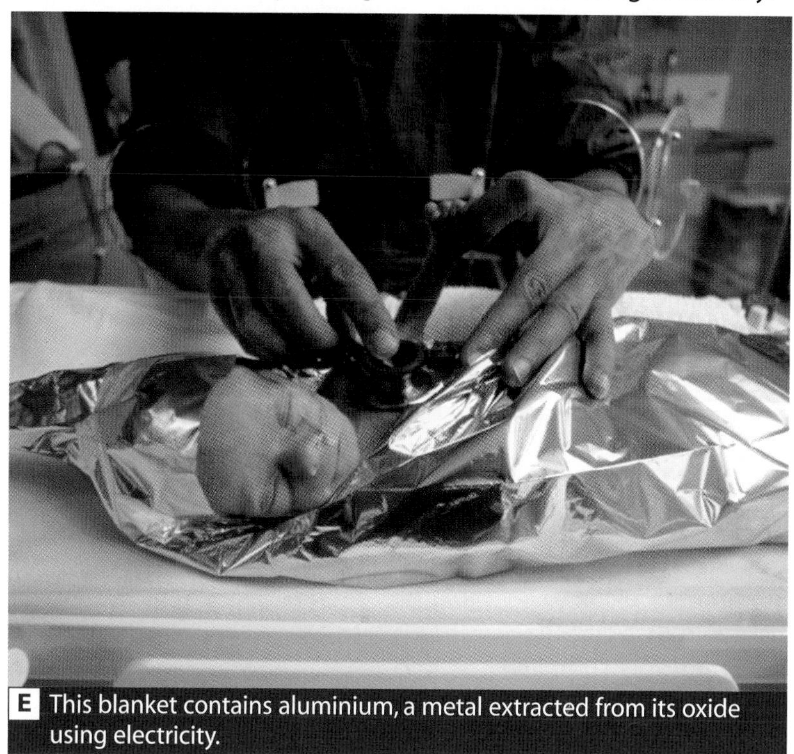

E This blanket contains aluminium, a metal extracted from its oxide using electricity.

C Iron is extracted from iron oxide in a blast furnace.

	Metal	
most reactive	potassium sodium calcium magnesium aluminium	Extracted using electricity
	carbon	
	zinc iron tin lead copper silver	Extracted by reacting with carbon
least reactive	gold platinum	Found uncombined in Earth's crust

D The reactivity series indicates how a metal is extracted from its oxide.

6 Explain whether copper or iron has the most stable oxide.

7 Why was silver discovered before potassium?

8 When iron is extracted from iron oxide, is the iron oxide reduced or oxidised? Explain your answer.

Summary Exercise

Higher Questions

Neutralisation

By the end of these two pages you should:

- be able to describe the reactions of dilute sulphuric acid and hydrochloric acid with metal oxides and metal hydroxides.

Remember that metals react with acids to produce salts and hydrogen. Metal oxides and metal hydroxides also react with acids. These reactions still produce salts, but water is formed instead of hydrogen. Reactions like these are called **neutralisation** reactions. Here are the equations for two examples:

copper oxide + sulphuric acid → copper sulphate + water

$$CuO + H_2SO_4 \rightarrow CuSO_4 + H_2O$$

sodium hydroxide + hydrochloric acid → sodium chloride + water

$$NaOH + HCl \rightarrow NaCl + H_2O$$

Neutralisation reactions are useful ways to make salts, especially if the metal itself is too reactive to use safely.

This is the word equation for the reaction of magnesium with hydrochloric acid, the formula equation is below it:

magnesium + hydrochloric acid → magnesium chloride + hydrogen

$$Mg + 2HCl \rightarrow MgCl_2 + H_2$$

In a balanced formula equation, you have the same number of each atom on both sides of the arrow. A large 2 is written next to the formula HCl to balance this equation. Without it, you would have two hydrogen atoms and two chlorine atoms on the right, but only one of each on the left. You must not change a formula to balance an equation: so it is 2HCl and not H_2Cl_2.

No bubbles of gas are produced in the reaction between a metal oxide or a metal hydroxide and an acid. So how can you tell that the reaction has happened? You could look for a colour change but often all the solutions are colourless. You can see if the acid has been neutralised using universal indicator paper. Unfortunately, this is not very accurate.

However, if the metal oxide is insoluble, you can add an **excess** to the acid. This means adding more than enough metal oxide to neutralise the acid. The extra solid can be filtered off, leaving the neutral salt solution. This is a useful way to make copper sulphate.

1 Copy and complete these word equations:
a metal + acid → salt + …
b metal oxide + acid → salt + …
c metal hydroxide + acid → salt + …

2 Why it is safer to make sodium chloride from sodium hydroxide and hydrochloric acid than from sodium and hydrochloric acid?

A Sodium chloride crystals.

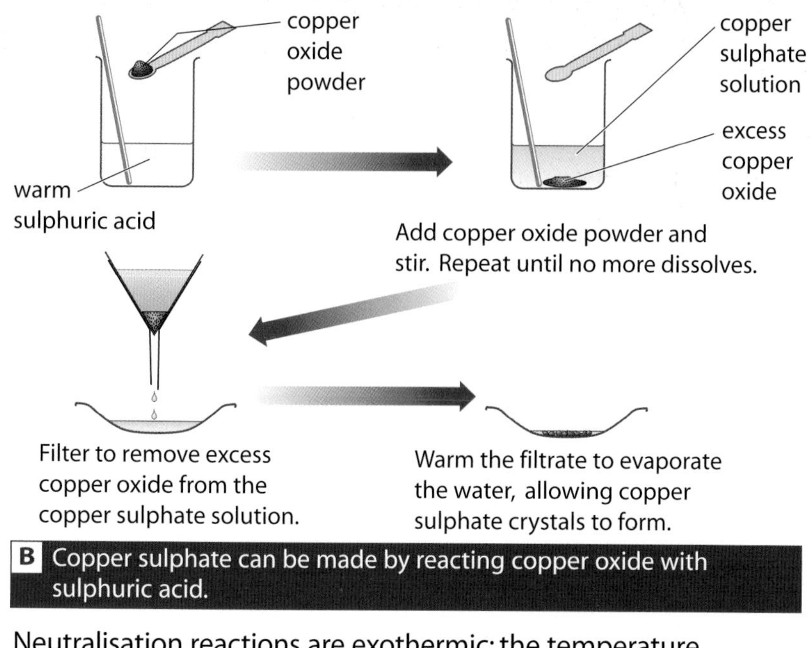

B Copper sulphate can be made by reacting copper oxide with sulphuric acid.

Neutralisation reactions are exothermic: the temperature increases as the reaction happens. The temperature is highest when the acid is exactly neutralised, and it starts to go down again as excess metal oxide or metal hydroxide is added. The reaction between sodium hydroxide solution and hydrochloric acid is often used to practise this method of following a reaction.

Sodium hydroxide solution is put into a polystyrene cup, which reduces heat loss to the surroundings. A small volume of hydrochloric acid is added, the mixture stirred briefly and the temperature recorded. This is repeated until the temperature has reached its maximum and then gone down again.

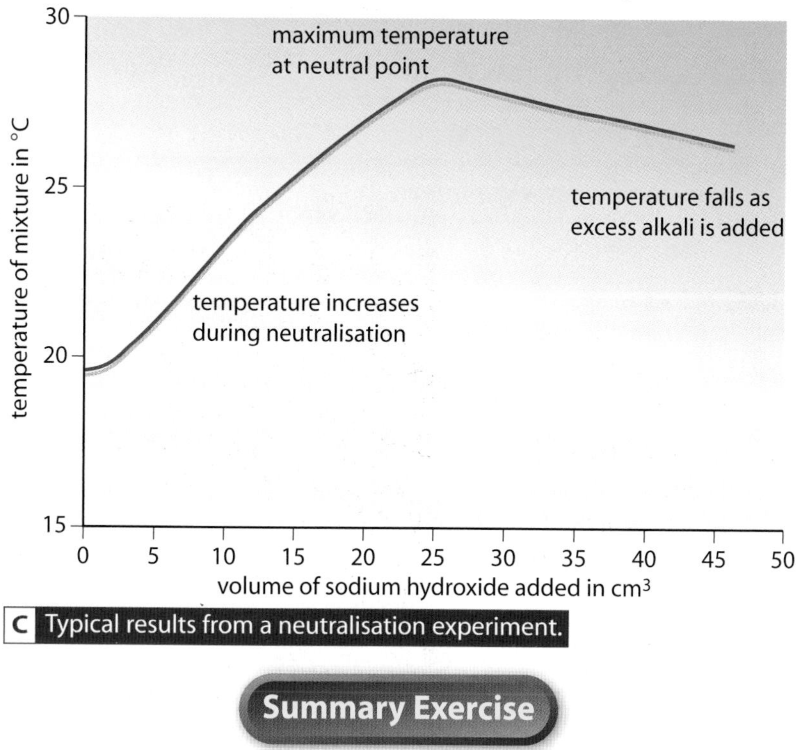

C Typical results from a neutralisation experiment.

Summary Exercise

3 Which salt will be made when copper oxide reacts with hydrochloric acid?

4 Suggest how the temperature could be measured in the experiment described on the left.

5 Why is it important to reduce the loss of heat in the experiment?

6 What does neutralisation mean? Give an example in your answer.

7 'Milk of Magnesia' contains magnesium hydroxide. It helps to settle an upset stomach by neutralising excess hydrochloric acid in the stomach.

a Which salt is formed when Milk of Magnesia reacts with stomach acid?

b What other product is formed in the reaction?

Higher Questions

Making useful salts

By the end of these two pages you should:

- be able to describe how to make pure, dry samples of insoluble salts from solutions of soluble salts
- know that some salts are used in fertilisers, in fireworks, as colouring agents and to help combustion of fuels
- be able to describe the use of hazard labels in the chemistry laboratory.

Soluble salts are compounds that dissolve in water. **Insoluble salts** are compounds that do not dissolve in water. The table lists some soluble and insoluble substances.

Soluble	Insoluble
all nitrates	none
most sulphates	lead sulphate, barium sulphate
most chlorides, bromides and iodides	silver chloride, silver bromide, silver iodide, lead chloride, lead bromide, lead iodide
sodium carbonate, potassium carbonate	most other carbonates
sodium hydroxide, potassium hydroxide	most other hydroxides

A Some soluble and insoluble substances.

1 Which substances in this list are soluble in water, and which are insoluble?
Sodium hydroxide, copper hydroxide, barium sulphate, copper sulphate, silver nitrate, lead chloride.

Lead nitrate and potassium iodide are soluble. Their solutions react to produce lead iodide and potassium nitrate. The equations are:

lead nitrate + potassium iodide → lead iodide + potassium nitrate

$Pb(NO_3)_2$ (soluble) + $2KI$ (soluble) → PbI_2 (insoluble) + $2KNO_3$ (soluble)

Lead iodide is insoluble. It forms a bright yellow **precipitate** – tiny particles of lead iodide suspended in the liquid.

Precipitation reactions are useful for preparing insoluble salts from solutions of soluble salts. When an insoluble salt has been produced by a precipitation reaction, it is filtered, washed on the filter paper and dried in a warm oven.

B Lead iodide forms a bright yellow precipitate.

2 **a** Which insoluble salt is formed when sodium chloride and silver nitrate react?
b Write the word equation for this reaction.

3 Explain how a precipitated salt can be washed and dried.

Many different salts are very useful. Nitrogen, phosphorus and potassium are very important for the healthy growth of plants. Fertilisers contain salts such as ammonium phosphate, potassium sulphate and ammonium nitrate.

C Fertilisers contain salts that help crops grow well.

4 **a** Which acid reacts with potassium hydroxide to make potassium sulphate?
b Write the equation for the reaction.

Have you ever wondered:

How do you make a firework?

Coloured salts are useful as colouring agents in oil paints and pottery glazes. Barium chromate makes lemon yellow paint, whereas cobalt phosphate makes violet paint. Fireworks contain salts to produce their wonderful colours when they burn. Copper chloride produces blue, barium chloride produces green and lithium carbonate produces red.

The fuel in fireworks must burn very quickly to shoot rockets into the sky and explode. **Oxidising agents** provide oxygen to make fuel burn quickly. Salts like potassium nitrate and potassium chlorate are used as oxidising agents in fireworks. Without them, fireworks would burn too slowly.

5 a Name the insoluble salt, used in bright red distress flares, made from strontium nitrate and sodium carbonate.
b Write the equation for the reaction.

D Pellets containing different salts used to make fireworks.

It would be dangerous and unwise to use chemicals without knowing their hazards. Containers of hazardous chemicals carry hazard warning symbols. The chart shows some of the common hazard symbols and their meanings.

Toxic

These substances can cause death.

Highly flammable

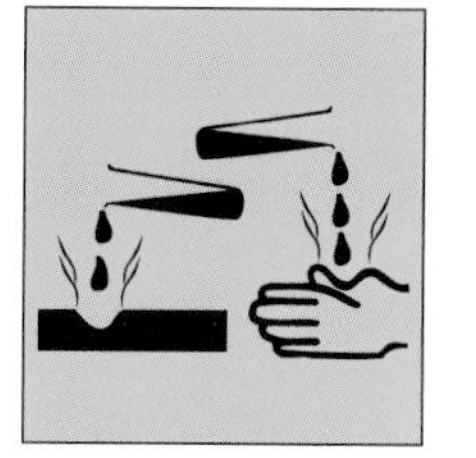

These substances catch fire easily.

Corrosive

These substances destroy living tissue.

Harmful

Similar to toxic substances but less dangerous.

Oxidising

These provide oxygen for other substances to burn.

Irritant

Not corrosive but may cause blisters or red skin.

F Some common hazard symbols.

E Salts are used in fireworks.

6 a What does an oxidising agent do?
b Sketch the hazard symbol that you would see on a bottle containing an oxidising agent.

7 Describe two uses of salts. In each case, name a salt that is used in that way.

Higher Questions

Baking with bubbles

A These cliffs are made from chalk, a form of calcium carbonate.

By the end of these two pages you should:

- be able to describe the reactions of dilute sulphuric acid and hydrochloric acid with metal carbonates and metal hydrogen carbonates
- be able to describe the use of sodium hydrogen carbonate as baking powder
- know the use in the home of carbon dioxide
- know how to collect carbon dioxide produced in reactions
- know a simple laboratory test for carbon dioxide.

Have you ever wondered:

How do the bubbles, that make cakes so light, actually get there?

Limestone, chalk and marble are types of rock that are mostly made of calcium carbonate. Geologists can do a simple test to see if a rock is one of these. They add a few drops of hydrochloric acid and, if they see fizzing, they know that the rock contains calcium carbonate.

A salt, water and carbon dioxide are produced when metal carbonates or metal hydrogen carbonates react with acids. Carbon dioxide gas causes the fizzing when hydrochloric acid hits limestone, chalk or marble. Here are the equations for this reaction:

calcium carbonate + hydrochloric acid → calcium chloride + water + carbon dioxide

$$CaCO_3 + 2HCl \rightarrow CaCl_2 + H_2O + CO_2$$

The same products are made if calcium hydrogen carbonate is used instead of calcium carbonate. Also, sulphuric acid reacts just as well as hydrochloric acid, but this would produce calcium sulphate.

1 Write the word equation for the reaction between calcium hydrogen carbonate and hydrochloric acid.

2 Which salt is produced when:
a sodium carbonate reacts with hydrochloric acid?
b sodium hydrogen carbonate reacts with sulphuric acid?

Sodium hydrogen carbonate, or bicarbonate of soda, is used in baking powder. The baking powder also contains a dry acid called tartaric acid. Moisture in the cake mixture allows the sodium hydrogen carbonate and tartaric acid to react, releasing carbon dioxide. The bubbles of carbon dioxide in the mixture make the cake rise in the oven.

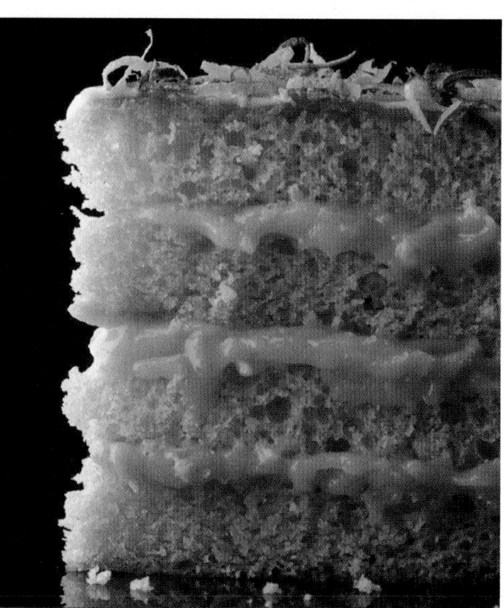
B Cakes rise because of carbon dioxide from sodium hydrogen carbonate.

Carbon dioxide is a colourless gas with no smell. It is denser than air, and it dissolves slightly in water to make a weak acid called carbonic acid. Carbon dioxide puts the fizz into fizzy drinks.

3 Explain why lemonade is slightly acidic.

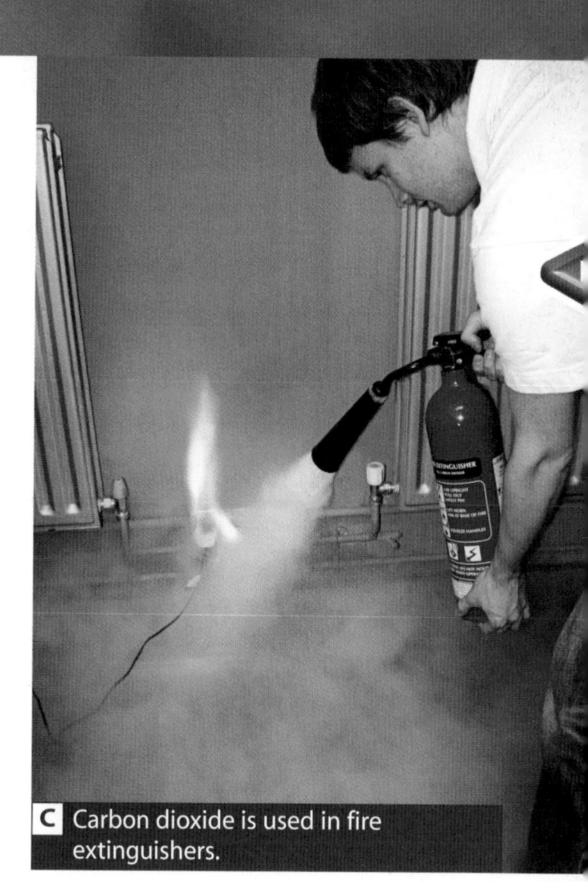

C Carbon dioxide is used in fire extinguishers.

Carbon dioxide is easily made in the laboratory by reacting marble chips with hydrochloric acid. Carbon dioxide can be collected in a gas syringe, over water or by **downward delivery**. This is the opposite of the upward delivery used for hydrogen. Carbon dioxide is denser than air, so it sinks into the container and pushes the air out.

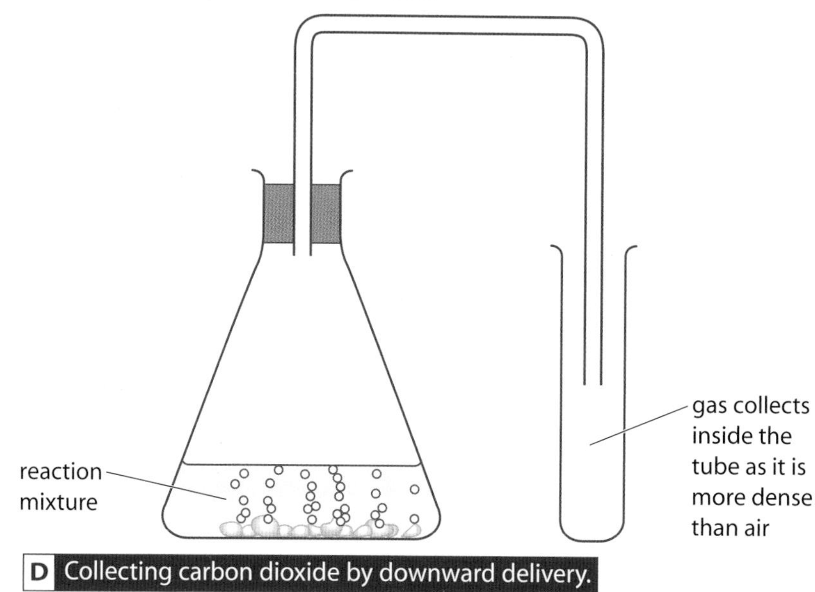

D Collecting carbon dioxide by downward delivery.

4 Suggest a problem with collecting carbon dioxide over water.

5 Carbon dioxide is colourless. Suggest a problem with collecting it by downward delivery.

Carbon dioxide does not allow fuels to burn. One test for carbon dioxide is to see if it will put out a lighted wooden splint, but some other gases will also do this, such as nitrogen. To check that a gas is carbon dioxide, it is bubbled through limewater. Carbon dioxide turns limewater cloudy white.

6 Describe the test for carbon dioxide.

7 Explain how sodium hydrogen carbonate is used in baking powder.

8 Complete these word equations:
a calcium carbonate + sulphuric acid →
b magnesium hydrogen carbonate + hydrochloric acid →

E Carbon dioxide turns limewater cloudy white.

Summary Exercise

Higher Questions

Breaking down in the heat

By the end of these two pages you should be able to:

- recall that when carbonates and hydrogen carbonates are heated they release carbon dioxide gas, and that this is called thermal decomposition
- write equations for these reactions.

Copper carbonate powder is green. When heated in a test tube it very quickly turns to black powder. It is no longer copper carbonate. Heat makes the copper carbonate break down to form copper oxide and carbon dioxide. The black powder is the copper oxide formed in the reaction. These are the equations for the reaction:

$$\text{copper carbonate} \rightarrow \text{copper oxide} + \text{carbon dioxide}$$

$$\underset{\text{(green)}}{CuCO_3} \rightarrow \underset{\text{(black)}}{CuO} + \underset{\text{(colourless gas)}}{CO_2}$$

Reactions in which a substance breaks down to form two or more other substances when heated are called **thermal decomposition**. *Thermal* means that heat is needed, and *decomposition* means that something decomposes or breaks down.

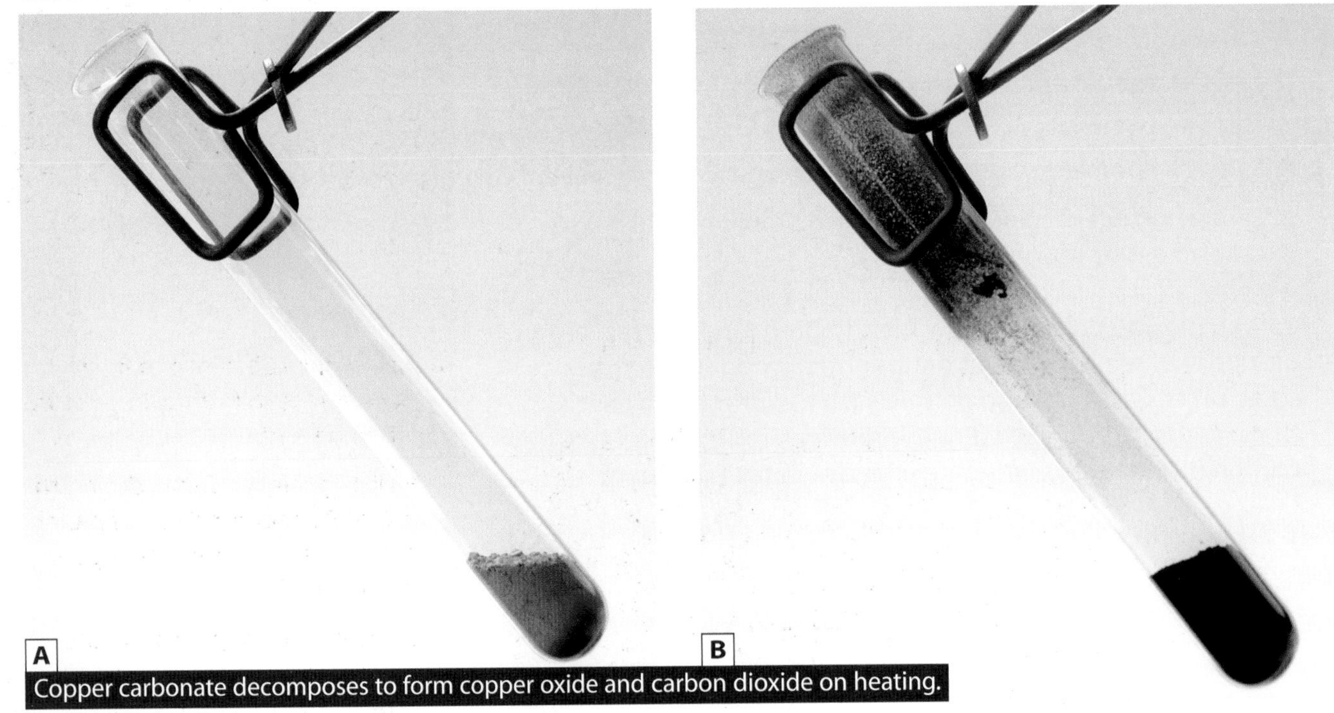

Copper carbonate decomposes to form copper oxide and carbon dioxide on heating.

1 What is a thermal decomposition reaction?

2 What happens to the carbon dioxide produced when copper carbonate is heated?

It is not only copper carbonate that decomposes to form a metal oxide and carbon dioxide when heated. Other metal carbonates do this too, but some are more easily decomposed than others. To predict which are easily decomposed and which are not, you need to look at the reactivity series again.

Metals at the top have very stable compounds. A lot of heat energy is needed to decompose them. As you go down the series, the metals become less reactive and their compounds less stable. Relatively little heat is needed to decompose the carbonate of a metal near the bottom of the reactivity series. For example, strong heating from a Bunsen burner for several minutes is needed to decompose calcium carbonate, but only a few seconds of heating is needed to decompose copper carbonate.

3 Write the word equation for the thermal decomposition of calcium carbonate.

4 Which carbonate will need the most heat energy to decompose it, magnesium carbonate or silver carbonate? Explain your answer.

most reactive metals | most stable carbonates | lots of heat needed

potassium
sodium
calcium
magnesium
aluminium
zinc
iron
tin
lead
copper
silver
gold
platinum

least reactive metals | least stable carbonates | little heat needed

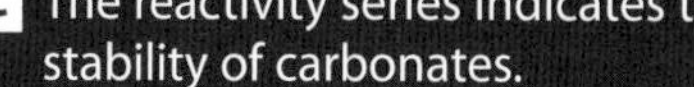

C The reactivity series indicates the stability of carbonates.

Metal carbonates are important **raw materials**. Calcium carbonate is used in the manufacture of iron. Lithium carbonate is used in the manufacture of heat-resistant glass for kitchens and laboratories.

D Lithium carbonate is a raw material in the manufacture of heat-resistant glass.

Metal hydrogen carbonates also decompose when heated. They release carbon dioxide gas, just like metal carbonates, but the other products are different. A metal hydrogen carbonate forms a metal carbonate and water when it decomposes. These are the equations for the thermal decomposition of sodium hydrogen carbonate:

sodium hydrogen carbonate → sodium carbonate + water + carbon dioxide

$$2NaHCO_3 \rightarrow Na_2CO_3 + H_2O + CO_2$$

5 Write the word equation for the thermal decomposition of calcium hydrogen carbonate.

E Stalagmites and stalactites consist of calcium carbonate, produced when calcium hydrogen carbonate in water decomposes.

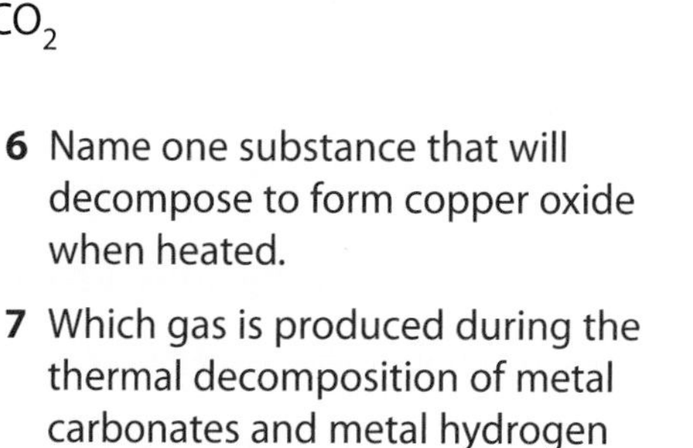

6 Name one substance that will decompose to form copper oxide when heated.

7 Which gas is produced during the thermal decomposition of metal carbonates and metal hydrogen carbonates?

8 Explain why copper carbonate, rather than potassium carbonate, is often used to demonstrate thermal decomposition in the laboratory.

Summary Exercise

Higher Questions

Chemicals in food

By the end of these two pages you should be able to:

- recognise that cooking involves chemical changes that lead to new products
- discuss the differences between 'natural' and artificial substances, including whether they are the same or chemically different, and any impacts on health
- H interpret data linking a chemical in food with a health impact.

A Baking bread involves chemical reactions such as browning.

The brown crust on a loaf of bread is caused by chemical reactions during baking, such as the decomposition of sugar to form carbon and water. Another browning reaction happens when sugars and proteins in the bread dough react together to form new substances. Similar reactions happen in other foods when they are cooked.

1 Describe two chemical reactions that cause bread to turn brown when it is baked.

Have you ever wondered:

Can you get cancer from eating too many food additives?

Chemical changes during cooking may lead to harmful products being made. Overcooking food on a barbecue produces chemicals called PAHs. Fried potato crisps contain tiny amounts of acrylamide. These chemicals have been shown to cause cancer in animals, although no link with cancer in people has been found.

B It is easy to burn food on a barbecue.

2 Suggest why some food scientists think that we should not eat too much burnt barbecue food or potato crisps.

Table salt is sodium chloride. It is found naturally in most food, and it is often added to improve flavour. However, if you have too much salt in your diet you risk high blood pressure. This can lead to heart disease and strokes. Food scientists recommend that you have no more than 6 g of salt a day in your diet.

3 Bill likes a lot of salt on his food. Give two possible health problems that this might cause.

C Common table salt is sodium chloride.

Have you ever wondered:

Could you tell the difference between ice-cream made with artificial vanilla and natural vanilla?

Vanilla is used to flavour ice-cream, yogurt and chocolate. Natural vanilla is extracted from vanilla beans, but natural vanilla is expensive. Ethyl vanillin is a chemical with a much stronger vanilla flavour than natural vanilla. It is cheap to make, so it is widely used as an artificial vanilla flavouring.

4 Why are food manufacturers more likely to use artificial flavourings than natural flavourings?

D Natural vanilla is extracted from vanilla beans.

Have you ever wondered:

Are artificial sweeteners good for you?

The bacteria in your mouth can produce acids from sugar, which then attack your teeth and cause tooth decay. Artificial sweeteners replace natural sugar in some types of chewing gum and other 'sugar-free' food. Mouth bacteria cannot produce acids from an artificial sweetener called xylitol, reducing the chance of tooth decay. Sorbitol, another artificial sweetener, can cause diarrhoea if eaten in large amounts.

Have you ever wondered:

How can sweeteners taste like sugar but have no 'calories'?

Sucralose tastes like sugar but has fewer 'calories' than the real thing. It is made from sugar that has been altered slightly. Sucralose still tastes sweet, but this small change is enough to stop your body digesting it.

5 Give one advantage and one disadvantage of artificial sweeteners.

6 Name two artificial sweeteners and discuss their possible impact on human health.

Liquids and gases in the home

A Ammonia is used in household cleaning products.

By the end of these two pages you should know:

- how to collect ammonia produced in reactions
- a simple laboratory test for ammonia and for chlorine
- the use in the home of ammonia, water, hydrochloric acid, phosphoric acid, ethanoic acid and citric acid.

Ammonia is a gas with a sharp, unpleasant smell. It is used in fertilisers and in home cleaning products. Ammonia is usually prepared in the laboratory by heating ammonium chloride and calcium hydroxide together:

ammonium chloride + calcium hydroxide → water + calcium chloride + ammonia

$$2NH_4Cl + Ca(OH)_2 \rightarrow 2H_2O + CaCl_2 + 2NH_3$$

You cannot collect ammonia over water because it is very soluble. As it is less dense than air, it is collected by upward delivery, just like hydrogen. But, unlike hydrogen, it must be prepared in a fume cupboard.

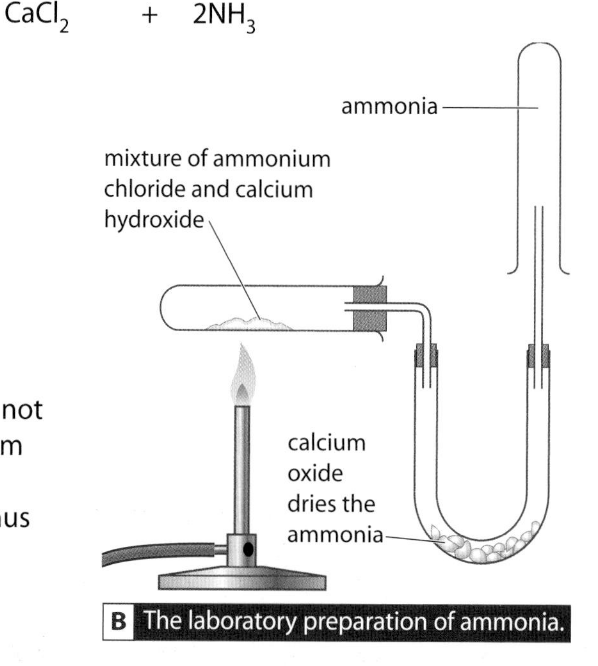

B The laboratory preparation of ammonia.

1 Why should ammonia be prepared in a fume cupboard?

2 Why is ammonia collected by upward delivery and not over water?

How can you be sure that you have collected ammonia and not a different gas? Ammonia forms a white smoke of ammonium chloride when hydrogen chloride gas, from concentrated hydrochloric acid, is held near it. It also makes damp red litmus paper turn blue.

3 Describe a test for ammonia.

Ammonia is not the only gas that can be tested using litmus paper. Chlorine is used to sterilise drinking water. It makes damp blue litmus paper turn red and then white. Chlorine also makes damp starch-iodide paper turn blue-black.

4 Look at the changes that ammonia and chlorine make to damp litmus paper.
Which gas is acidic, and which is alkaline?

C Chlorine kills harmful bacteria in drinking water and swimming pools.

Water is a very good solvent. Many substances dissolve in it, including solids, liquids and gases. Many household products contain water. It is usually listed as 'aqua' on liquid soap and shampoo labels.

5 Name at least three substances that are soluble in water. Try to name solids, liquids and gases.

Laboratory acids such as sulphuric acid and hydrochloric acid must be handled carefully. You should wear eye protection and clean up spills with plenty of water. You might be surprised to learn that acids are found around the home. However, these are weak acids or very **dilute** acids, so they are safer to use.

Several acids are used to give a tangy taste to food. **Citric acid** is found naturally in oranges and lemons. It is an ingredient in sherbet, fizzy drinks, sweets and ice cream. Vinegar for your chips is dilute ethanoic acid (also called acetic acid). Dilute means mixed with water. Ethanoic acid is also used in crisps and sauces. Phosphoric acid is a strong acid but is diluted with water for use in cola. It is also found in jam, jelly and cheese.

Acids are ingredients in household cleaning products. Citric acid is found in limescale removers, ethanoic acid is in window cleaning liquids, and hydrochloric acid is used in toilet cleaners. Concentrated phosphoric acid is an ingredient in rust removers.

6 Why should you wear rubber gloves when using household cleaners?

D Citric acid is used in limescale removers.

7 How is ammonia made in the laboratory?

8 Apart from their smells, describe how you could tell ammonia and chlorine apart.

Summary Exercise

Higher Questions

Solids in the home

By the end of these two pages you should:

- be able to describe hydration and dehydration
- know the use in the home of carbohydrates, caustic soda and common table salt.

What do doctors and walls have in common? The answer is gypsum, a soft rock consisting of **hydrated** calcium sulphate. **Hydration** means that water has been chemically joined to the calcium sulphate. **Dehydration** happens if the gypsum is heated to around 180°C. Most of the water is driven out and the gypsum is **dehydrated** to become the substance we know as plaster of Paris. Plaster of Paris is used by doctors. It also resists fire and is used to line walls.

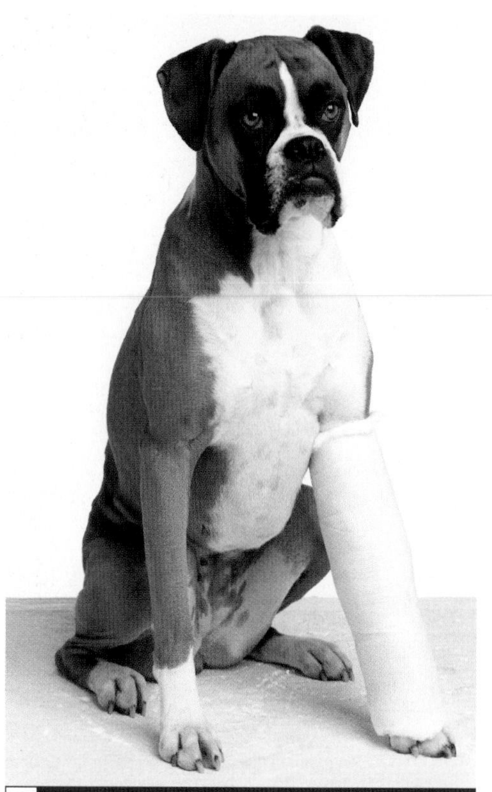

A Plaster of Paris casts are used to support broken bones.

You have seen how sulphuric acid reacts with metal oxides or metal hydroxides in neutralisation reactions to produce salts and water. However, concentrated sulphuric acid also has the ability to dehydrate other substances. Sugar molecules are made from carbon, hydrogen and oxygen atoms. Sulphuric acid is so good at dehydrating, it can even remove the hydrogen and oxygen atoms from sugar **molecules** as water, leaving the carbon behind. The reaction produces a lot of heat, and a hot column of black carbon emerges from the container. Here are the word and formulae equations for the reaction:

$$\text{sugar} \xrightarrow{\text{concentrated } H_2SO_4} \text{carbon} + \text{water}$$

$$C_6H_{12}O_6 \xrightarrow{\text{concentrated } H_2SO_4} 6C + 6H_2O$$

1 What do hydration and dehydration mean?

Sugar, wood and cotton are linked in an unexpected way. They all contain **carbohydrates**, which contain carbon, hydrogen and oxygen atoms only. Ordinary table sugar contains a small carbohydrate molecule called sucrose. You are able to digest sucrose so that your body can use it. However, the molecules found in wood and cotton each consist of thousands of glucose molecules, chemically joined end to end. This means that you cannot digest these carbohydrates, and it also gives them different properties to table sugar. They form tough fibres that are useful for making paper, building materials and clothing.

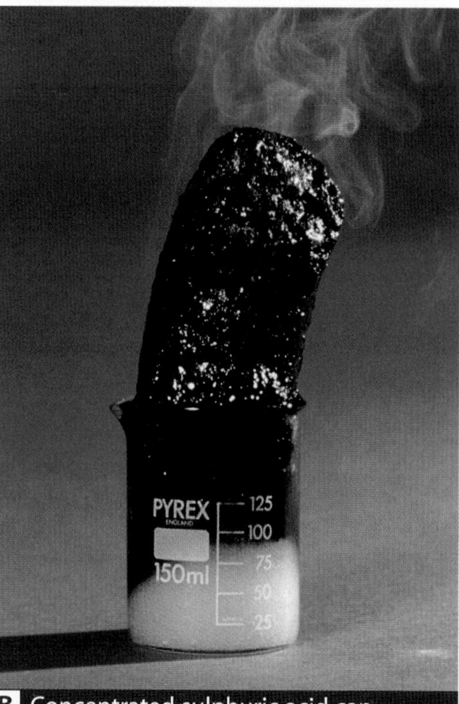

B Concentrated sulphuric acid can dehydrate sugar, leaving carbon behind.

C Paper, cotton and wood contain carbohydrates.

2 What is a carbohydrate?

3 Name three items in your home that contain carbohydrates.

Common table salt is sodium chloride. Dishwashers need very pure sodium chloride to keep their built-in water softener working. Without it, the dishwasher would gradually fail to clean the cups and plates properly. Dishwasher salt is in large pieces so that water can pass between them easily.

4 Table salt also contains other substances. Some of these are needed to stop the tiny grains sticking together while others provide us with minerals needed for health.
   a Suggest why table salt must not be used in a dishwasher.
   b Suggest why dishwasher salt must not be eaten.

D Dishwasher salt is very pure sodium chloride.

Caustic soda is the common name for sodium hydroxide. It is an alkali that is used in drain cleaners. When the drain cleaner is poured into the drain, the caustic soda reacts with the greasy substances from cooking and washing. These break down and the drain is unblocked.

5 The chemical formula of caustic soda is NaOH. Name the elements it contains.

E Caustic soda is used in drain cleaners.

Summary Exercise

6 White anhydrous copper sulphate contains no water molecules. How could you change it to blue hydrated copper sulphate?

7 a What is the chemical name for caustic soda?
   b Describe a household use of caustic soda.

Higher Questions

Questions

Multiple choice questions

1 Which of the following is a laboratory test for oxygen?
- **A** A lighted splint goes pop.
- **B** A glowing splint ignites.
- **C** A lighted splint goes out.
- **D** Damp litmus paper turns blue.

2 Which gas is produced when a metal reacts with an acid?
- **A** ammonia
- **B** carbon dioxide
- **C** hydrogen
- **D** oxygen

3 Which of the four metals listed below is the least reactive?
- **A** potassium
- **B** platinum
- **C** iron
- **D** copper

4 Why can iron be extracted by heating iron oxide with carbon?
- **A** Iron is less reactive than carbon.
- **B** Iron is more reactive than carbon.
- **C** Iron is found in its native state.
- **D** Iron ore contains iron oxide.

5 What are oxidation and reduction?
- **A** The gain of oxygen by a substance.
- **B** The loss of oxygen from a substance.
- **C** Oxidation is the loss of oxygen and reduction is the gain of oxygen.
- **D** Oxidation is the gain of oxygen and reduction is the loss of oxygen.

6 Which general equation is correct?
- **A** acid + metal → salt + water
- **B** acid + metal oxide → salt + hydrogen
- **C** acid + metal oxide → salt + water
- **D** acid + metal carbonate → salt + carbon dioxide

7 What does this hazard symbol mean?

- **A** flammable
- **B** oxidising
- **C** explosive
- **D** harmful

8 Which two substances would react to produce an insoluble salt?
- **A** Hydrochloric acid and sodium hydroxide.
- **B** Potassium nitrate and sodium hydroxide.
- **C** Silver nitrate and sodium nitrate.
- **D** Lead nitrate and potassium iodide.

9 The correct word equation for a decomposition reaction is:
- **A** copper oxide → copper carbonate + carbon dioxide
- **B** sodium carbonate → sodium hydrogen carbonate + oxygen
- **C** sodium hydrogen carbonate → sodium carbonate + water + carbon dioxide
- **D** calcium carbonate → calcium + carbonate

10 During the cooking of any food, which change will happen?
- **A** A reversible chemical change.
- **B** An irreversible chemical change.
- **C** No chemical change.
- **D** A change of state.

11 A simple laboratory test for ammonia is:
- **A** damp litmus paper turns blue.
- **B** damp litmus paper turns red.
- **C** damp litmus paper turns red, then bleaches white.
- **D** limewater turns cloudy white.

12 Which two methods are used to collect ammonia in the laboratory?
- **A** Over water and by gas syringe.
- **B** Over water and by downward delivery.
- **C** By downward delivery and by gas syringe.
- **D** By upward delivery and by gas syringe.

13 A simple laboratory test for chlorine is:
A damp litmus paper turns blue then white.
B damp litmus paper turns white then blue.
C damp litmus paper turns red then white.
D damp litmus paper turns white then red.

14 Name the chemical shown by the formula $NaHCO_3$.
A sodium hydrogen carbonate
B sodium carbon oxide
C sodium carbohydrate
D sodium hydrogen carbon trioxide

15 Which substance is the main component of common table salt?
A sodium hydroxide
B potassium iodide
C sodium chlorate
D sodium chloride

Short-answer questions

1 Use the reactivity series below to help you answer these questions about metals.

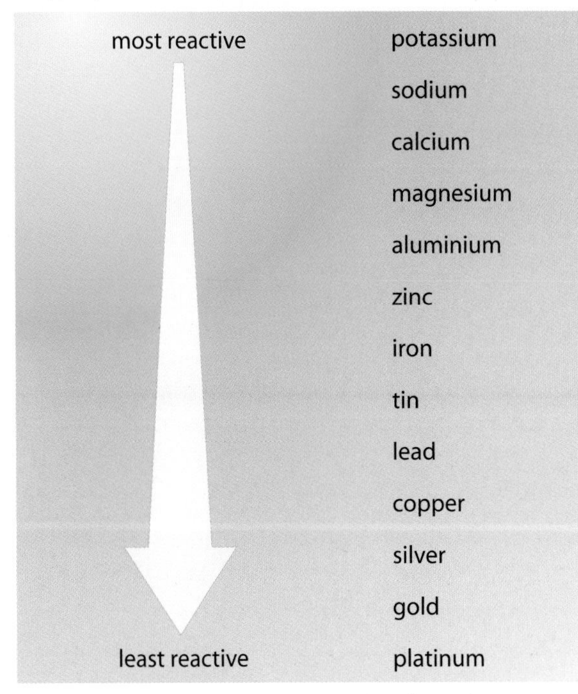

a Name two metals that can be found uncombined in the Earth's crust.
b Name two metals that cannot be extracted from their oxides by heating with carbon.
c Explain why sodium was not discovered until 1807, even though it is the sixth most abundant element in the Earth's crust and the most abundant metal in seawater.

2 David wanted to prepare some dry, insoluble barium sulphate crystals from barium nitrate solution and another chemical.
a Which other chemical (sulphuric acid, hydrochloric acid or nitric acid) does David need?
b Outline the method that David needs to follow to make his barium sulphate crystals. Include in your answer the names of suitable types of apparatus.

3 Draw the hazard symbol or symbols that you are likely to see on containers of the following substances, and explain why you would expect to see them there.
a Dilute ethanoic acid.
b Concentrated sulphuric acid.
c Potassium nitrate, which is a salt that helps other substances to burn.
d Barium chloride, which is a salt used in rat poison.

D

4 You have just won a contract to explain some science to children at a primary school. There are three topics to choose from: the chemistry of fireworks, chemistry in the home or the chemistry of food. You have been given enough money to make a television programme, a booklet or a poster. Decide which topic you would like to explain to the children. Write a short script for a television programme, produce a small booklet or draw a poster.

P

Glossary

D

carbohydrates Substances that are compounds of carbon, hydrogen and oxygen.
caustic soda The common name for sodium hydroxide.
citric acid A weak acid found naturally in oranges and lemons.
combustion Burning.
crust The rocky outer layer of the Earth.
decomposition A reaction in which a substance breaks down into two or more new substances.
dehydrated A dehydrated substance has had its water removed.
dehydration A reaction in which water is removed from a substance.
dilute A dilute substance is mixed with water to make it less concentrated.
downward delivery The method of collecting dense gases in a container.
excess Enough of a substance to react completely and have some left over at the end of the reaction.
flammable Able to burn.
gas syringe A syringe, usually made of glass, used to collect gases in the laboratory.
hydrated A hydrated substance contains water molecules.
hydration A reaction in which water is added to a substance.
insoluble Does not dissolve in water.
insoluble salt A salt that does not dissolve in water.
molecule A particle made from atoms joined together by chemical bonds.
neutralisation The reaction between an acid and an alkali (or other base) to form a neutral solution.
ore A compound of a metal found naturally from which the metal can be extracted.
oxidation A chemical reaction in which oxygen is added to a substance.
oxidised A substance that has had oxygen added to it in a chemical reaction is said to be oxidised.
oxidising agent A chemical that can cause an oxidation reaction.
precipitate The insoluble solid formed in a precipitation reaction.
precipitation reaction A reaction in which a solid is formed when two solutions are mixed together.
raw material A starting substance for the manufacture of a particular chemical.
reactive The more reactive a substance is the more likely it is to react with other substances.
reactivity series A list of metals or non-metals, in order of their reactivity.
reduced A substance that has had oxygen taken from it in a chemical reaction is said to be reduced.
reduction A chemical reaction in which oxygen is taken away from a substance.
salt One of the substances produced in a neutralisation reaction.
soluble Dissolves in water to make a solution.
soluble salt A salt that dissolves in water.
stable A stable compound needs a lot of energy to break it down into simpler substances.
thermal decomposition A reaction in which a substance is broken down using heat energy.
uncombined Not joined together.
upward delivery The method of collecting gases, that are less dense than air, in an upside-down container.

b.7.0

There's only one Earth

A Cars can pollute our environment.

B Rubbish as sculpture: a lifetime's worth of electronic waste.

Modern science and technology give us a fantastic standard of living. But some problems go with this. Traffic congests our streets. Vehicles pump choking gases into the air. Globally, our use of fossil fuels could be changing the Earth's climate in dangerous and unpredictable ways.

The Earth's resources are being used alarmingly fast. Oil may run out in your lifetime and many other raw materials may become scarce. Ways must be found to use resources more efficiently, or more mines and factories will be needed. More materials must be recycled, if only to avoid us being buried in rubbish!

In this topic you will learn that:

- all substances are obtained or made from substances in the Earth's crust, sea or atmosphere
- many natural resources are mixtures of substances
- products obtained from crude oil are essential to modern life
- production and disposal of substances have environmental impacts.

Look at these statements and sort them into the following categories:

I agree, I disagree, I want to find out more

- The Earth has plenty of resources. We just need to open more mines.
- When the oil runs out we can run our cars on alcohol, like they do in Brazil.
- Careful recycling won't solve all our problems, but it will give us more time to find new scientific solutions to our problems.
- Global warming is wrecking the Earth and it's all our fault!

Getting energy from fuels

By the end of these two pages you should know:

- how to recognise a good fuel
- that you get carbon dioxide and water when you burn most fuels.

Crude oil, natural gas, coal, wood, wax and paper are all **fuels**. Fuels contain stored chemical energy. When they burn they react with oxygen in the air and give out heat energy. You can use this energy to cook, move a car, heat a house or make electricity. The general reaction looks like this:

fuel + oxygen → waste gases (+ energy)

Reactions that give out energy like this are called **combustion** reactions.

1 Which of these could be used as a fuel? Household rubbish, scrap iron, cooking oil, concrete, waste plastic, cowpats.

2 What other kind of energy do combustion reactions often give out alongside heat energy? (*Hint*: think of a candle.)

A Natural gas is useful for cooking.

We use fuels to get heat energy, so a useful fuel must contain a lot of stored energy that is released when you burn it. However, some fuels also make black sooty smoke. This can cause pollution. It also means that you are not getting as much energy out as you could. **Sootiness** is something to avoid when choosing a fuel.

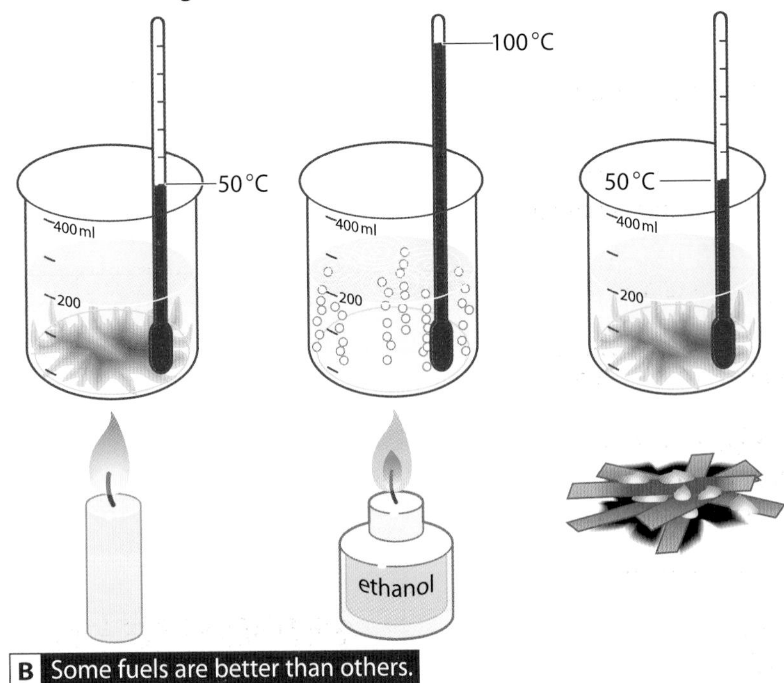

B Some fuels are better than others.

Candles burn with a yellow flame. Candles do not give out much heat and they make a lot of sooty smoke. Some of the fuel is not burning and giving out its energy. **Ethanol** (a type of alcohol) burns with a clear, clean, pollution-free flame. You get all the energy out so it can be a very good fuel. Brazilians use it in their cars instead of petrol.

Fuels such as wood and coal burn with sooty, yellow flames, like wax. They also leave a lot of ash behind as a **residue** when they burn.

Many fuels such as oil and natural gas are chemicals called **hydrocarbons**. These compounds are made from carbon and hydrogen atoms only. When they burn, the carbon atoms react with oxygen to form carbon dioxide. The hydrogen atoms react with oxygen to make water.

3 Suggest two reasons why ethanol is a better fuel than candlewax.

4 Suggest two reasons why very few people heat their homes using coal today.

C Burning natural gas: a methane molecule (CH_4) reacts with two oxygen molecules ($2O_2$) to make one carbon dioxide (CO_2) and two water ($2H_2O$) molecules.

You can show this reaction as a word equation.

hydrocarbon + oxygen → carbon dioxide + water (+ energy)

Most other fuels contain carbon and hydrogen atoms, so they too produce carbon dioxide and water when they burn.

5 Which gas in air is used in combustion?

6 Copy and complete the word equation for the combustion reaction of methane in air.
methane + __________ → carbon dioxide + __________

7 Coal is mostly made of carbon atoms. What gas will you get if you burn coal in lots of air?

8 Match the 'ends' to the 'beginnings' to make sensible sentences. Copy these into your book.

All fuels release their stored energy…	…when hydrocarbons react with oxygen.
Carbon dioxide and water form…	…when it burns with a clear, smoke-free flame.
You know you have a good fuel…	…when you burn them.

Summary Exercise

The dangers of incomplete combustion

By the end of these two pages you should know that:

- if a fuel burns without enough oxygen it can produce carbon monoxide gas
- carbon monoxide is a toxic gas that reduces the ability of the blood to carry oxygen
- faulty gas heating appliances can give off carbon monoxide.

A Carbon monoxide is dangerous.

Have you ever wondered:

Did you know carbon monoxide can suffocate you to death before you realise it?

Newspapers sometimes carry stories about people being killed by gas from 'faulty' gas boilers. The victim might have felt sick or had a headache, and fallen asleep. They never woke up. They are victims of silent, odourless carbon monoxide.

If natural gas (**methane**) burns with all the oxygen it needs, the products will be carbon dioxide and water. This is **complete combustion**. If there is not enough oxygen around, the hydrogen atoms react to form water as usual, but the carbon atoms can be left on their own. You see the carbon as soot. Sooty air can cause breathing problems. The big problem comes if there is *not quite enough* oxygen. The carbon atoms can only get one oxygen atom each instead of two, forming **toxic** carbon monoxide. This is **incomplete combustion**.

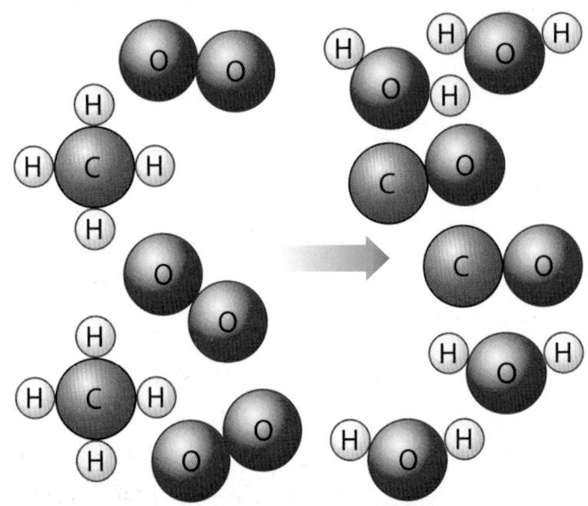

B Incomplete combustion of natural gas: two methane molecules ($2CH_4$) react with three oxygen molecules ($3O_2$) to make two carbon monoxide ($2CO$) and four water ($4H_2O$) molecules.

1 Copy and complete the word equation for the incomplete combustion of methane (CH_4).
methane + insufficient oxygen → water + carbon __________ (and/or carbon)

2 Copy and complete this balanced formula equation for the side reaction that forms the 'soot' (carbon) when methane burns in insufficient oxygen.
$CH_4 + O_2 \rightarrow C + 2$____

3 A yellow Bunsen burner flame leaves a sooty residue if you heat a test tube with it. Is this complete combustion or incomplete combustion? Explain your answer.

Gas central heating boilers, water heaters, cookers and fires need enough air to ensure complete combustion. The boiler flame should be pale mauve with a central blue cone, just like the roaring flame of a Bunsen burner. If the air-intake vent gets blocked, incomplete combustion will turn the flame yellow and soot will form around the gas jet.

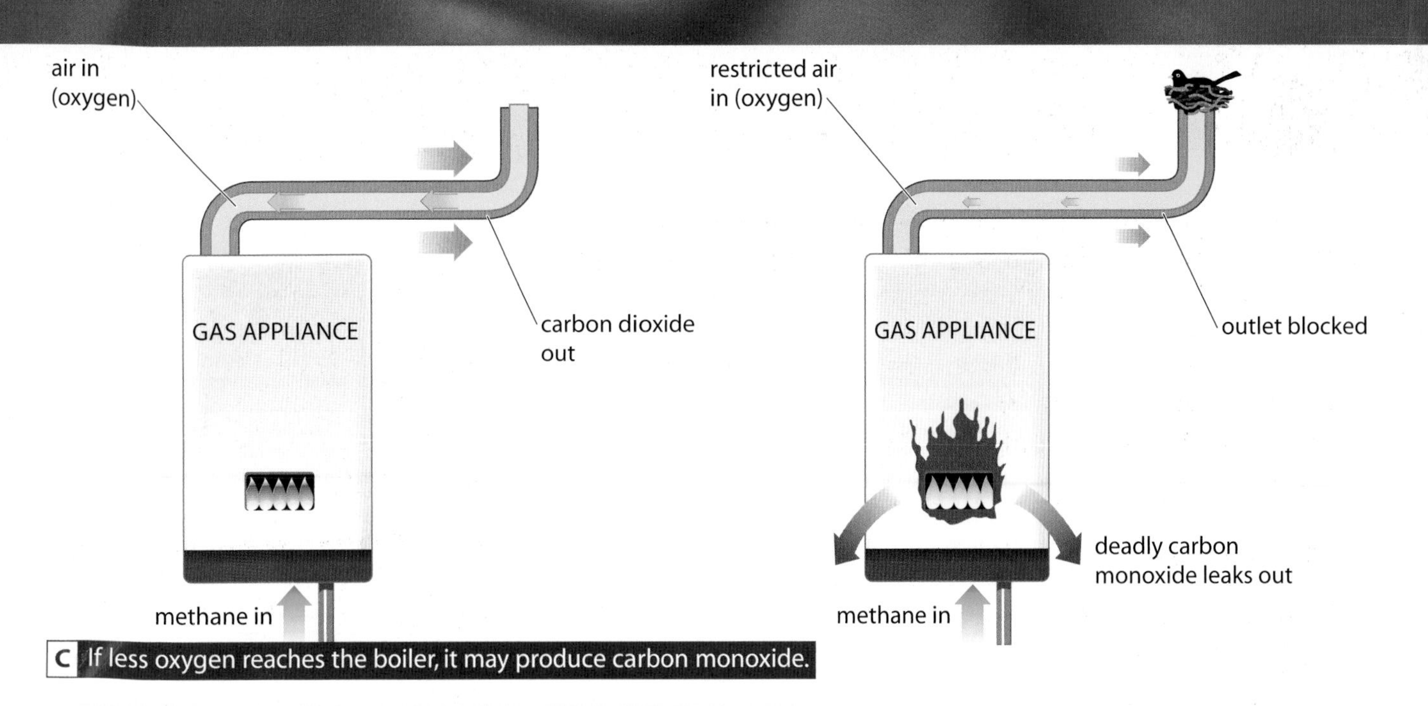

C If less oxygen reaches the boiler, it may produce carbon monoxide.

4 What's reducing the air intake to the boiler in this picture?

5 What clue might warn you that there is a problem here?

Your blood transports oxygen and carbon dioxide around your body. Red blood cells carry oxygen atoms from your lungs to where your body needs them and then release them. Your blood then carries carbon dioxide back to your lungs to be breathed out.

Carbon monoxide molecules hijack this system. They attach to your red blood cells and don't let go. They stop your red blood cells carrying oxygen. You suffocate as no oxygen is getting to your body, particularly your brain.

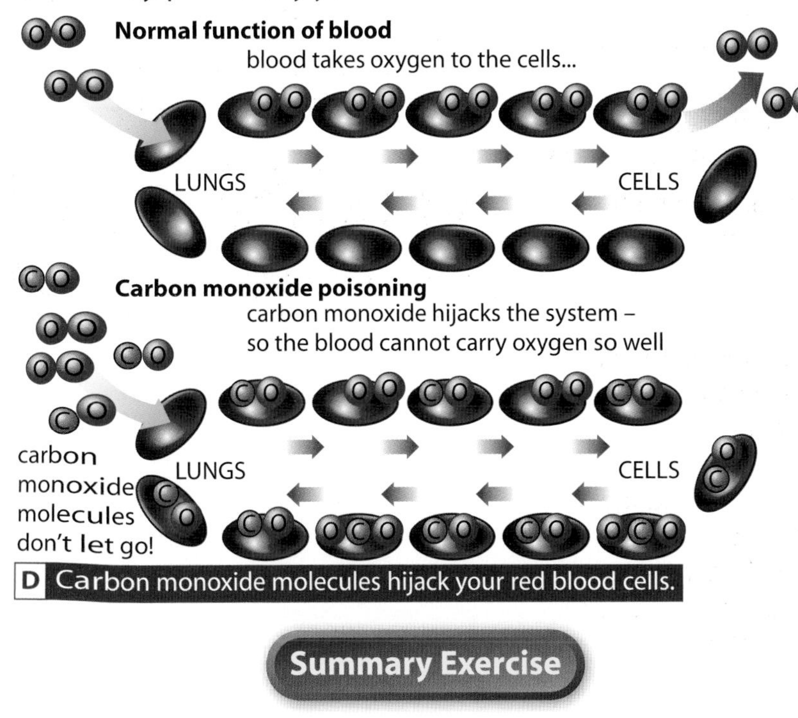

D Carbon monoxide molecules hijack your red blood cells.

Summary Exercise

6 Car exhaust contains small amounts of carbon monoxide. Why would it be dangerous to run a car engine in a closed garage? Explain your answer.

7 Design a 'public information' leaflet about the risks from faulty boilers. Explain how carbon monoxide forms and why it is so toxic.

Higher Questions

Global warming and fossil fuels

By the end of these two pages you should know that:

- burning fossil fuels is increasing the amount of carbon dioxide in the atmosphere
- the increased carbon dioxide may be causing global warming
- the Earth's climate has changed dramatically over time
- the idea of global warming, first put forward nearly 50 years ago, is now accepted by most scientists.

A Does burning fossil fuels cause this?

Have you ever wondered:

Will the UK freeze over one day, like in the film The Day After Tomorrow?

You will have heard a lot about **global warming**. The Earth is warming up, changing our **climate** and leading to droughts or violent storms. Some people say that it is all our fault. They say it is because we burn **fossil fuels** and pump carbon dioxide into the **atmosphere**. This idea was first suggested in the 1950s, but was not taken seriously. Today most scientists agree that there is a problem.

By day, the Sun's energy warms up the Earth. The warm Earth then **radiates** energy back into space. When the energy flow is balanced, the Earth has a steady average temperature.

Our atmosphere acts like a blanket, trapping energy and keeping the planet warm. Without it, the Earth would freeze over. Carbon dioxide in the atmosphere is particularly good at trapping energy. The amount of carbon dioxide in the atmosphere is rising. Many people think this causes global warming, making the average temperature of the Earth slowly increase.

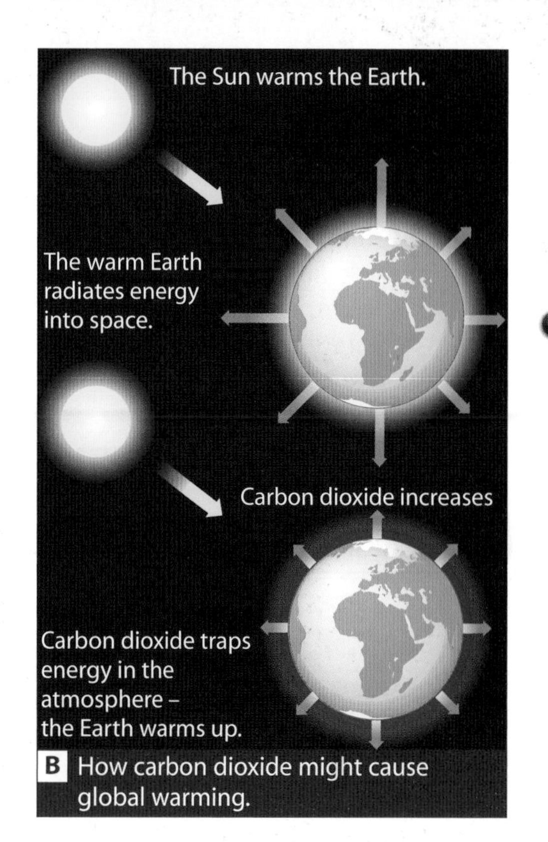

B How carbon dioxide might cause global warming.

1 What would happen to the Earth if it didn't radiate energy into space?

2 Mars is a cold planet with a thin atmosphere. In some science fiction stories, carbon dioxide is pumped into the Martian atmosphere to warm the planet up so that people could live there. How would that work?

There has always been carbon dioxide in the atmosphere. You breathe it out and plants take it in. But carbon dioxide in the atmosphere is rising rapidly because we burn vast amounts of coal, oil and gas. These fossil fuels took hundreds of millions of years to form. We are burning them up a million times faster than it took them to form.

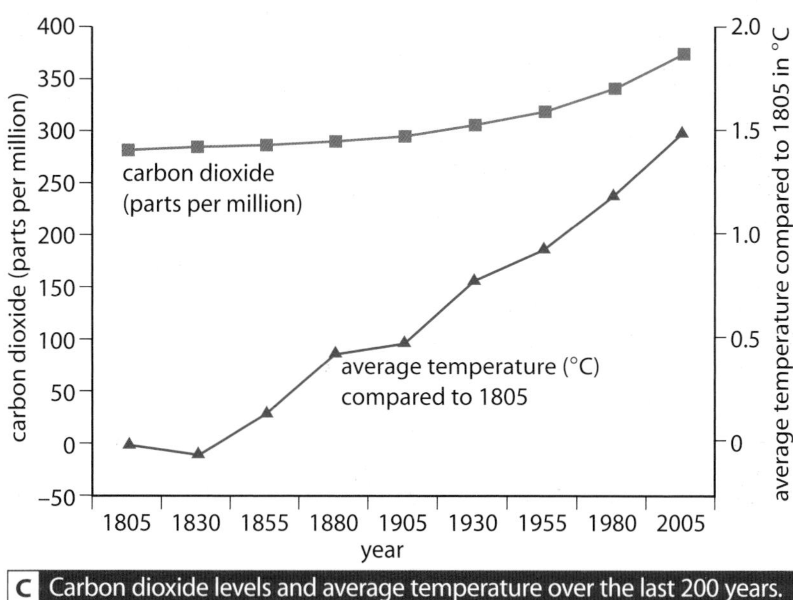

C Carbon dioxide levels and average temperature over the last 200 years.

Because carbon dioxide is not toxic, nobody really worried about the increase at first. But over the last 30 years scientists have raised concerns that we could be upsetting the balance of our climate. Summer heat waves and droughts, often followed by violent and destructive storms, have made people aware of these concerns.

In 1997, world leaders met in Kyoto, Japan, and agreed to reduce the amount of carbon dioxide their countries produce. Much needs to be done to stop global warming getting out of control.

3 How do *you* put carbon dioxide directly into the atmosphere?

4 What weather features over the last 30 years have made people start to think more about the problems of global warming?

5 From figure C, describe how the Earth's average temperature and the amount of carbon dioxide in the atmosphere have changed over the last 200 years.

Higher Questions

The Earth's changing climate

By the end of these two pages you should know that:

- the Earth's climate has a long history of change
- our predictions for climate change are based on complex computer models
- the Internet is a good source of information, but you need to check that it is reliable.

Have you ever wondered:

Why do scientists need to do their work in exotic locations like Hawaii and Antarctica?

When the Earth first formed 4 billion years ago, the atmosphere was very different from today. It was formed from the gases that came out of volcanoes. This atmosphere was mostly carbon dioxide. After plants appeared the atmosphere changed. The plants removed most of the carbon dioxide and replaced some of it with oxygen by a process called **photosynthesis**.

A Scientists in Antarctica.

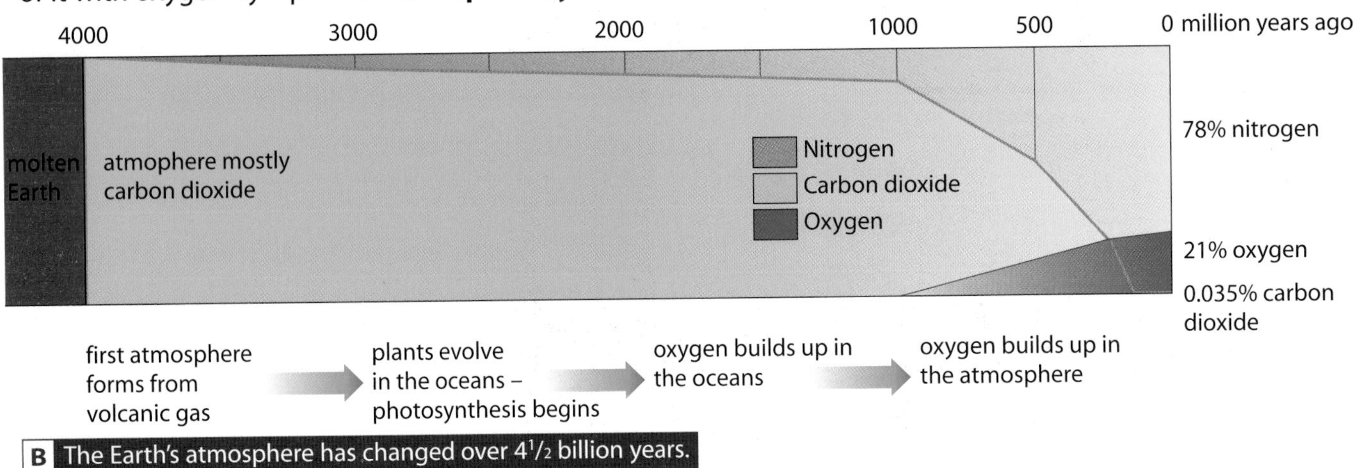

B The Earth's atmosphere has changed over $4^1/_2$ billion years.

1 Why do scientists study volcanoes in Hawaii?

2 Which important 'plant process' removes carbon dioxide from the air and replaces it with oxygen?

As the atmosphere changed, the climate changed. Over the last 2 million years or more, the climate has been flipping between ice ages and warm periods. We can find evidence for this by studying the gases trapped in the ice of Antarctica.

3 Describe what has happened to the average temperature of the Earth over the last 500,000 years.

4 How has the amount of carbon dioxide in the atmosphere changed compared to this?

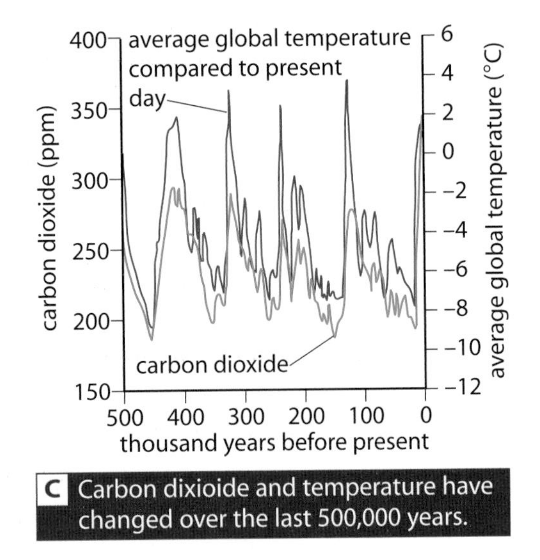

C Carbon dixioide and temperature have changed over the last 500,000 years.

Scientists use complex computer models to predict how the Earth's climate will change if the amount of carbon dioxide in the atmosphere increases. They are similar to the models used for weather forecasting. They are based on:

- what we know about our climate today
- what we know about the climate and atmosphere in the past
- scientific laws that govern how energy from the Sun is absorbed and radiated by the Earth and its atmosphere.

Climate models are very sophisticated but not perfect. We do not fully understand how our climate works today. There are always uncertainties and a range of 'possible outcomes'. There is much debate about what will happen and what we should do.

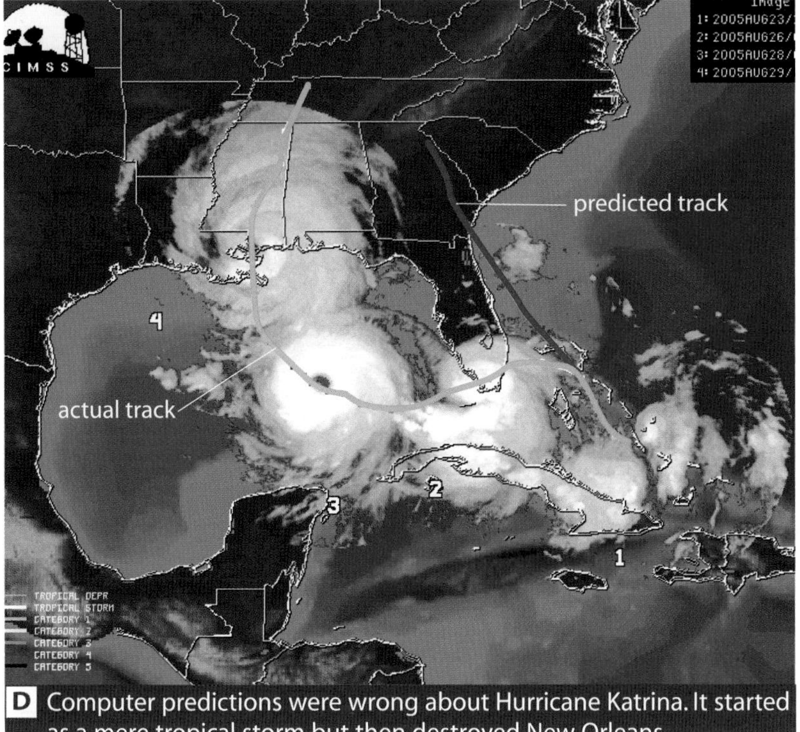

D Computer predictions were wrong about Hurricane Katrina. It started as a mere tropical storm but then destroyed New Orleans.

The Internet is a great place to research global warming and other environmental issues like **acid rain**. But be careful: there is no 'quality control' on the Internet. You could be reading someone's personal opinion, which might be completely wrong. Use sites from universities, the Government or well-known sources like TV companies.

Beware of sites giving only one side of the argument. Environmental pressure groups may push 'environmental concerns', whereas oil companies will give a less worrying view of the future. These sites can be useful if you are aware of the **bias** in the way they present the information and balance it out yourself by looking at the other side of the argument.

5 If you were researching global warming on the Internet, what sort of bias would you expect from

a an environmental pressure group?

b a multi-national oil company?

6 a How do scientists work out what the climate will be like in the future?

b What do they base these predictions on?

c Why are they not 100% reliable?

Higher Questions

Tackling the problem of fossil fuels

H ***By the end of these two pages you should be aware of:***

- what could be done now to tackle global warming
- other pollution issues associated with burning fossil fuels.

A link between fossil fuels and climate change has not been proved. However, many people agree that the climate appears to be changing. Climate change may or may not be *caused* by burning fossil fuels, but either way it is best to be on the safe side and reduce our carbon dioxide output. This 'playing safe' is called the **precautionary principle**.

A good way to reduce carbon dioxide production is to use less energy. As individuals we can use less energy at home, reduce car use and **recycle** rubbish instead of throwing it away.

The big companies that generate electricity must invest in new technologies that don't rely on fossil fuels, such as wind farms. Governments must also do what they can, rewarding energy efficiency with lower taxes and educating the public about reducing energy consumption.

1 List all the ways in which you could help to save energy and reduce the amount of carbon dioxide produced.

2 Cars with small engines pay less road tax than cars with larger engines. How does that help to reduce carbon dioxide production?

Have you ever wondered:

Could we stop global warming by capturing the CO_2 we generate instead of letting it escape into the atmosphere?

Capturing the carbon dioxide we generate before it got into the air would stop global warming. This would be a great solution if we could find an easy and economical way to do it. Pilot schemes are being set up in the USA and China, but we won't know if they work for a few years.

Have you ever wondered:

Is there really enough pollution in the air to kill people?

In the first half of the last century, coal was king. Coal was used to heat houses, in power stations and to make steam trains run. The smoke and soot turned everything black and grimy. It got into the lungs and killed elderly and ill people. In big cities autumn fog became thick, black, deadly **smog**. Thousands died from bronchitis and other lung diseases.

A We can do our bit to reduce global warming.

B City smog many decades ago.

3 Why were steam trains bad for your health in the 1940s?

C A modern-day photochemical smog.

Coal has mostly been replaced, but smogs have changed rather than vanished. Diesel engines produce very fine soot, called carbon **particulates**. These tiny particles react with other exhaust gases to form **photochemical smog**. This is the brown haze seen hanging over cities on sunny days.

Over the last 50 years, the percentage of people with asthma has increased steadily. Many people think that this has been caused by the increase in pollution due to carbon particulates from diesel, which could get deeper into the lungs than coal soot did. However, this has not been proved.

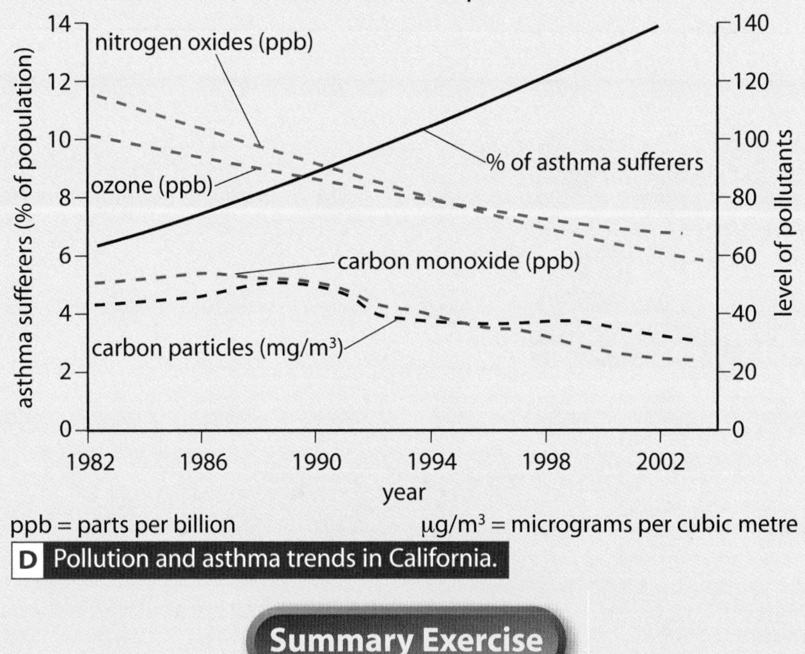

ppb = parts per billion   μg/m^3 = micrograms per cubic metre

D Pollution and asthma trends in California.

4 The suggested link between asthma and pollution has been studied extensively in California.

a Describe how the percentage of children suffering from asthma has changed since 1982.

b Describe how levels of the major pollutants have changed over that period.

c Does this study support the suggestion of a link between pollution and asthma or not? Explain your answer.

5 Explain why we should try to reduce the amount of fossil fuels we burn, even though we cannot prove that this will stop our climate from changing.

Summary Exercise

Higher Questions

Is there an alternative to oil?

By the end of these two pages you should know that:

- hydrogen is the cleanest fuel, but it is not easy to use
- bio-fuels such as ethanol are promising alternatives to oil.

Have you ever wondered:

When oil starts running out, will petrol cost as much as gold?

Every time there's a natural disaster or major global political problem oil prices skyrocket. So just imagine what will happen to the price of oil and oil products such as petrol when the oil really starts to run out. No wonder scientists are searching for alternatives.

A Buses can be powered by hydrogen.

We can burn hydrogen with oxygen to get energy. Spacecraft have been fuelled by hydrogen for decades. It gives out a lot of energy when it burns and is completely pollution-free. The only product is water.

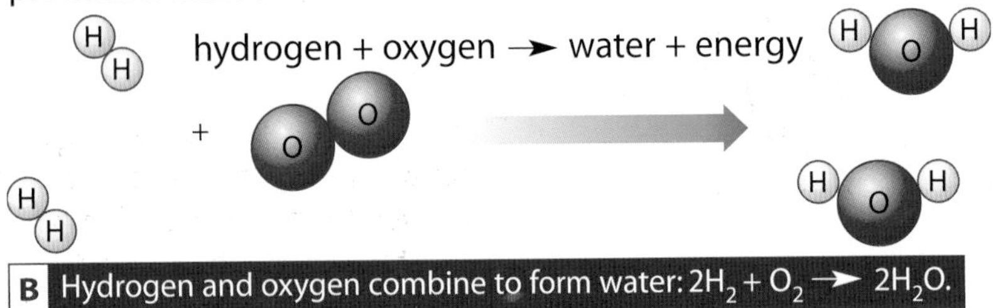

B Hydrogen and oxygen combine to form water: $2H_2 + O_2 \rightarrow 2H_2O$.

There are already trials using hydrogen-fuelled buses in California. The buses work well, but hydrogen is not easy to use. It has to be compressed and stored in very strong cylinders. If handled badly, it can be very dangerous.

There are also production problems. The most common way to make hydrogen is to remove it from water using electricity. But most electricity is made by burning fossil fuels – so the pollution simply moves from the bus to the power station.

1 Why does burning hydrogen produce less pollution than burning oil?

2 Hydrogen-powered buses produce no pollution from their exhausts, but using hydrogen is not a completely pollution-free option at the moment. Explain why.

Have you ever wondered:

What is the cleanest, greenest fuel for a car?

Bio-fuels are fuels made from living things. Plants can be grown as crops to make bio-fuel. As the plants grow, they take carbon dioxide out of the air. When they are burnt, they put the same amount of carbon dioxide back into the air. So overall they are carbon-neutral. They are also renewable.

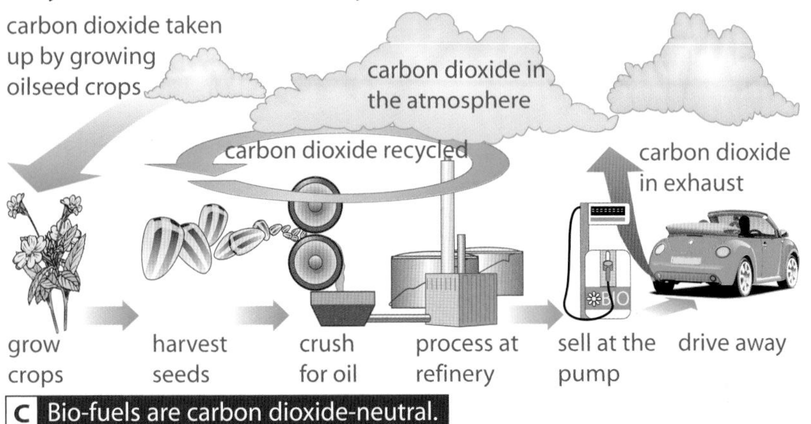

C Bio-fuels are carbon dioxide-neutral.

3 What two advantages do bio-fuels have over oil?

Sugar cane stores a lot of its energy as sugar, which can be fermented with yeast to make ethanol. That's the 'alcohol' in beer and wine. In Brazil, they make ethanol from the sugar and distil it. They then use the pure ethanol to run their cars, instead of petrol. Ethanol makes less pollution than petrol and, because it is a bio-fuel, it is carbon-neutral.

D Ethanol can be used as a fuel for cars.

The disadvantage of ethanol and other bio-fuels is that you need a lot of land to grow the crops. In Brazil, areas of rainforest have been lost to sugar cane farming. It's too cold to grow sugar cane in Britain, but we can grow sugar beet instead. But to replace all of our petrol we'd have to turn half the country over to beet farming! Still, growing even a small amount would replace a fraction of our petrol with alcohol. Some oil companies have already started to do this.

4 The formula for ethanol is C_2H_5OH. What two new products do you get when you burn it? (*Hint*: what will you get when the carbon and hydrogen atoms react with oxygen?)

5 Compare hydrogen and ethanol as 'fuels for the future'. What are their advantages and disadvantages? Which do you think will be more important?

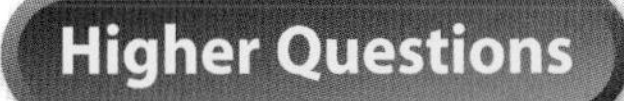

Waste not, want not…

By the end of these two pages you should know:

- why it is important that we recycle as much as we can
- why sustainable development is the best way forward for us all.

We live in a throw-away society, but if we don't do something soon we'll all be buried in our own rubbish! It would be a start if we could recycle as much as possible.

Have you ever wondered:

Why do we recycle so little of our rubbish in this country?

Packaging makes up a large proportion of the rubbish we produce in Britain – roughly half a tonne of it from every family, every year and as much again from industry. The chart shows that our record for recycling in Britain is not very good.

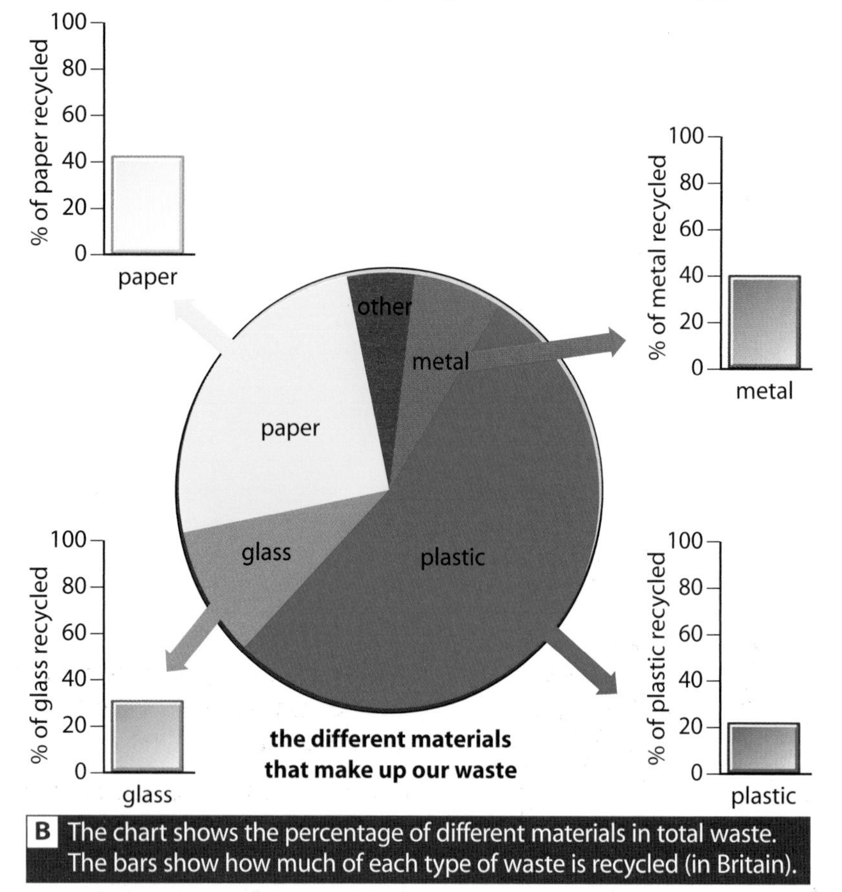

B The chart shows the percentage of different materials in total waste. The bars show how much of each type of waste is recycled (in Britain).

We recycle just a third of our glass in Britain but Europe as a whole recycles more than twice that. We used to recycle more glass as you got a deposit back on bottles when you returned them. But glass is made from sand, which is quite cheap and in plentiful supply.

A Recycle – or be buried by rubbish.

Paper is quite easy to recycle for cardboard or toilet paper, but it is too expensive to clean it up for books or newspapers. Some people think it is better to burn it as a fuel and grow trees for new paper. Paper is biodegradeable, so you could use it to make compost for your garden!

Metals are well-worth recycling. This is because some metals are rare. Other metals are common, such as aluminium, but recycling saves vast amounts of energy. That saves burning fossil fuels.

Plastics make up most of our rubbish. Different types have to be separated for recycling, which is not easy. Waste plastic can be burned as a fuel for power stations. Plastic contains as much energy as oil, but you have to be careful when you burn it or it can cause pollution.

Fresh water is recycled from sea water in **desalination** plants in hot, dry countries like Saudi Arabia.

Our technological society lets us enjoy a high standard of living. Developing countries want our lifestyle too. China and India are developing rapidly and may soon catch up with us. The developed world is using up natural resources rapidly, damaging the environment in the process. We must develop ways that are more in tune with the Earth, so as not to damage the environment or squander our resources. This is called **sustainable development**. **Sustainability** is very important for the future, and recycling can help us develop in this way.

C Alex Hill from the Meteorological Office says we must reduce our impact on the environment.

1 Suggest why we recycle less glass now than in the past.

2 List three things you could do with waste paper.

3 Why is it so important to recycle metals such as aluminium?

4 Why are plastics so difficult to recycle? What are the alternatives?

5 In Germany many people reuse the plastic bags from the supermarket. Why is this a good thing?

6 Make a leaflet for shoppers at your local supermarket, encouraging them to use the recycling banks. Explain in simple terms why recycling is important.

C1b.7.8

What do we get from crude oil?

By the end of these two pages you should know:

- how the fractional distillation of crude oil works
- the uses for the main fractions of crude oil
- H where the main fractions are produced on the fractionating column and how this relates to their properties.

A From this…

B …to this, and many other things.

Crude oil is a mixture of hydrocarbon molecules. Hydrocarbons are chains of carbon atoms with hydrogen atoms attached. The chains can be very short or very long. These different molecules have different properties. They must be separated out before they are used.

One property difference is that short-chain molecules boil at lower temperatures than long-chain molecules. Because of this, crude oil can be separated by a special type of **distillation**.

1 Why can't you use crude oil directly in cars?

To distil salty water, you heat it and the water boils off at 100°C. You cool and condense it to get pure water. You would have to heat the salt to nearly 1500°C to make that boil, so it is left behind.

The molecules in crude oil boil over a much closer range of temperatures, from 40 to 400°C. If you distil crude oil, you cannot get such a clean separation. What you get is a series of liquids containing molecules of *similar* chain length and *similar* properties, called fractions. This is called **fractional distillation**. These fractions include petrol and diesel, and they are very useful.

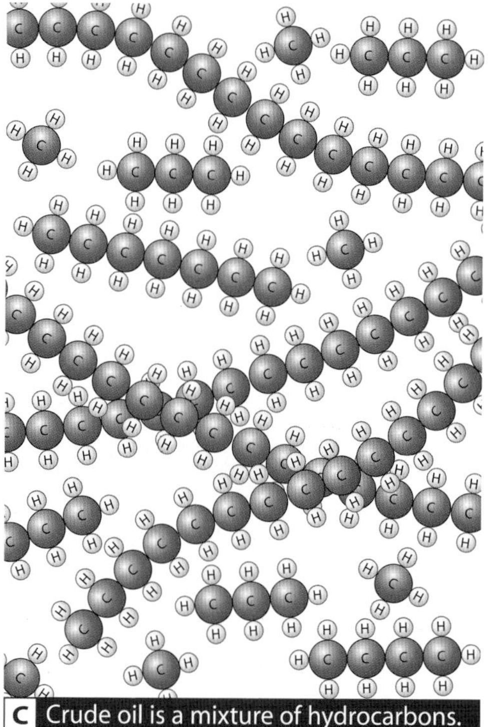

C Crude oil is a mixture of hydrocarbons.

In an oil refinery, the oil is heated and completely vaporised. This hot vapour then passes up the **fractionating column**. The column is hot at the bottom but cooler towards the top. As each type of hydrocarbon in the vapour cools to its boiling point, it starts to condense out. The liquid fractions are trapped in trays and can be piped off. This process is continuous and very economical.

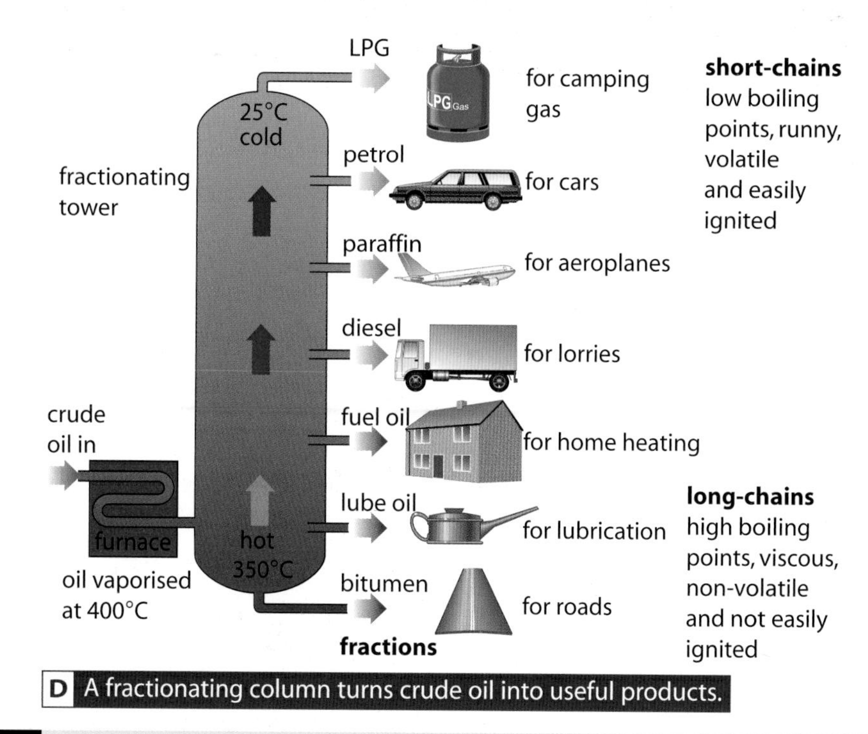

D A fractionating column turns crude oil into useful products.

2 Where does petrol collect in the fractionating column?

3 Why does engine oil collect near the bottom of the column?

H **Viscosity** – how thick or runny a liquid is – changes with carbon chain length. Short-chain molecules make very runny liquids like petrol (low viscosity). Longer chains make thicker liquids like engine oil (high viscosity).

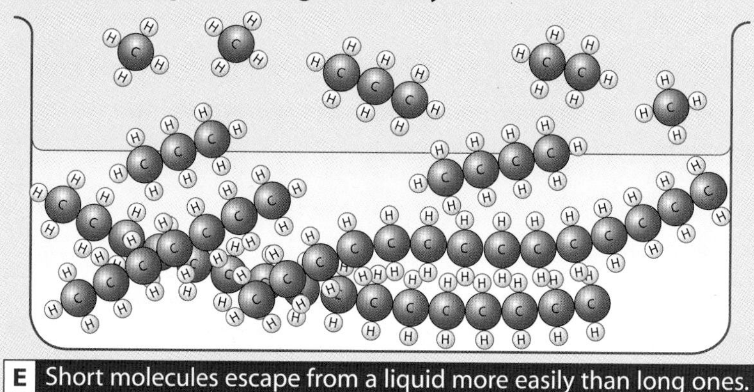

E Short molecules escape from a liquid more easily than long ones.

Ease of **ignition** also changes with carbon chain length. It is the hydrocarbon vapours that catch fire, so the short-chain molecules, which vaporise easily, also catch fire easily. Long-chain molecules do not vaporise or catch fire easily.

These properties control the uses of the main fractions. Petrol ignites easily with the spark in a petrol engine but is dangerous to store. Diesel has to be heated to get it to vaporise in a diesel engine, but is much safer to store than petrol. Engine oil doesn't vaporise easily at all. It isn't used as a fuel. It is used to **lubricate** car engines. Its high viscosity keeps it in place between the moving parts of the engine.

4 Choose the correct word from each pair given.
As carbon chain length increases
a viscosity (increases/decreases).
b ease of ignition (increases/decreases).

5 Why can't you buy petrol in an open container like a bucket?

6 Explain how the different fractions are obtained from crude oil.

What can we get from air?

By the end of these two pages you should know:

- how fractional distillation can be used to get oxygen and nitrogen from air
- about the main uses of oxygen and nitrogen.

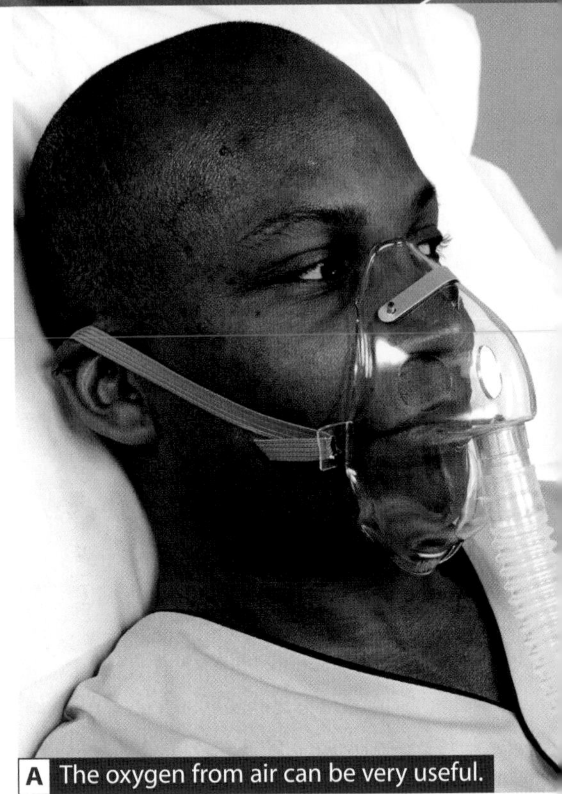

A The oxygen from air can be very useful.

The air is a mixture of gases. Dry air is made from approximately one-fifth oxygen and four-fifths nitrogen. Our lungs and blood take out the oxygen we need. Plants get the carbon dioxide they need. But for other uses the gases must be separated.

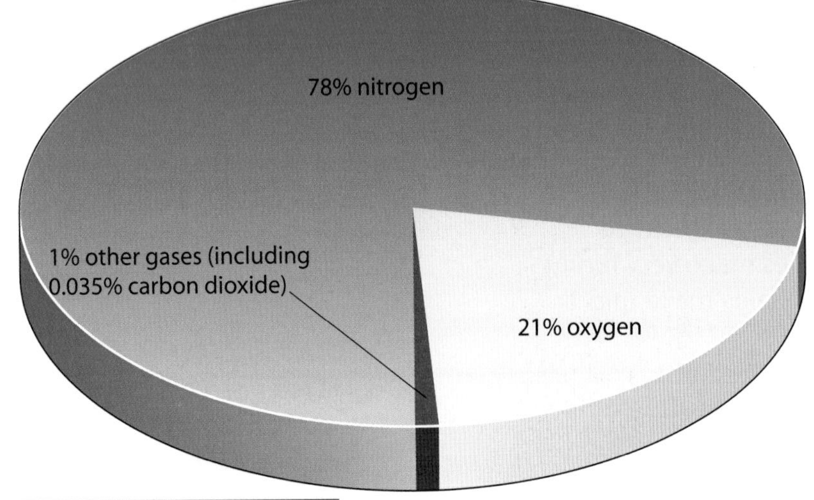

B The different gases in air.

1 What other common substance is usually found in air? (*Hint*: is it always dry?)

2 Carbon dioxide is only a tiny percentage of air, yet plants can get all they need. Why is this?

Air is separated by fractional distillation. The process is similar to the fractional distillation of oil, but the air has to be **liquefied** first. To do this, the air has to be cooled down to –200°C. At this temperature, any water and carbon dioxide turn solid and can be removed easily.

water	freezes at 0°C
carbon dioxide	freezes at –79°C
oxygen	liquefies at –183°C
nitrogen	liquefies at –196°C

C Freezing and liquefaction points of the gases in air.

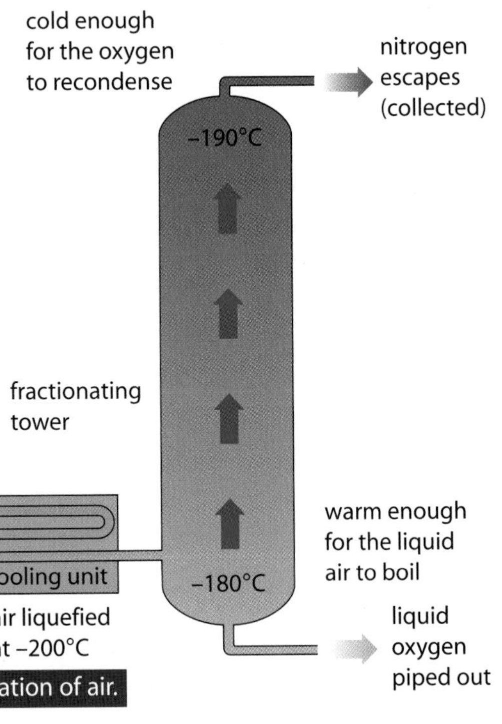

D The distillation of air.

The liquid air is then allowed to boil gently and passes up a fractionating column which is warmer at the bottom and cooler at the top. The temperature at the top is just cold enough to make the oxygen condense and fall back down the column, but just warm enough to let the nitrogen gas flow out from the top. The liquid oxygen can be tapped off and stored in heavy metal cylinders. The nitrogen has to be cooled down and liquefied again before it can be stored.

Pure oxygen is used in hospitals for people who are having breathing difficulties and to keep premature babies alive. Pure oxygen is also used when a fuel needs to burn really fiercely. Spacecraft burn hydrogen fuel with oxygen to give the power they need. Welders (and safe-crackers) burn special gas mixed with pure oxygen to give a flame hot enough to melt steel.

E Oxygen helps other fuels to burn.

Most of the nitrogen produced is used to make fertilisers for plants. Plants need nitrogen to grow, but can't get it straight from the air. In industry, nitrogen is used to make ammonia (NH_3) and nitric acid (HNO_3). These are combined to make ammonium nitrate (NH_4NO_3).

ammonia + nitric acid → ammonium nitrate

Ammonium nitrate is an excellent fertiliser that has helped farmers increase the yields of their crops for 100 years, though it can cause problems if it washes into rivers or lakes.

Some ammonia is used in household cleaners, and for making nylon. Some nitric acid is used in dyemaking, and for making nitroglycerine for dynamite.

3 The top of the column is at about −190°C. Is that above or below the boiling point of oxygen?

4 How does nitrogen help to feed the world?

5 The formula of the simplest unit of ammonium nitrate is NH_4NO_3.

a How many nitrogen (N) atoms are there?

b How many oxygen (O) atoms are there?

c How many hydrogen (H) atoms are there?

d How many atoms altogether?

6 Describe the stages you need to go through to turn nitrogen from the air into a bag of fertiliser.

What can we get from sea water?

By the end of these two pages you should know:

- how products such as chlorine, hydrogen and sodium hydroxide can be made from sea water
- why these chemicals are so useful
- how you can get fresh drinking water from sea water.

A Sodium chloride in solution.

B Solid sodium chloride – a dried-up sea, Bonneville Salt Flats, Utah, USA.

When sea water evaporates you get white cubic crystals of sodium chloride, common salt. This is happening today in the Great Salt Lake, Utah, USA, and it happened 200 million years ago in Britain. Thick rock salt deposits are mined from the rocks beneath Cheshire.

Every litre of sea water contains about 25 g of common salt, plus some other salts of sodium, calcium, potassium and magnesium. If you pass electricity through sea water, the elements inside the salt and water get separated, and produce the useful products chlorine, hydrogen and sodium hydroxide. This process is called **electrolysis**.

1 How did the rock salt that is mined in Britain get in the rocks?

2 What form of energy is used to break sodium chloride solution apart?

3 Copy and complete this word equation:
sodium chloride + water → (electrolysis) _____ + _____ + sodium hydroxide

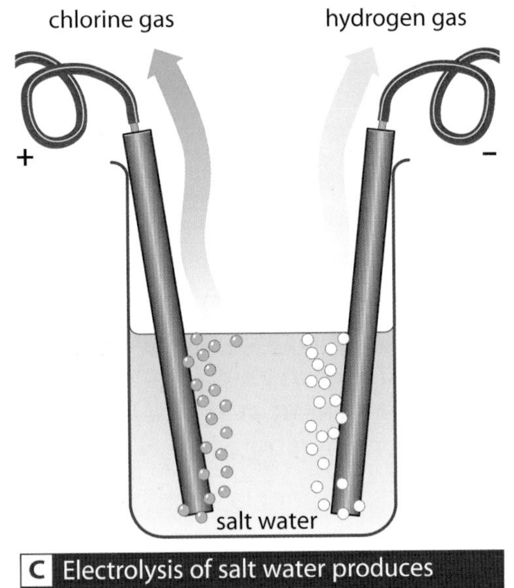

C Electrolysis of salt water produces hydrogen and chlorine.

Hydrogen is a highly flammable gas that can be used as a fuel for spacecraft. It may also become one of the 'green' fuels of the future. It is used in industry to make ammonia for fertilisers – and margarine!

Chlorine is deadly poisonous. It was used as a poison gas in World War I. Today it is used to kill bacteria. It is used to purify drinking water and keep swimming baths clean. It is also used to make PVC and solvents such as dry-cleaning fluid.

Sodium hydroxide is a very strong alkali. It will burn your skin. It is used as an oven cleaner and to make soap and detergents. In industry it is used in fabric and paper production.

Sodium hydroxide and chlorine can also be reacted together to make bleach. Bleach is used around the home to kill germs. Be careful not to spill it on your clothes, or it will bleach out the colour!

D Everyday items made using the chemicals in sea water.

4 How do we put the 'killing power' of chlorine to good use today?

5 Sodium hydroxide turns fat into soap. Why would a person's fingers feel soapy if they got caustic soda on them? What should they do, very quickly, if that happens?

Sea water contains a lot of water too! For countries that are desperately short of clean, fresh, drinking water one solution is to recycle sea water by distillation. Saudi Arabia now has several desalination plants to do this. Their plentiful oil supplies are used to boil the water. Desalination means taking out the salt. A more sustainable option is to use energy from sunlight. The purification process may be slower, but it will work long after all the oil has gone!

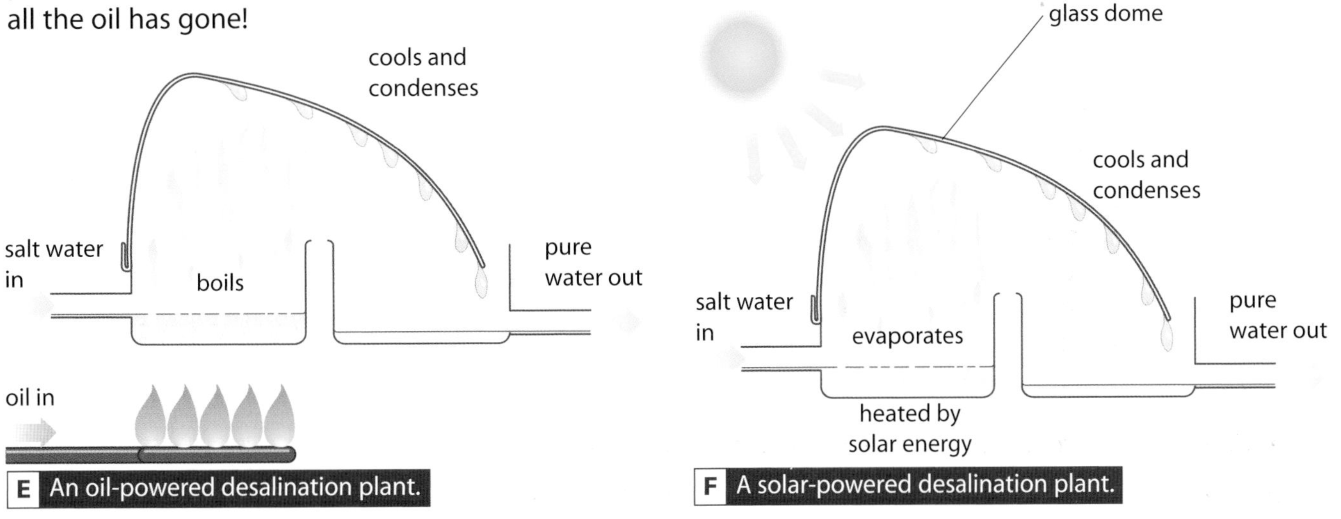

E An oil-powered desalination plant.

F A solar-powered desalination plant.

6 Explain why a solar-powered desalination plant is a good example of sustainable development.

7 The chlorine and sodium atoms in a bottle of bleach used to be in the sea 200 million years ago. List the stages on their journey from sea to bottle.

Higher Questions

Questions

Multiple choice questions

1 A good fuel must be

A easy to light and burn with a clean flame.
B easy to light and burn with a yellow flame.
C hard to store and burn with a yellow flame.
D hard to light and burn with a yellow flame.

2 When hydrocarbon fuels burn, the products are

A methane and water.
B methane and carbon dioxide.
C water and carbon dioxide.
D water and ammonia.

3 A faulty gas boiler can kill because

A carbon dioxide forms if there is too much air getting to the flame.
B carbon dioxide forms if there is not enough air getting to the flame.
C carbon monoxide forms if there is too much air getting to the flame.
D carbon monoxide forms if there is not enough air getting to the flame.

4 Burning fossil fuels may lead to global warming because

A burning fuels gives out a lot of heat energy.
B carbon monoxide is formed and traps heat in the atmosphere.
C carbon dioxide is formed and traps heat in the atmosphere.
D carbon dioxide makes the sun hotter.

5 Which of these statements is true?

A The Earth's climate has been the same for the last 2 million years.
B The Earth's climate has got steadily hotter over the last 2 million years.
C The Earth's climate has got steadily cooler over the last 2 million years.
D The Earth's climate has changed from hot to cold many times over the last 2 million years.

6 The predictions about global warming

A are made using powerful computers so they must be correct.
B may not be accurate as they are based on many assumptions and incomplete data.
C are based on very accurate data so they must be correct.
D may not be accurate because the computers are not reliable.

H **7** The number of asthma sufferers has increased steadily over the last 40 years. This may be due to

A there being no pollution before 1965.
B the soot particles from car exhausts having got bigger and bigger.
C the tiny sooty particulates and other chemicals from diesel and petrol engine exhausts getting deeper into the lungs than coal soot did.
D the radiation from mobile phones making the pollution much worse.

8 Hydrogen produced using energy from a wind turbine is a 'green' fuel because

A it makes no carbon dioxide from either its production or its combustion.
B it gives out no heat when you burn it.
C the carbon dioxide it gives out when it burns is taken in when the next batch is made.
D its only waste gas is methane.

9 Ethanol, made by fermentation, and biodiesel are 'greener' fuels than petrol or diesel because

A they do not give out water when they burn.
B they do not give out as much carbon dioxide when they burn.
C they do not give out any carbon dioxide when they burn.
D the carbon dioxide they give out when they burn is taken in when the next batch grows.

10 Aluminium cans ought to be recycled because

A aluminium is a very rare metal and so its ore is very expensive.
B aluminium is a very rare metal that uses up a lot of energy when you extract it.
C aluminium is a common metal but you use up a lot of energy when you extract it.
D aluminium is very rare and will run out in a few years.

11 In an oil refinery

A crude oil is heated in batches and the different fractions boil off one by one.
B crude oil is vaporised in a furnace and the gas is then cooled by water.
C crude oil is vaporised in a furnace and the gas is then passed up a tower where the long chains escape at the top.
D crude oil is vaporised in a furnace and the gas is then passed up a tower where the long chains condense at the bottom.

H **12** Petrol is a good fuel for a car engine because
- **A** it vaporises and ignites very easily.
- **B** it is very safe and easy to store.
- **C** it burns with a yellow, sooty flame.
- **D** it is very thick and viscous.

13 Pure oxygen from the air is
- **A** mixed with special gas to give a super-hot flame for a welding torch.
- **B** used to fill light bulbs to stop the filament burning out.
- **C** used to fill modern airships as it doesn't catch fire like hydrogen.
- **D** used to make bleach for use in the home.

14 There are beds of rock salt beneath Cheshire because
- **A** the salt came out of ancient volcanoes.
- **B** an ancient sea dried up there millions of years ago.
- **C** rain water soaks down through the rock carrying salt.
- **D** it was washed into the mines by a great flood thousands of years ago.

15 The main products you get when you electrolyse salt water are
- **A** chlorine, fluorine and bromine.
- **B** sodium, chlorine and water.
- **C** hydrogen, chlorine and sodium hydroxide.
- **D** chlorine, sodium and sodium hydroxide.

Short-answer questions

1 The table shows the boiling point of some hydrocarbons with different carbon-chain lengths.

Number of carbon atoms in the chain	Boiling point (°C)
5	36
6	69
7	98
8	missing
9	151
10	174
11	197

- **a** Plot a graph of boiling point (°C) against the number of carbon atoms.
- **b** Draw a line of best fit and use this to find the missing boiling point for the hydrocarbon chain of eight carbon atoms.

2 Jane tested some oil made by heating a type of rock called oil shale, to see if it could be used to make petrol. She fractionally distilled 100 ml of the oil and got three fractions.

Fraction	'Average' boiling temperature of the fraction (°C)	Amount collected (ml)
1st	69	40
2nd	126	30
3rd	197	30

- **a** Use the graph from question 1 to work out the 'average' carbon-chain length for each fraction based on their boiling point.
- **b** Petrol needs to vaporise easily, so it must have a boiling point below the boiling point of water. Which fraction could be used as petrol?
- **c** What percentage of petrol could you get from this oil, by volume?

3 Some people have discovered an interesting way to get rid of waste cooking oil. They run their diesel engine lorries on it!
- **a** Cooking oil is not made of simple hydrocarbons, but the molecules are made from the elements hydrogen and carbon (plus some oxygen). What waste gases will come out of the exhaust pipe?
- **H** **b** Cooking oil has a higher boiling point than diesel and does not vaporise so easily. It only works as a fuel if the engine is started on ordinary diesel, then switched to cooking oil once it has warmed up. Why is this?
- **c** Waste cooking oil has to be disposed of carefully. Suggest two reasons why using it as a fuel is a better solution.
- **d** Some oil companies are now starting to make biodiesel from vegetable oil. Why would switching from diesel to biodiesel help to reduce the risk of global warming?
- **e** You need an acre of land to grow the crops to make the biodiesel to run a car for a year. There are 25 million cars in Britain. Why is biodiesel unlikely to fully replace diesel or petrol from crude oil?

D

P

Glossary

acid rain Rain with dissolved pollutants such as sulphur dioxide that make the rain acidic.

atmosphere The layer of gas that surrounds the Earth.

bias (on the Internet) Where an unbalanced account of a problem is given, to push forward one point of view over another.

bio-fuel A fuel made from living things – usually from plants grown as a crop, such as sugar cane or sugar beet, which is used to produce ethanol, or from seed oils such as corn oil, which is used to produce biodiesel.

climate The general weather conditions.

combustion Another word for burning. A chemical reaction between a fuel and oxygen that gives out heat.

complete combustion When a fuel burns with a good supply of oxygen to form carbon dioxide and water.

crude oil The natural form of the fossil fuel oil. A mixture of hydrocarbons.

desalination Turning salt water into fresh water; 'taking out the salt'.

distillation Purifying by boiling a liquid, collecting and cooling the gas, and then condensing it back to a liquid.

electrolysis Tearing a compound apart using electricity.

ethanol The type of alcohol in alcoholic drinks. Can be used as a fuel.

fossil fuels Coal, oil and gas; fuels that formed millions of years ago from the remains of plants and tiny animals.

fractional distillation A form of distillation that does not give complete separation, but various fractions.

fractionating column A tall tower used in industrial fractional distillation. Different fractions condense at different levels.

fuel A substance which is burnt to get energy.

global warming A rise in the average temperature of the Earth that could lead to climate change, with more storms or droughts in different parts of the world.

hydrocarbon A compound made from carbon and hydrogen atoms only.

ignition Catching fire.

incomplete combustion When a fuel burns without a good enough supply of oxygen to form deadly carbon monoxide (and/or soot) and water.

liquefied Turned into a liquid (in the case of air, by cooling to very low temperatures).

lubricate Reduce friction; for example, oil in an engine or grease on a wheel.

methane The chemical name for natural gas.

particulate Tiny particles. Carbon particulates are tiny bits of carbon that get into the atmosphere from diesel exhaust.

photochemical smog A choking mixture of particulates, nitrogen oxides and ozone that forms in cities when car exhausts react under strong sunlight.

photosynthesis A process in plants that converts water and carbon dioxide from the air into glucose. Oxygen is given off as a waste gas. The reaction uses energy from the Sun.

precautionary principle Another way of saying 'playing safe'. With global warming, it means that it is safer to reduce the amount of fossil fuels we burn, even if we can't prove that this is the cause of the problem.

radiate To give off radiation, energy in the form of electromagnetic waves like light or infrared.

recycle Use the same material over and over again.

residue Something that is left behind. For example, when wood burns it leaves a residue of ash.

smog An unpleasant mixture of fog and smoke that was common in the 1940s and 1950s.

sootiness The amount of soot (carbon) produced by a burning fuel.

sustainability Being able to keep doing the same thing over and over again without causing harm to the environment.

sustainable development Development that involves balancing the need for economic development, standards of living and respect for the environment.

toxic Poisonous.

viscosity How thick or runny a liquid is. Low viscosity is very runny, high viscosity is thick.

Designer products

A This skier is using 'smart' skis that can sense and reduce vibrations.

In the past, we used materials because they had the properties we needed. We used glass because it is transparent, steel because it is strong and copper because it conducts electricity.

Today, designers think about the properties they would like to have, and scientists create new materials with just those properties. Sometimes they do this by combining old materials, sometimes they make completely new materials. And, increasingly, scientists are developing new 'smart' materials that can actually respond to conditions and change their properties.

The same processes are applied in designing food and drinks. Beer has been made for thousands of years but breweries now use modern chemistry to make the best product every time. And it takes a lot of careful chemistry to make mayonnaise stick to a lettuce like that!

In this topic you will learn that:

- materials differ in their properties and so are suitable for different purposes
- new materials are developed to meet specific requirements
- useful substances are made by chemical reactions
- chemical processes use energy and have environmental consequences.

Use these words to finish the sentences below with the most suitable properties:

elastic, flexible, hard, lightweight, shatter-proof, soft, stiff, strong, transparent, warm, waterproof.

- The material for my raincoat must be…
- The material for my tennis racquet must be…
- The material for the walls of my house must be…
- The material for my car windscreen must be…
- The material for my helium balloon must be…

Getting the right materials

By the end of these two pages you should know how:

- commercial materials such as Lycra™ and Thinsulate™ are designed to have special properties
- smart materials can change their properties to suit changing conditions.

Fast-moving athletes need clothing that is flexible and comfortable yet strong enough for extreme activity. It must also reduce air resistance and help improve performance times. So, many athletes now use clothes made from **Lycra**™.

Lycra™ is an artificial fibre that was invented nearly 50 years ago. It can stretch even more than rubber but it is much more resistant. It doesn't perish in the sun or when covered in sweat. It was originally used for making girdles to hold in flabby stomachs!

A Skin-tight clothing reduces air resistance.

B Extreme activities call for extremely well-designed materials.

Polar bears keep warm in two ways. First, their fine fur traps lots of air. Air is a very good insulator as it is a poor **conductor** of heat. Second, the white fur reflects body heat back into the body and does not **radiate** it outwards.

Modern **Thinsulate**™ fabric does the same trick in a similar way. It is made from polypropene fibres like some modern carpets. But in Thinsulate™ the fibres are very thin – just one-hundredth of the thickness of a human hair. Fabric made from these tiny fibres traps a lot of air, like polar bear fur. The dense mat of fibres is also good at reflecting body heat back in, and not radiating the energy out.

1 How does Lycra™ help great athletes beat world records?

2 How is a skier in a Thinsulate™ jacket like a polar bear?

C The special properties of carbon fibres can help to improve the performance of top players like Roger Federer.

Objects from helicopter rotor blades to tennis racquets need material that is as stiff as possible. If a material bends in use, it reduces efficiency. Even steel can be too bendy.

Carbon fibres are very stiff, but are not very easy to shape. But scientists have made a **composite** material by setting the carbon fibres in strong plastic. This can be moulded into any shape. These carbon-fibre composites are four times stiffer than the finest steel, weight for weight.

3 Why can't you make a tennis racquet frame out of pure carbon fibres?

Have you ever wondered:

How do those glasses that remember their shape work?

Smart materials can change their properties when conditions change. '**Shape-memory alloys**' are a group of smart materials. When you make something at a high temperature from this metal the atoms are fixed in a certain pattern. When it is cooled, the metal can be bent out of shape. But warm it up again and the atoms move back, restoring the original shape.

Superelastic glasses are made from a shape-memory **alloy** that is warm enough at room temperature for the atoms to snap back into position.

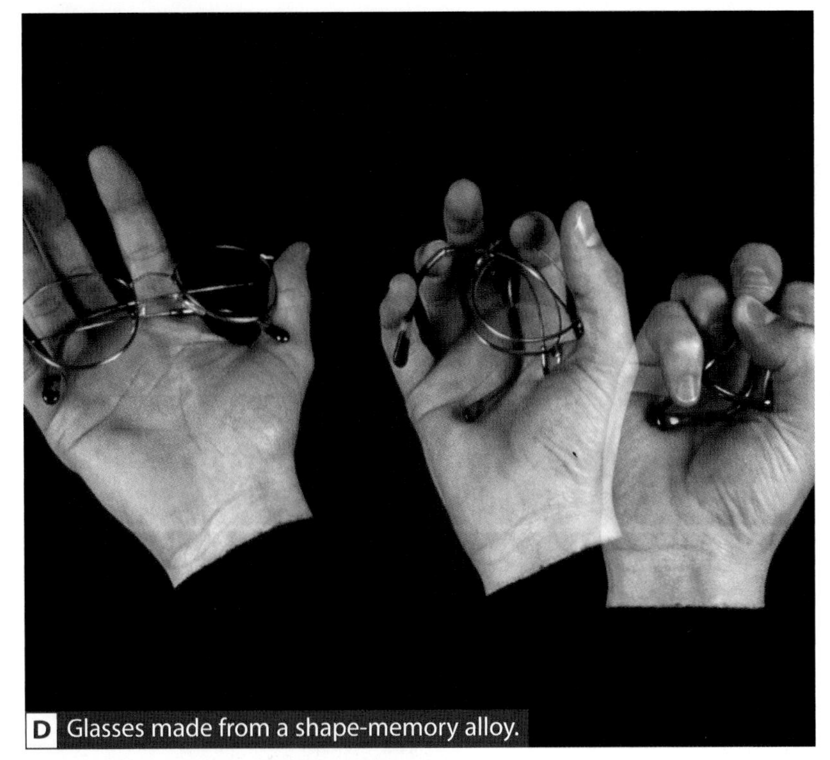

D Glasses made from a shape-memory alloy.

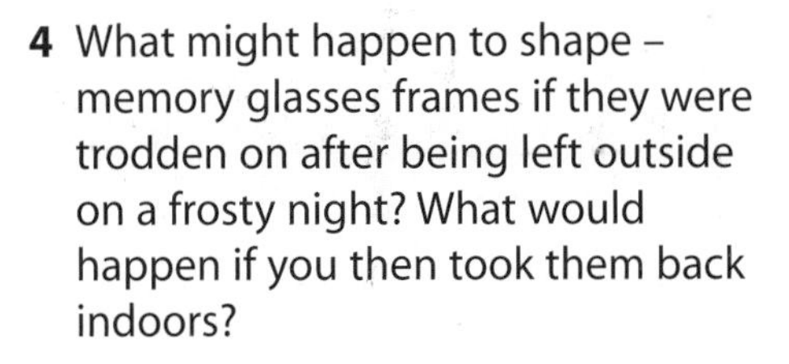

4 What might happen to shape – memory glasses frames if they were trodden on after being left outside on a frosty night? What would happen if you then took them back indoors?

5 a Suggest a use for
- **i** Lycra™
- **ii** Thinsulate™
- **iii** carbon-fibre composite.

b Explain why each is suitable for the task.

c The materials in i, ii and iii all keep their special properties whatever the conditions. In what way is a 'smart material' different?

Have you ever wondered:

Will scientists one day create toasters that feel 'cuddly' if you touch them gently?

Higher Questions

Making new materials

By the end of these two pages you should understand:

- why materials such as Kevlar™ and Gore-tex™ have special properties
- that new materials are sometimes discovered by accident.

Have you ever wondered:

How can modern body armour, made of soft clothing, stop bullets?

For centuries, soldiers had to rely on heavy and uncomfortable steel armour or chain mail to protect them. Then, fifty years ago, a chemist made a new polymer fibre called **Kevlar**™. Weight for weight it is five times stronger than steel. Yet Kevlar™ can be woven into cloth and made into comfortable and lightweight bullet-proof vests.

A Modern body armour must be strong yet light and flexible.

1 You can make strong body armour from either steel or Kevlar™. Give two advantages of Kevlar™.

Kevlar™ is strong because...

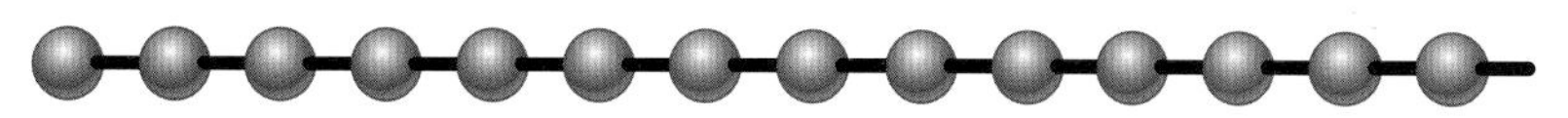

it has long molecules with strong chemical bonds...

the molecules stack together in crystal-like structures

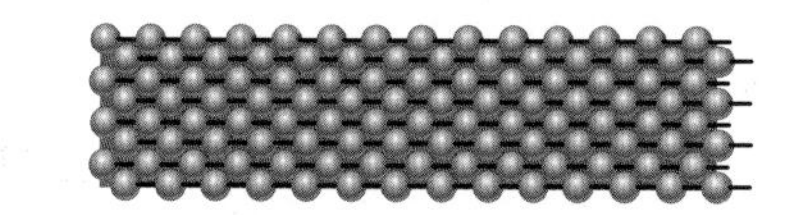

forces between the molecules hold the crystal-like stacks closely together

B Kevlar™ is very strong.

Kevlar™ is strong because it has long molecules – like sticks of dry spaghetti in a packet. They make chemical bonds between them. This gives the fibres a crystal-like structure. Today, Kevlar™ has a wide range of uses from skis to brake pads on cars.

Have you ever wondered:

Why is Gore-tex™ 'breathable'?

Gore-tex™ fabric was invented to overcome the problem of rain macs making you sweaty. It has a special plastic layer sandwiched between fabric. The plastic layer has millions of microscopic holes in it. The tiny water molecules in water vapour from your sweat are small enough to escape through these pores. This '**breathability**' stops you getting sweaty inside. But rain drops are too big to get through the microscopic holes, so you don't get wet from the rain either.

C A good overcoat needs to keep the rain out, but mustn't make you sweaty.

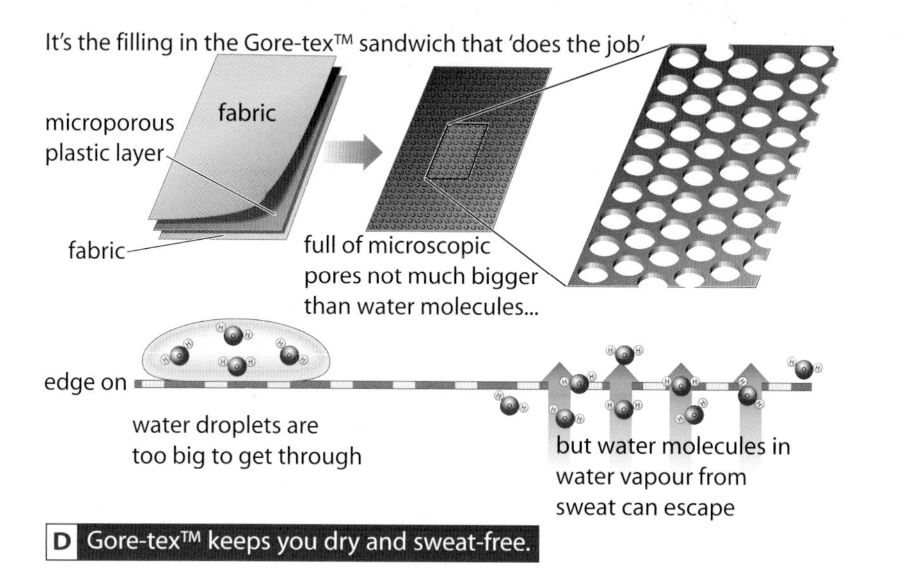

D Gore-tex™ keeps you dry and sweat-free.

Sometimes new materials are discovered by chance. It is only later that a use for the material is found. Fifty years ago a chemist called Silver was trying to produce a very strong adhesive. Instead he made a very weak one. You could stick things together with it, but then you could easily pull them apart. Then, ten years later a friend complained that he was always losing his place in books as the bookmarks he used just fell out. Silver thought of his 'weak' adhesive and 'Post-it™' notes were the result.

E Teflon™ – a very useful, accidental discovery.

Teflon™ is the 'non-stick' coating used for frying pans. This unusual plastic, made in 1938 by accident, is incredibly slippery. Since then, many more uses have been discovered. Teflon™ is now used to coat fibres for clothes so that dirt and stains can't stick to them. And it is also used to make the **microporous** plastic **membrane** for Gore-tex™.

2 Hikers often carry thin polythene macs for emergency use in the summer, in case it rains. Why wouldn't these be very good to wear for a long, strenuous hike?

3 How could a weak adhesive make a better bookmark?

4 Food doesn't stick to Teflon™, but unfortunately the Teflon™ layer doesn't stick to the metal pan very well either. What can happen to the Teflon™ layer if you scrape out a non-stick pan with a metal spoon?

5 Describe carefully how Gore-tex™ lets sweat escape but keeps rain out. What material is used to make the plastic membrane?

6 Make a list of the materials mentioned on this spread. For each briefly describe its special property and say whether you think it was discovered accidentally or on purpose.

Higher Questions

C1b.8.3

The future could be very small...

By the end of these two pages you should know:

- just how small a nanoparticle really is
- that nanoparticles have special properties
- that nanoparticles can alter and improve existing materials.

Just how small is a **nanometre**? Too small to see, that's certain. The best way to imagine the scale is to start from 1 metre and move down by a factor of 10 each time.

1 metre	1 m	The visible zone – things we can see unaided	child
1/10th or 0.1 metre	1×10^{-1} m		DVD
1/100th or 0.01 metre (1 centimetre)	1×10^{-2} m (1 cm)		marbles
1/1000th or 0.001 metre (1 millimetre)	1×10^{-3} m (1 mm)		salt crystals
1/10,000th or 0.0001 metre	1×10^{-4} m	The microscopic zone	thickness of human hair
1/100,000th or 0.00001 metre	1×10^{-5} m		plant cells
1/1,000,000th metre or 0.000001 (1 micrometre)	1×10^{-6} m (1 μm)	The electron microscopic zone	small bacterium
1/10,000,000th or 0.0000001 metre	1×10^{-7} m		virus
1/100,000,000th or 0.00000001 metre	1×10^{-8} m	The nanoparticle zone	large molecule
1/1,000,000,000th or 0.000000001 metre (1 nanometre)	1×10^{-9} m (1 nm)		atoms and small molecules

A You can get very small indeed in just a few 'power of ten' jumps...

So a nanometre is 1 billionth of a metre. **Nanoparticles** range in size from 10 to 100 nanometres. They are too small to see, even using the most powerful microscope. You cannot see them because they are smaller than light waves.

Being so small, not much bigger than atoms, nanoparticles have some very strange – but useful – properties.

1 How many nanometres are there in 1 metre?

2 How many nanometres are there in 1 millimetre?

Have you ever wondered:

Why do sunscreens now rub in better and no longer leave your skin white?

B Who's not wearing nanoparticles?

You may think that nanoparticle technology is the stuff of science fiction, but it is here already. If you used a new 'clear' sunblock last summer you were using **nanotechnology**.

Most sunblocks use titanium dioxide particles to trap the harmful ultraviolet (UV) rays. In 'ordinary' sunblock, these particles are up to 50,000 nanometres in diameter. That's too small to see the individual particles but big enough to reflect light. So sunblock looks like white face-paint. This is not surprising, as titanium dioxide is also used to make white house paint.

Nanoparticle sunblocks also use titanium dioxide, but this time the particles are only 50 nanometres in diameter – the old type were a thousand times bigger. These particles are too small to reflect light. They are 'invisible', as visible light simply passes them by. However, they still block the UV rays.

3 What advantage do nanoparticle sunblocks have over the old type?

4 Why do we not use nanoparticle titanium dioxide for house paint?

5 Powdered substances have been used for many products in the past. How are nanoparticle products different and why are scientists interested in them?

Summary Exercise

Higher Questions

The nanoparticle balance sheet

By the end of these two pages you should be able to:

- compare sizes of nanoparticles with conventional materials
- relate the sizes of nanoparticles to their uses
- H discuss some of the potential dangers of nanotechnology.

A Nanotechnology could come to the rescue in cases like this.

Some nanoparticles can help speed up chemical reactions. Iron nanoparticles can speed up the breakdown of oils or solvents in polluted areas. Silver nanoparticle suspensions can now be used to clean surfaces in kitchens or hospitals. In Japan you can now buy babies' bottles with silver nanoparticles in the teat. Special nanoparticles on the surface of a new type of 'self-cleaning' glass break down dirt and grime, so that rain can simply wash it away.

Nanoparticles can also improve the properties of ordinary materials. Nylon fibres with embedded clay nanoparticles are much stronger than ordinary fibres. Nanoparticles used in combination with other materials like this are called **nanocomposites**. In the future, other nanocomposites could revolutionise electronics.

Sheets of carbon atoms can be rolled up to make **nanotubes**. These could be used as electrodes for new electronic devices, or as pipes or sieves to move molecules around.

As the technology improves, scientists are finding ways to move even single atoms about. Nano-sized machines called **nanobots** could soon be produced in their millions and set to work on tasks such as cleaning dust particles in your home or crawling through the body to clean fat from arteries or to destroy bacteria.

1 Dry cleaning fluid is difficult to dispose of. It is harmful if it gets into the environment. How can iron nanoparticles help?

2 Why might it be a good idea to have silver nanoparticles in babies' toys or dummies?

B Nanotubes are made from rolled-up sheets of atoms.

Have you ever wondered:

Are the new sunscreens that contain nanoparticles safe?

H Nanoparticles often show new and unexpected properties, which worries some people. They are concerned about what might happen if these particles spread widely in the environment. This is of course possible, as new technologies often bring unexpected problems alongside their undoubted benefits. There is always a balance between what we gain and what we might upset. Scientists need to be vigilant and test their new products thoroughly before they are widely used.

C In the future, you might be injected with millions of nanobots to kill bacteria or clean your arteries.

Some scientists have suggested that nanobots could become self-replicating, like living things. You could get the millions needed to do a job without having to build them yourself. But they might keep multiplying until the whole world was covered in them! However, most scientists think this would be impossible.

Many newspapers have run 'scare stories' about the possible dangers of nanobots, but they have not balanced this out by explaining the potential benefits. Nanobots could be very helpful to us. Every new technology brings potential benefits and potential dangers. What counts is the way we humans choose to use it.

Summary Exercise

3 Why would you need millions of nanobots to clean your house? (*Hint*: How big are they compared to the size of a house?)

4 Why are some people concerned about the use of nanoparticles?

5 Write a balanced newspaper article about nanobots. You should include one paragraph about the possible benefits of nanobots and another about the possible dangers.

Higher Questions

Making beer and wine

By the end of these two pages you should know:

- how beer and wine are made
- how yeast can be used to ferment any sugar solution to make alcohol.

Have you ever wondered:

How do you make beer?

Beer starts its life as a weak **sugar** solution made by boiling up barley with water. Yeast is added to this solution, which is kept warm at about 'body temperature' (37°C). Yeast is a living organism; it is a simple, single-celled fungus. If yeast is put in a sugar solution it starts to feed on the sugar, grow and reproduce. It can get its energy without using oxygen. It does this by breaking the sugar down into **alcohol** and carbon dioxide.

sugar → alcohol + carbon dioxide

This process is called **fermentation**. If grape juice is used instead of barley in water, it gives wine instead of beer, but the process is the same.

1 Bitter hops are used to make beer but do not take part in fermentation. What are they used for?

2 Home-made wine is often fermented in large glass jars. If you looked at the jar, how would you know that carbon dioxide was being released?

3 Why is fermentation usually carried out at 'body temperature'? (Would fermentation happen as fast in the fridge?)

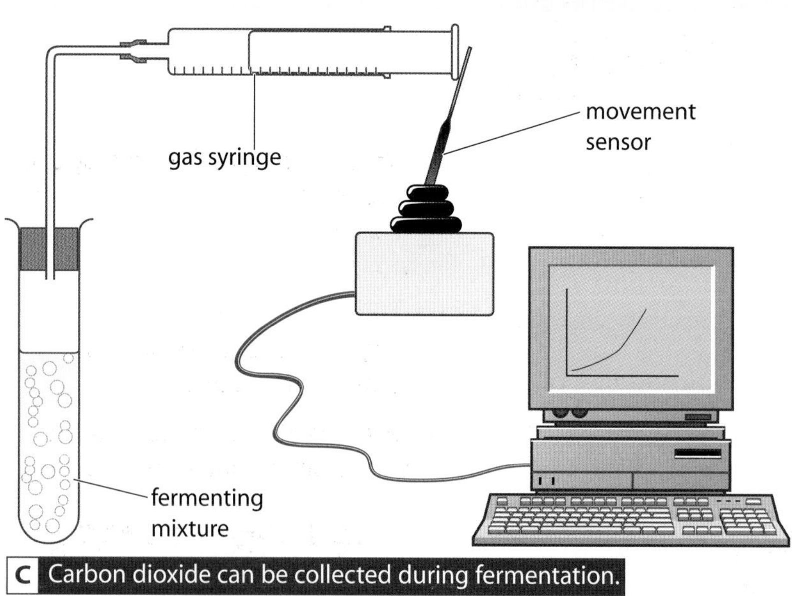

C Carbon dioxide can be collected during fermentation.

A How do we get from this …

B … to this?

Yeast splits sugar up into roughly equal amounts of alcohol and carbon dioxide by weight. The alcohol remains in solution but most of the carbon dioxide gas bubbles out. The gas has a surprisingly large volume. One kilogram of sugar would give roughly half a kilogram of carbon dioxide, with a volume of over 250 litres – that's about 50 balloon-fulls!

If you monitor the amount of carbon dioxide produced you can see how the fermentation is proceeding. The graph shows three phases. At first things start slowly, but start to speed up as the yeast reproduces. Then the gas bubbles out quickly when fermentation is in full swing. Towards the end, gas production slows before finally stopping. Fermentation has come to its natural end.

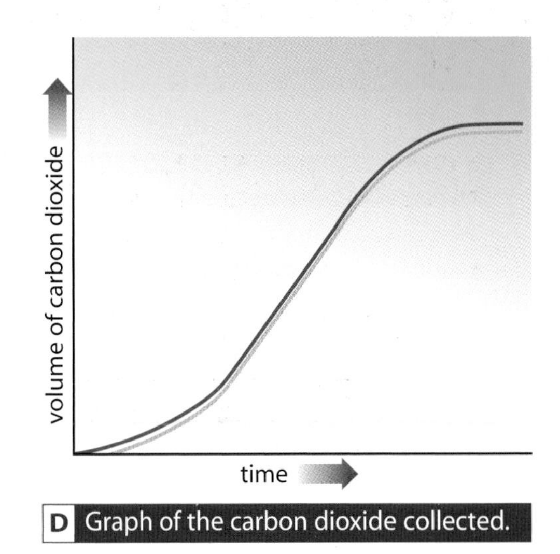

D Graph of the carbon dioxide collected.

4 Suggest a reason why fermentation stops eventually.

E Wine, beer and spirits have different strengths.

Wine is stronger than beer because grape juice has more sugar than barley in water, so wine ferments for a longer time and you get a much stronger alcohol solution, often as high as 13%.

It is not easy to get a stronger solution of alcohol by adding more sugar. Alcohol is toxic and affects living organisms. By about 15% alcohol the yeast cells are killed by their own waste.

To get stronger alcohol, the alcohol must be concentrated by **distillation**. Spirit drinks are usually about 40% alcohol. Anything above this would be dangerous to drink. To run a car on alcohol instead of petrol it would need to be 100% pure.

5 What is the typical percentage of alcohol in

a beer

b wine

c spirits?

6 Why can't vodka be made by fermentation alone?

7 Describe how to make a glass of wine from a bunch of grapes.

How alcohol affects the human body

By the end of these two pages you should understand:

- what alcohol does to the human body.

The alcohol in alcoholic drinks is called **ethanol**. In small amounts, ethanol makes people less self-conscious and more chatty. Medical research even suggests that adults who drink small amounts of alcohol are actually more likely to stay healthy than people who abstain completely.

But ethanol is a drug that affects both brain and body. In large amounts the effects become more negative and harmful. Users may become emotional, aggressive or violent. Ethanol also affects the way the body works. Balance and coordination are lost. The digestive system loses control and vomiting is common. In larger amounts, blood pressure suffers, leading to fainting. And, if the intake is sufficiently high, ethanol can kill you.

1 People can often be a little shy and self-conscious at parties. How can a little alcohol help to overcome that?

2 How does increased alcohol intake affect the brain and body?

A Alcohol clouds the higher thought processes…

B Five different ways to get one unit of alcohol.

The government and health agencies have issued guidelines in terms of 'units' of alcohol. One unit of alcohol is the same as

- half a pint of beer
- a small glass of wine
- a single 'pub' measure of spirits
- two-thirds of a bottle of alcopop.

The recommended daily limits for alcohol consumption for adults are

- 3–4 units for men
- 2–3 units for women.

Below this level, there are no associated health risks. Above this level, the health risks start to multiply. And you cannot abstain on some days and drink excessively on others. **Binge drinking** puts an excessive strain on the body and can cause permanent damage.

Despite clear guidelines about alcohol it can be difficult to judge your intake as drinks come in different strengths and different-sized cans and bottles. Drinking even small amounts clouds judgement. Once people have had a couple of drinks they can't think as clearly about whether they have had too much.

But payback time comes in the morning – with a hangover. One of the main causes is **dehydration**. Alcohol makes the body pass more urine – more than the water taken in when drinking. The brain shrinks away from the skull slightly, causing a thumping headache. Excessive urination also leads to loss of important salts of potassium and magnesium. This can cause irregular heart beat. The liver breaks down the alcohol and makes waste products that upset body chemistry.

Long-term damage from excessive drinking is far worse and the NHS in Britain spends almost £2 billion pounds a year treating alcohol-related conditions. Excessive drinking increases the risk of heart disease and strokes, as well as some types of cancer. It also slowly destroys the liver. Without a liver transplant, this will cause death.

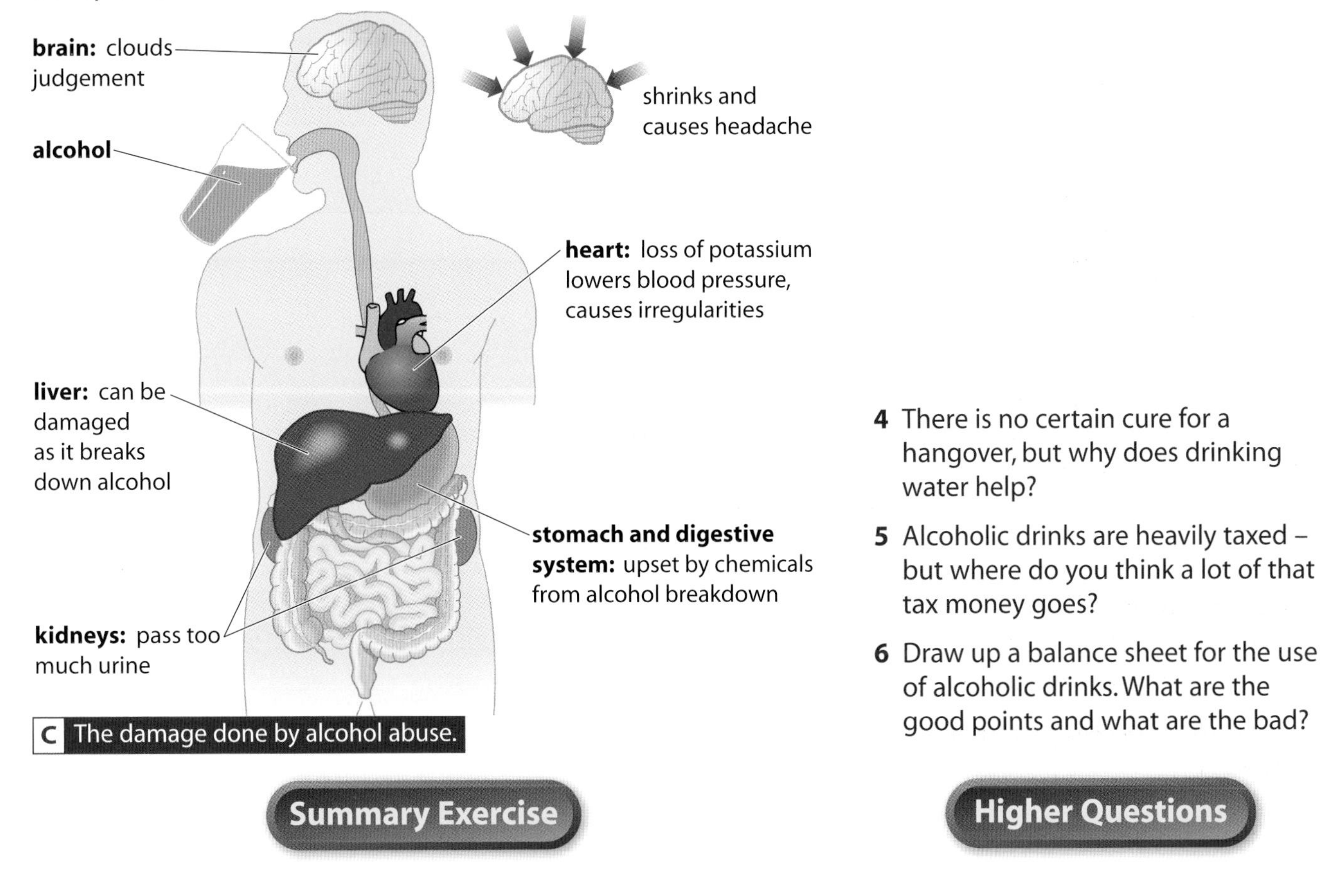

C The damage done by alcohol abuse.

Summary Exercise

3 Suggest reasons why the level for women might be lower than the level for men.

4 There is no certain cure for a hangover, but why does drinking water help?

5 Alcoholic drinks are heavily taxed – but where do you think a lot of that tax money goes?

6 Draw up a balance sheet for the use of alcoholic drinks. What are the good points and what are the bad?

Higher Questions

What alcohol does to society

By the end of these two pages you should understand:

- that excessive use of alcohol can cause harm to society.

Alcohol-related problems are not a new phenomenon. Alcohol was considered so bad by the government of the USA in the 1920s that they passed a Prohibition law. For a few years it was completely illegal to make, sell or drink alcohol. But Prohibition did not work in the long run.

Drinking alcohol clouds the higher brain functions, leading to a loss of inhibitions. For some people this means acting foolishly, for others it can lead to aggression and violence. The more alcohol that is drunk, the more this effect increases.

Binge drinking is now a serious problem. It is a health issue for those involved, as they will be doing long-term damage to their bodies, but it is also a problem for our society. City centres are littered and vandalised. A large police presence is needed on a nightly basis to keep order. People are assaulted and hospital casualty departments are kept busy. And of course all of this costs money, which is provided by the taxpayer.

A Alcohol was illegal in the USA in the 1920s.

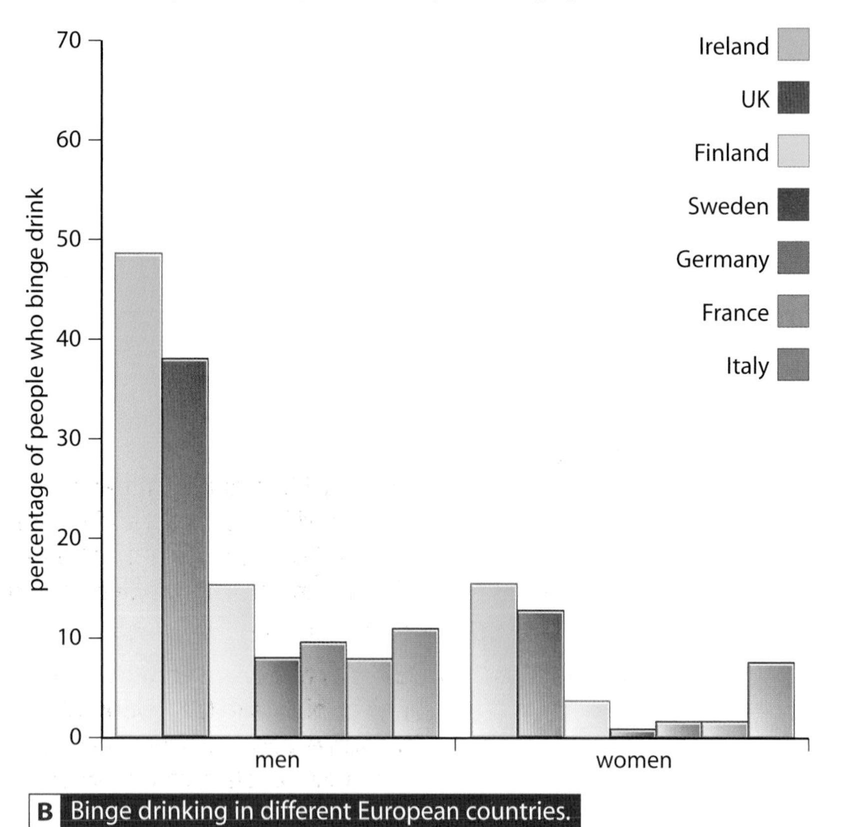

B Binge drinking in different European countries.

Many people see this as a 'British disease' as it is far worse in this country than in many others.

1 Look at figure B. Roughly how many times worse is the problem in the UK, compared to Germany (for men)?

2 Suggest some reasons why you think people might binge drink.

Figure B might suggest that this is a male problem, but there are worrying trends that women are catching up fast. Statistics show that women are increasingly drinking more than the recommended amounts. Some people blame the media. Others blame the advertising industry. Advertisers claim that sweet alcopops offer a wider choice. Others consider that they were designed specifically to 'get girls drunk'.

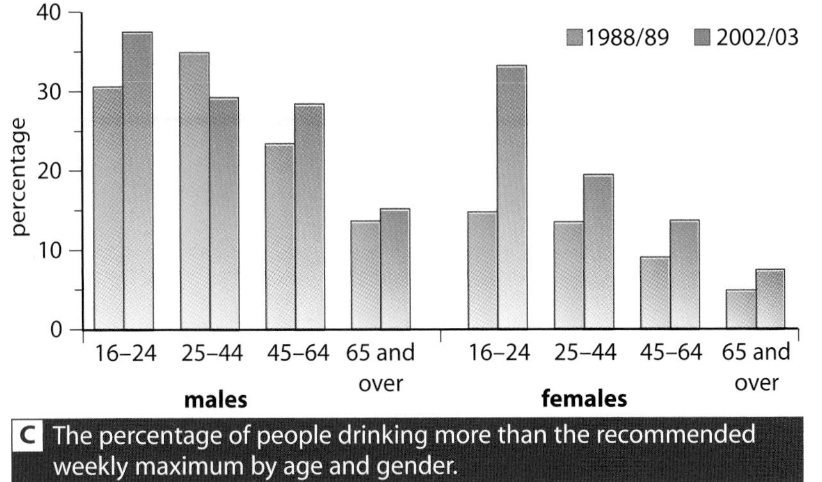

C The percentage of people drinking more than the recommended weekly maximum by age and gender.

3 Look at figure C. Which group has the greatest rise in alcohol consumption over the decade shown?

4 Suggest reasons why women are now drinking much more than they used to.

Driving is particularly dangerous when under the influence of alcohol. Judgement is impaired and reaction times are slowed. A drunk driver is far more likely to have an accident than a sober one. Advertising campaigns use images of death and misery to discourage drink-driving.

Motorists can be 'breathalysed' at the side of the road by the police. If their breath has more than 35 micrograms of alcohol in 100 ml of breath they will be arrested and are likely to be banned from driving. This two-pronged approach – public education enforced by law – has resulted in the number of injuries caused by drunk-driving falling substantially over the years.

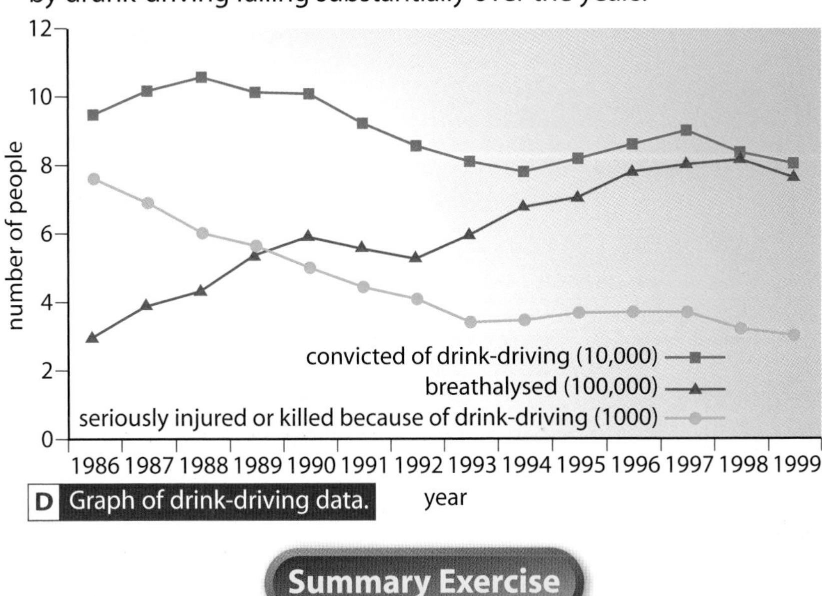

D Graph of drink-driving data.

5 The roadside breathalyser was introduced in the 1980s and its use has steadily increased. Describe its effect on the number of deaths and serious injuries.

6 How much of a problem do you think alcohol is in this country? Should we ban it, as they once did in the USA? Explain your answer.

Summary Exercise

Higher Questions

Intelligent packaging

By the end of these two pages you should know:

- how intelligent packaging is used to keep food fresh
- how intelligent packaging in the future may 'talk' to your kitchen.

A Supermarket fish is packaged so that it can be safely stored for longer than fresh fish.

Have you ever wondered:

How does 'intelligent packaging' keep food fresh?

Food goes bad because microbes grow in it. Some microbes might just make the food taste nasty, but others can cause illness or even death. Wrapped or covered food will keep fresh longer than 'open' food because flies or airborne microbes can't get to it. But modern packaging goes a step further. Like you, most microbes need water and oxygen to live and grow. Keep out either or both of these, and the food will last much longer. Modern packaging is designed to do this. It is often called '**intelligent packaging**' as it actively controls the way the food is kept.

1 Why is it a good idea to keep flies off food?

2 Which two chemicals do microbes need to live and grow?

B Just add boiling water…

Perhaps the simplest way to stop food going bad is to dry it. This method has been used for thousands of years. Dried fish or meat may not look or smell very appetising but without water the microbes that cause decay cannot grow. Modern methods allow foods to be dried out without changing their smell or flavour quite so much and modern packaging keeps the moisture out. Food packed like this can last for months. 'Soup in a cup' is made like this. Pour on boiling water and the food absorbs it and swells up.

In intelligent packaging, air is removed and replaced by other gases such as carbon dioxide or nitrogen. With no oxygen most microbes cannot grow and so the food keeps for longer. There are other advantages. Lack of oxygen stops fats going **rancid** and stops cut fruit and vegetables turning brown. The contents of sealed packs do not get squashed during transport.

3 Why does bread go mouldy faster than dry crackers?

4 Crisp bags are often filled with nitrogen. Suggest two reasons for this.

5 **a** Why does it matter if cut apples in a fruit salad turn brown?
b How can this discolouration be prevented?

C Nitrogen fresh…

The labelling of foods is also changing. Old-style barcodes give basic information such as price, but new labels will contain much more interactive information held on microchips. They might be able to give details of a packet's supply chain. They could also check that the storage temperature is correct and monitor fruit and vegetables for ripeness.

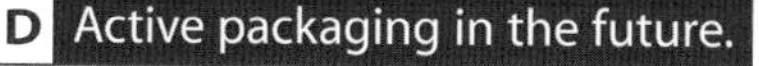
D Active packaging in the future.

In the future, these microchips might even be able to talk to kitchen appliances. Freezers might be able to check the sell-by dates of their contents. And ovens might know how long to cook food and give nutritional details of the food.

6 Why would it matter if a chilled fresh lasagne spent an hour in the hot sun before it went into the fridge?

7 Explain why food goes bad and describe two ways in which modern packaging can help to prevent this.

Higher Questions

How does mayonnaise work?

By the end of these two pages you should know:

- how emulsions are formed
- how they are used in the food industry.

Both mayonnaise and salad dressing are made from oil and vinegar – but they are very different.

Have you ever wondered:

How do they keep the oil and vinegar in mayonnaise from separating?

Ethanoic acid dissolves in water. Vinegar is a weak solution of ethanoic acid in water. Oil does not mix with water. If you shake oil and vinegar the two will mix as they break up into tiny droplets. But when you stop shaking the droplets of oil float up through the vinegar and pop back together. That's fine for a salad dressing, but not for mayonnaise.

Mayonnaise is made from oil and vinegar, but the two have been thoroughly mixed and do not separate out. Mayonnaise under a microscope has tiny droplets of vinegar trapped in oil. Something has been added to the mixture to stop the tiny droplets popping back together and forming a layer. In mayonnaise it is egg yolk, which contains a chemical called **lecithin**.

Mayonnaise has very different properties from 'runny' oil or water. It is 'thick' and sticks to the food that is dipped in it. Mixtures of oil and water like this are called **emulsions**. Chemicals that stop the oil and water parts in an emulsion from separating are called **emulsifiers**.

1 Which weighs more for the same volume, oil or water? Explain your answer.

2 Why can't you make mayonnaise from vinegar and oil alone?

3 What advantages do foods that are thick, like mayonnaise, have over foods that are runny, like oil and vinegar?

Many common foods are emulsions. Milk, cream and butter are natural emulsions.

- Milk is an emulsion of a few per cent of oil droplets in a watery liquid.
- Cream is an emulsion of a larger percentage of oil droplets in a watery liquid.
- Butter is an emulsion of a few per cent of water droplets in fat (fat is just solid oil).

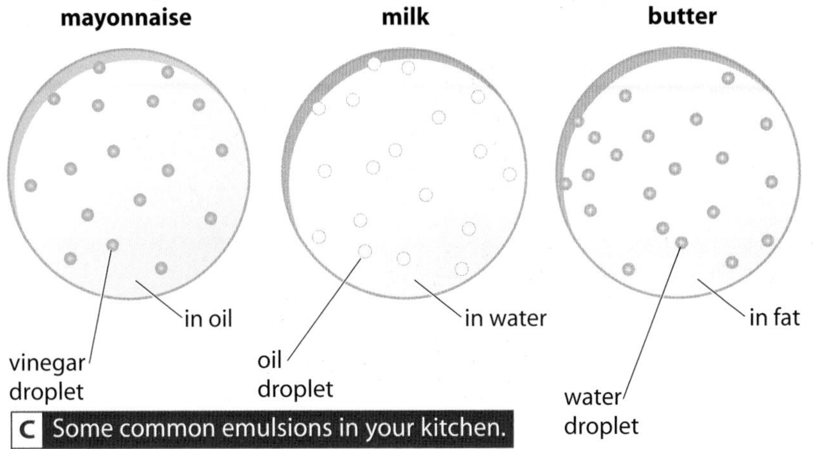

C Some common emulsions in your kitchen.

Emulsifiers like lecithin have very special molecules. At one end the molecule is **hydrophilic** (water-loving). This end of the molecule is attracted to the watery droplets. At the other end, the molecule is **hydrophobic** (water-hating). This end is attracted towards the oil. The molecules arrange themselves so that they 'stitch' the oil and water parts together, holding them in place and stopping them from separating out.

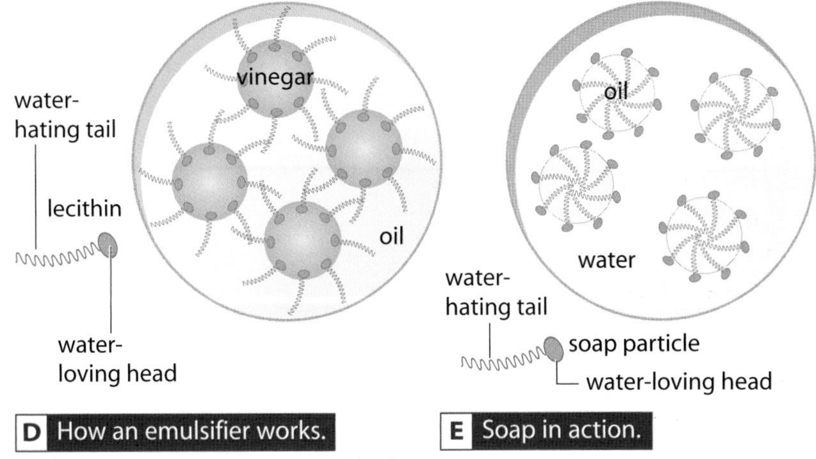

D How an emulsifier works.

E Soap in action.

Soap and detergent molecules work in a similar way. Most 'difficult' dirt is stuck in place with grease or oil. The hydrophobic ends of the detergent molecules attach themselves to the oil, leaving the hydrophilic ends sticking up into the water. The little droplets of oil are then pulled up into the water, forming an emulsion – and leaving your clothes clean.

4 What do lecithin and soap molecules have in common?

5 Emulsion paint is an emulsion of tiny droplets of oil paint in water. Draw a diagram to show what this looks like under a microscope. What else must be in the mixture, apart from the oil, water and colour?

Design and test

H ***By the end of these two pages you should be able to:***

- suggest the properties that would be needed for a given product.

A What do people want from toothpaste?

The main purpose of a toothpaste is to get your teeth clean when you brush them. Toothpaste can also perform other useful functions such as making your breath fresh. But sometimes people are swayed by factors that are not really important.

Have you ever wondered:

What would the properties of a perfect hair gel be?

1 Draw up a table with the headings 'essential for a toothpaste', 'not essential but a good addition' and 'not important'. Discuss each of the comments in figure A with your group and put them into one of your three columns.

B Two types of hair gel.

2 Draw up list of the properties you think would be important for
a suntan lotion
b eye make-up
c shower gel.

If you were part of a product design team it would be essential that you thought very carefully about the properties or features your product needed to have. Advertisers often make extravagant claims about their products to try to make people buy them. But customers will not buy something twice if the product doesn't do the job properly.

As a customer, you need to look at it from the other direction. Advertisers are trying to make you buy, but cannot guarantee that their product is the best. You usually have to pay more for the famous brand names, but you might be better off with a cheaper version if it does the job just as well. If you understand the key properties needed for a product, you will be better able to make a good decision.

The Advertising Standards Authority can ban adverts if they are incorrect or misleading. It is the duty of all manufacturers to test out their products and to be able to back up any claims they make with evidence.

P

3 For each of the advertisements below, list the points made under the headings 'important' and 'not important'. In a third column, add anything else you think you ought to know about the product.

a 'Dazzle' ultra-white paint
- Tough gloss finish.
- Contains 'Ultradazzel™'.
- Non-drip.
- Washes off the brush easily.
- Your friends will be amazed!

b 'Jammy Fruit Dippers'
- Crunchy finger biscuit.
- Real fruit jam.
- Attractive packaging.
- Brings a smile to your face.
- No artificial colours or flavours.

c 'Biofuel Ecodiesel'
- Use in ordinary diesel engines.
- 'I love it' says Jeremy Trucker, lorry driver.
- Gives the same miles per litre as ordinary diesel.
- Reduces engine wear and tear.
- Made from soya beans.
- 'Help save the planet' says Josiah Crusten, environmentalist.

C A selection of products.

Questions

Multiple choice questions

1 Kevlar™ is used to make bullet-proof vests because
- **A** it is cheap and easy to manufacture.
- **B** weight for weight it is five times as strong as steel.
- **C** you can easily mould it into solid sheets.
- **D** it contains chemicals that dissolve metal.

2 Thinsulate™ material is such a good insulator because
- **A** it has no air in it.
- **B** its very thick fibres trap more air.
- **C** its very thin fibres trap more air.
- **D** it contains nitrogen which stops radiation.

3 Gore-tex™ fabric is good for raincoats because
- **A** it has a solid plastic film in it to keep the rain out.
- **B** it has holes between the fibres so that it can 'breathe'.
- **C** it has a plastic film in it that keeps the rain out but has tiny holes so it can 'breathe'.
- **D** it has a porous film in it that can 'breathe' but the holes close up when it rains.

4 Sunblock with nanoparticles does not leave white streaks because
- **A** you only put it on very thinly.
- **B** it uses different chemicals to normal sunblock.
- **C** the nanoparticles are too small to reflect light.
- **D** the nanoparticles go through the pores in your skin.

5 Nanoparticles of iron can be used to clean up oil spills because
- **A** they speed up the natural breakdown of oil-based chemicals.
- **B** nanoparticles are easier and cheaper to make than iron filings.
- **C** iron reacts with oil.
- **D** iron makes the oil molecules clump together.

6 Alcohol is made by growing yeast in a sugary solution. The process is called
- **A** fermentation.
- **B** distillation.
- **C** catalytic conversion.
- **D** thermal decomposition.

7 The gas produced during the production of alcohol by yeast is
- **A** oxygen.
- **B** nitrogen.
- **C** carbon monoxide.
- **D** carbon dioxide.

8 Long-term alcohol abuse can lead to death because it damages the
- **A** liver.
- **B** lungs.
- **C** kidneys.
- **D** stomach.

9 Alcohol causes increased urination. One of the main causes of a hangover is
- **A** a build-up of toxins in the kidneys.
- **B** the brain shrinking inside the skull.
- **C** a build-up of water in the lungs.
- **D** a build-up of water in the brain.

10 Excessive drinking can lead to public disorder because
- **A** alcohol is a drug that makes everybody violent.
- **B** alcohol is a drug that makes people lose their inhibitions.
- **C** alcohol is a drug that always makes people fall asleep.
- **D** alcohol is a drug that always gives people a headache.

11 Fruit salad packed in an atmosphere of nitrogen does not go brown because:
- **A** there is no oxygen to make it go brown.
- **B** there is no carbon dioxide to make it go brown.
- **C** the nitrogen kills off the bacteria.
- **D** the nitrogen 'bleaches' the fruit.

12 Smart packaging could tell you if your fruit is ready to eat by
- **A** changing colour after one week.
- **B** getting softer if you squeeze it.
- **C** sensing the flow of electricity inside the package.
- **D** sensing the appearance of 'ripening' chemicals inside the package.

13 Mayonnaise is
- **A** a solution of oil in vinegar.
- **B** a solution of vinegar in oil.
- **C** an emulsion of vinegar in oil.
- **D** an emulsion of egg yolk in vinegar.

14 Chemicals like lecithin that, for example, stop the different parts of mayonnaise from separating are called
- **A** gelling agents.
- **B** acidifiers.
- **C** emulsifiers.
- **D** plasticisers.

H **15** Which of these is the most important property for a toothpaste?
- **A** It gets your teeth really clean.
- **B** It has coloured stripes.
- **C** It has a minty flavour.
- **D** It comes in a really snazzy tube.

Short-answer questions

1 Read the following passage and answer the questions below.

> One important growth area for 'smart materials' involves producing 'plastic electronics'. In a simple example, traffic lights now use 150 W light bulbs. In the future they may be replaced by light-emitting plastic devices which would use only 15 W. Meanwhile 'plastic computers' are being developed. These will have a great advantage over silicon-based computers as they will be flexible. One day you might be able to have a pocket-sized computer with a large roll-out screen. These new materials are expensive now, but will become cheaper once they are manufactured on a large scale.

- **a** What would be the advantage of using 'light-emitting plastic', instead of light bulbs, for traffic lights?
- **b** What potential saving could you make on every £100 currently spent?
- **c** Give two big advantages that the plastic 'pocket-computer' would have over current silicon-based equivalents.
- **d** Why will 'plastic electronic' devices become cheaper in the future?

2 The graph shows the strength of the alcohol solution produced from different strengths of sugar solution during fermentation. The experiments were conducted in a warm room.

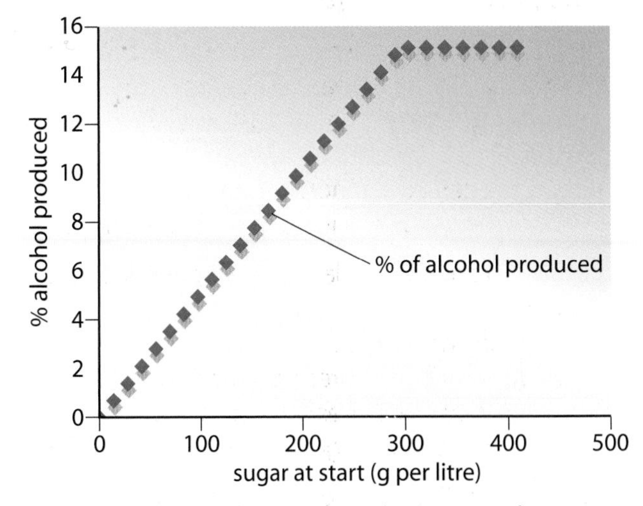

- **a** What needs to be put into the sugar solution to make the sugar turn to alcohol?
- **b** Describe the pattern shown by the graph.
- **c** What is the relationship between the amount of sugar you start with and the amount of alcohol produced, up to 300 g of sugar per litre?
- **d** What is the relationship after 300 g of sugar per litre?
- **e** Suggest a reason why the pattern changes at this point.
- **f** This fermentation process takes about a month to convert 300 g of sugar to alcohol in a warm room. How would this time change if the experiment was conducted in a cold room?
- **g** In another experiment, the different solutions were fermented in a cold room and a graph of sugar against alcohol was plotted. How do you think this graph would look, compared to the one shown above?

D

P

Glossary

alcohol A family of chemicals made by fermenting sugars.
alloy A metal mixture, often with special properties.
binge drinking Drinking excessive amounts of alcohol over a short space of time with the intention of getting completely drunk.
breathability Clothing that lets the water vapour from sweat escape has this.
carbon fibre Fibres made from carbon atoms only. They are stiffer than steel.
composite Something made from more than one type of material, carefully arranged together.
conductor Something that lets energy (such as heat or electricity) pass through it easily.
dehydration (of the body) Not having enough water in your body.
distillation Boiling a liquid and then cooling and condensing the gas; used to concentrate or purify.
emulsifier A chemical that stops the oil and water parts from separating in an emulsion.
emulsion Mixtures of water and oil that do not separate out.
ethanoic acid The acid in vinegar. You may hear it called acetic acid, which is the old name for it.
ethanol The chemical name of the 'alcohol' in alcoholic drinks.
fermentation The process in which yeast turns sugar into alcohol and carbon dioxide. Used to make beer and wine.
Gore-tex™ A special fabric that can 'breathe' but keeps liquid water out.
hydrophilic 'Water loving' – chemicals that dissolve or mix completely with water.
hydrophobic 'Water hating' – chemicals like oil that do not 'mix' with water.
intelligent packaging Modern packaging that actively controls the environment around the food and stops it going bad.
Kevlar™ A very strong plastic fibre that can be used for bullet-proof vests.
lecithin The chemical in egg yolk that acts as an emulsifier in mayonnaise.
Lycra™ A brand of very stretchy polymer (plastic) fibre.
membrane A very thin layer, or sheet, of material.
microporous Something with millions of microscopic holes (pores).
nanobot A nano-sized machine or robot.
nanocomposite A material which has nanoparticles used in combination with other materials.
nanometre One billionth of a metre (10^{-9} m) – the scale of atoms.
nanoparticle A particle with a diameter of between 10 and 100 nanometres.
nanotechnology New technology built around the use of nanoparticles.
nanotube A nano-sized tube made by rolling up a sheet of carbon atoms.
radiate Spread out from a source. For example, light from a bulb or infrared radiation from a hot object.
rancid Oils and fats go rancid when they react with oxygen. It makes them taste bad.
shape-memory alloy A special alloy that goes back to its original shape on warming if you bend it.
smart material A material that can change its properties as conditions change.
sugar A sweet-tasting chemical found naturally in fruits.
Teflon™ The material used to coat 'non-stick' pans.
Thinsulate™ A brand of insulating material that is made from very fine polypropene fibres. Polypropene is a type of plastic.

Producing and measuring electricity

A B Chris Little develops new mobile phones for Motorola.

Less than 10 years ago hardly anyone had used a mobile phone or sent an e-mail. Technology has developed at an incredible rate, making all kind of things possible. Many of these advances have come as a result of electricity.

In many ways electricity runs our lives. All around us there are many things that use electricity to entertain us, make our lives easier and allow us to communicate with others. All these technologies work because we can measure, control and use electricity very accurately.

Electricity is still very cheap and easily portable, with cells and batteries being very convenient to use. But each year over 20,000 tonnes of used batteries are thrown away in the UK alone. There are alternatives – rechargeable and solar cells – that can save money and reduce waste.

In this topic you will learn that:

- there are a variety of ways we can produce electricity
- electricity can be measured
- the voltage, current and resistance in a circuit are related
- the change in resistance of electrical devices is used in a variety of applications.

Sort these statements into three groups:

Agree, Disagree, or It depends on…

- Computers will get smaller and faster in the future.
- There are different types of current.
- Electricity is still – it does not flow.
- Rechargeable batteries are better for the environment than disposable batteries.
- Only data collected by a computer are reliable.

Telecommunications

By the end of these two pages you should know:

- how telephones and electricity have helped shape the modern world
- that there are two types of electrical current
- the difference between AC and DC.

Have you ever wondered:

Which invention changed the world the most?

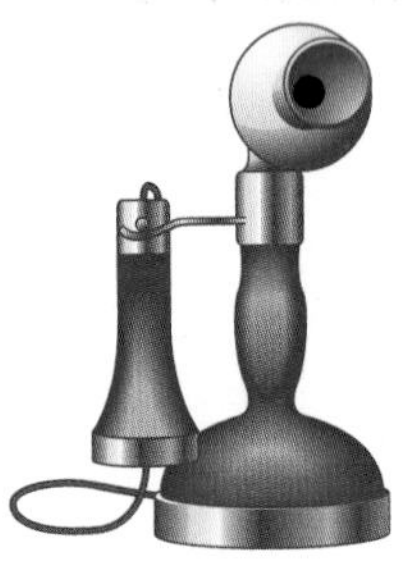

1875
Telephone.

1928
First transatlantic television broadcast.

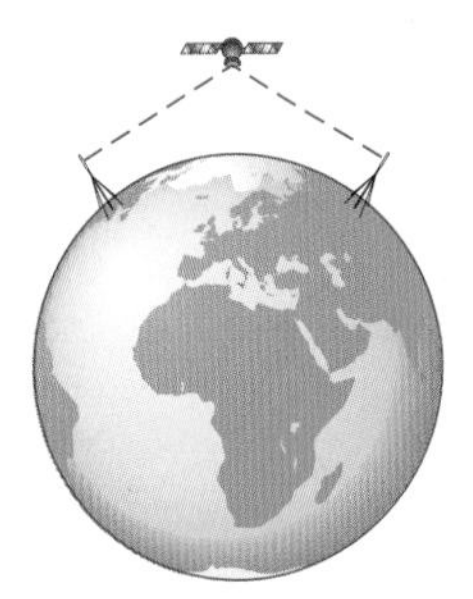

1967
First satellite TV network.

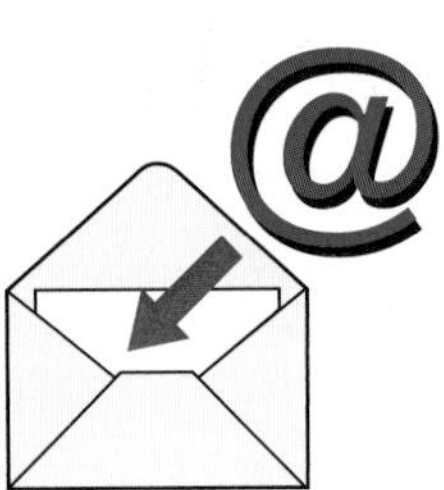

1971
First e-mail with an @ sign in it sent.

1983
Hand held mobile phones become available.

A Developments in technology have revolutionised the way we communicate.

People had known about electricity for a long time but it wasn't until the electric light bulb was developed around the 1860s that people's everyday life was changed. Try and imagine what your life would be like without electric lights.

The next big step forward was in 1875 when a telephone was first used to transmit a voice from one place to another. Before then messages were sent by telegraph using Morse code, or needed to be carried from one place to another, by hand, by rail or even by pigeon. Once the telephone had been developed and we started to use electricity to send voice messages, communication changed forever.

Today, live sound and moving pictures are beamed worldwide. Telecommunications and electricity together have shrunk the whole planet and transformed our lives.

1. When was the first voice transmission by telephone?
2. How were messages sent before then?
3. Name three benefits that the invention of the telephone gave to communications.

The first ever e-mail as we know it (with an @ sign in it) was sent in 1971 and although e-mail did not become widespread for another 10 or 15 years it was the start of something that has changed many lives.

When mobile phones were first developed the idea of SMS/text messaging was added on as an afterthought. Those who worked on it did not imagine that it would be very popular. In 2004 an estimated 20 billion text messages were sent in the UK alone – another dramatic change in how we communicate with each other.

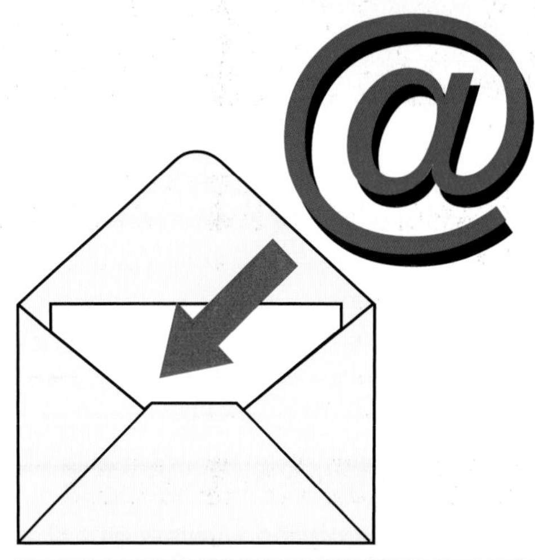

B E-mail is an instant and worldwide way of communicating.

4 How many text messages were sent in 2004 in the UK?

5 Write down three advantages of e-mail communication over previous methods.

Have you ever wondered:

Why did people believe electricity could cure all your aches and pains?

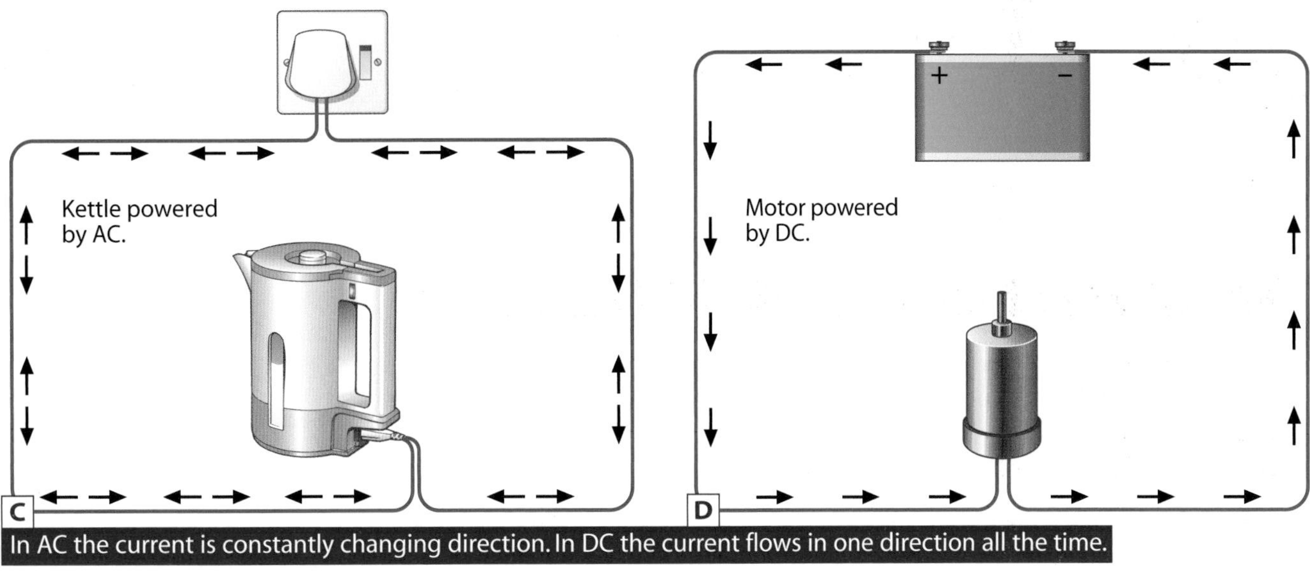

In AC the current is constantly changing direction. In DC the current flows in one direction all the time.

The electricity that has transformed our lives so much comes in two types, alternating current and direct current (usually known as AC and DC). Electricity from the mains sockets in your home, dynamos and generators is AC and the type from batteries is DC.

The difference between AC and DC is the way that the **current** flows around the **circuit**. In AC the current is constantly changing direction whereas in DC the current always flows in the same direction. Lights plugged into AC don't flicker because the current changes direction so quickly we cannot see it, about 50 times every second.

6 What are the two types of electrical current?

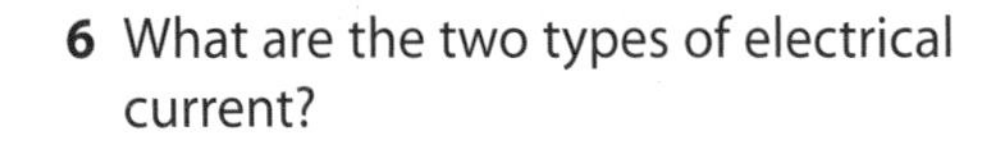

7 What type of current do we get from the mains and dynamos?

8 What type of current do we get from batteries?

Summary Exercise

Higher Questions

Direct current

By the end of these two pages you should:

- know about different types of cell
- know which way the conventional current flows
- understand what electric current really is
- know how to measure current using an ammeter.

A Mobile phones are powered by a rechargeable battery.

Have you ever wondered:

Why is my phone wireless, but I have to plug my hairdryer into the wall?

Appliances that don't plug into the wall usually run on **batteries**. Batteries are a useful, portable and generally safe way of carrying electricity around. They are made up of individual **cells** connected together. The most common type of cell is the non-rechargeable type called a **dry-cell**. Inside a dry-cell a chemical reaction creates the current. It is also possible to get **rechargeable** batteries such as the ones used in mobile phones.

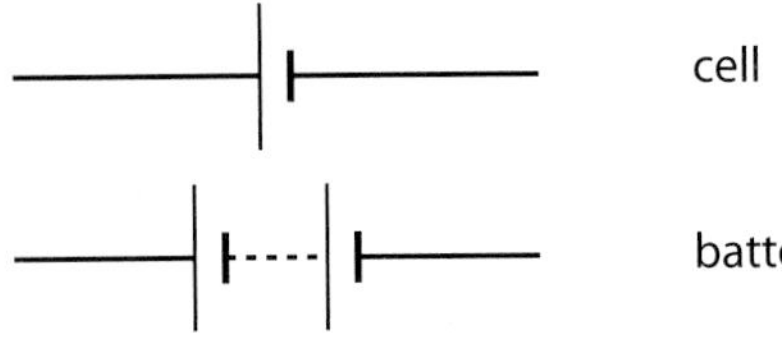

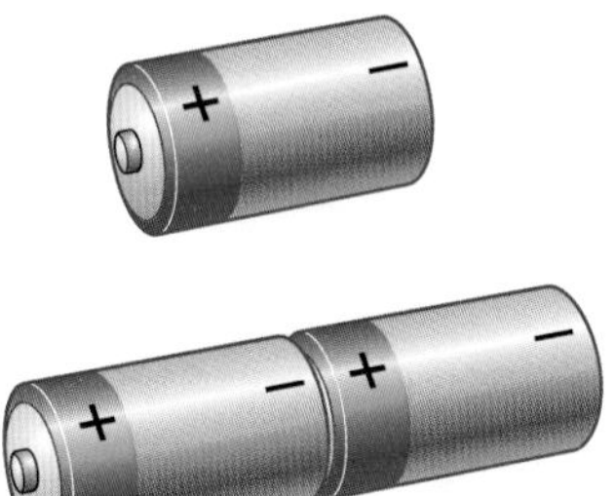

B The circuit symbols for a cell and a battery are not the same.

The electricity that a cell produces is always DC. It is possible to have a cell that generates DC that does not use a chemical reaction. **Solar cells** are the most common example of this. They are made of thin sheets of material that has been treated so that when sunlight falls on them they create an electric current. They have the advantage of never running out as long as the sun is shining.

1 Give two advantages of using cells instead of mains electricity.

2 What is another name for a normal, non-rechargeable cell?

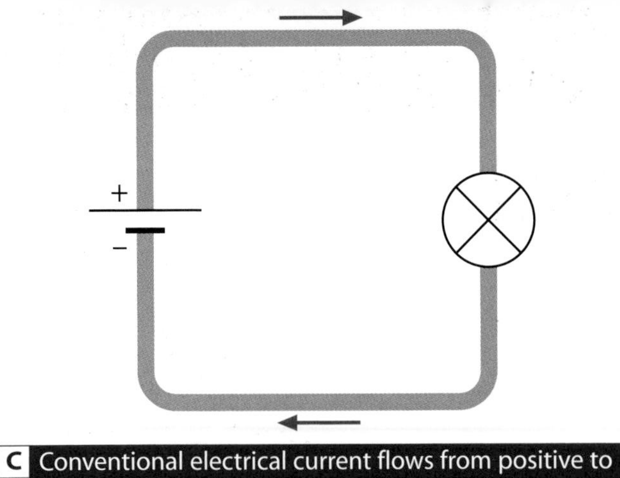

C Conventional electrical current flows from positive to negative.

D When a current flows the electrons move from negative to positive.

When electricity was first discovered, people said that the current flows around a circuit from the positive to the negative terminals of a cell. This is called **conventional current**. However, if you were able to look incredibly closely at a wire with a current passing through it then you would see lots of tiny particles flowing along the wire, all in the same direction. This is because electrical current is a flow of negatively charged particles called **electrons**. The electrons actually move around the circuit in the *other* direction to the conventional current. This is all a bit confusing but happened because electricity was discovered long before they knew what electrons were and what was actually going on.

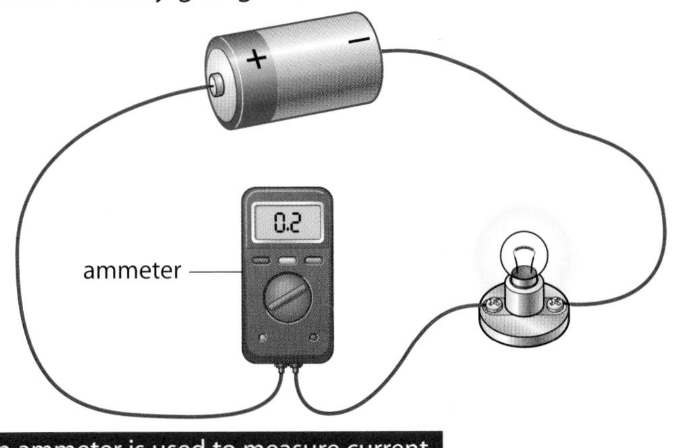

E An ammeter is used to measure current.

Current is a flow of electrons. When current is measured in amperes (usually known as amps, written as A) what is actually being measured is the number of electrons flowing per second. In order to measure this flow you need to put your measuring device in the flow of current. An **ammeter**, placed in **series** in the circuit, is used to measure current. It is normally measured in amps but if the current is very small then it is sometimes measured in milliamps (mA). 1 mA is one-thousandth of an amp (0.001 A).

3 Describe conventional current.

4 When an electrical current flows, what is actually moving?

5 Name three different types of cell.

6 What piece of equipment is used to measure current?

7 Should it be connected in series or in parallel for it to work?

8 How many milliamps are in 1 amp?

Producing electric current

By the end of these two pages you should know:

- how to produce an electric current using a magnet and a coil of wire
- what affects the size and direction of induced voltage
- how a dynamo works.

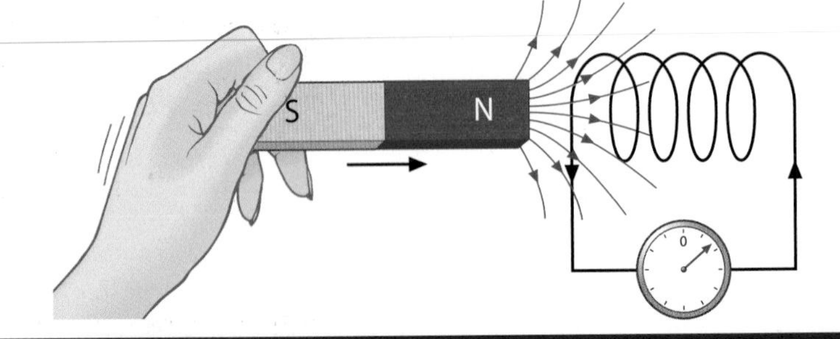

A When a magnet is moved near a wire, an electric current is generated.

Michael Faraday discovered that electricity can be made with a **magnet** and a coil of wire. Around every magnet is a **magnetic field** and if a wire is moved through that magnetic field then a **voltage** is made. Voltage is the amount of electrical 'push' that makes a current flow around a circuit. Moving the magnet or the wire will create a voltage and the current will flow. This is called **electromagnetic induction**, or the **dynamo** effect.

1 What two things are needed to generate electricity?

2 A magnet is placed inside a coil of wire and does not move. Will it create a current?

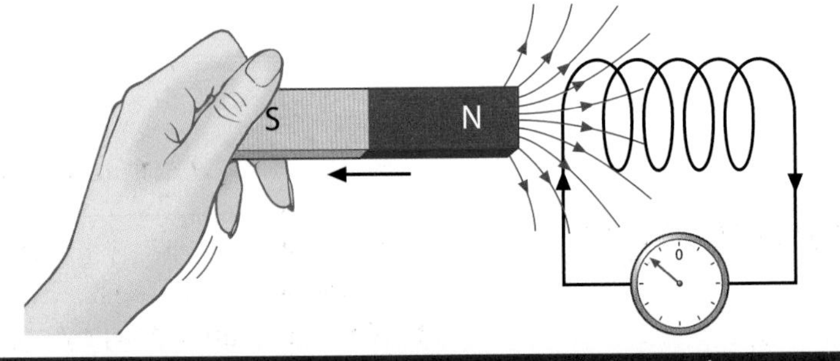

B Moving the magnet in the other direction will change the current's direction.

If you move the magnet or the wire in the opposite direction then the voltage will also change direction. The current will still flow, but in the *other direction*. If you turn the magnet around then the current will also change direction.

The amount of electricity produced with a single wire and a magnet is very small. But there are several ways to increase the induced voltage and the current produced. You can

- move the magnet closer to the wire
- move the magnet faster
- use a stronger magnet
- use more coils of wire.

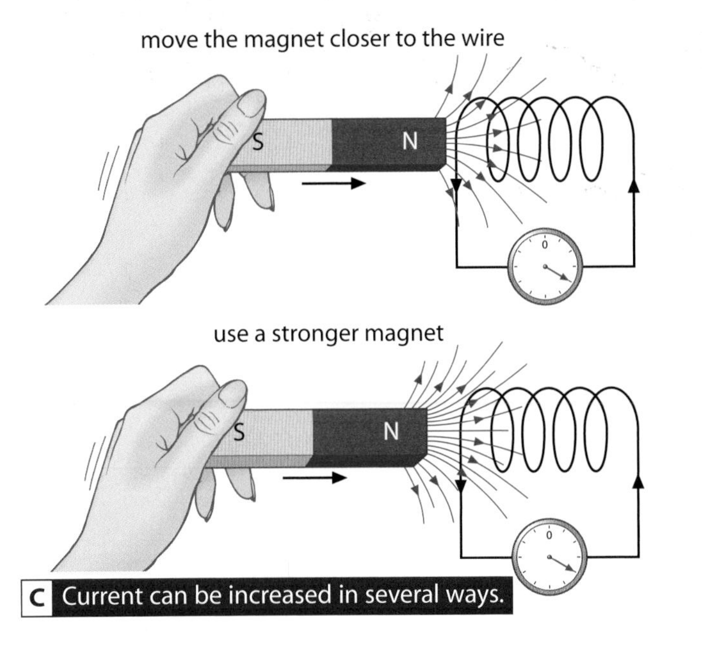

C Current can be increased in several ways.

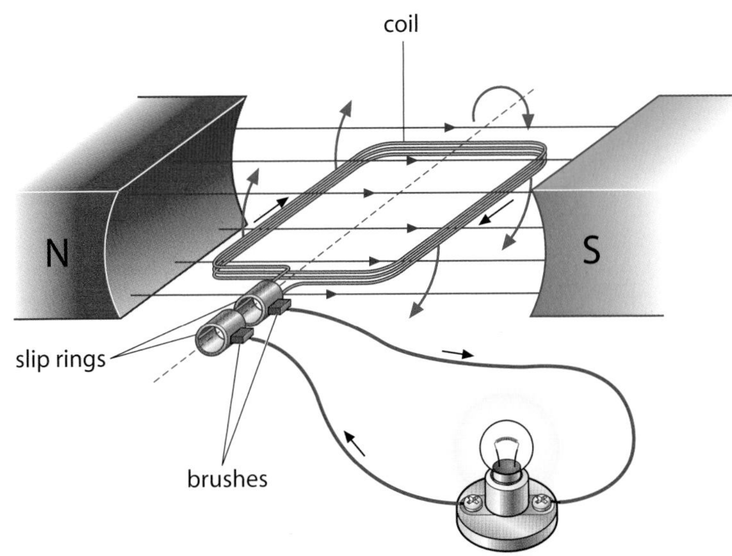

D A generator creates current by rotating a coil of wire in a magnetic field.

Moving a magnet near a coil of wire creates an electric current. Generators use this principle to generate electricity by rotating coils of wire *inside* a magnetic field. Rotating the coils means that they can be moved much faster and so produce a greater current. This principle is used on a small scale to power bicycle lights with a dynamo.

There is a rubber wheel on the dynamo that touches the bicycle wheel. This is connected to a magnet inside the main body of the dynamo. As the bicycle moves, the rubber wheel turns and so the magnet is rotated. This induces a voltage in the coil, which generates a current, powering the lights. The faster the wheel turns, the brighter the light.

3 In a competition to produce the most electricity using a magnet and a coil of wire, what four things could you do to ensure you win?

4 When you connect everything up, you find the current is going in the wrong direction. What can you do to make it flow in the other direction?

5 A cyclist uses a dynamo to power their lights.
  a If the bike stops moving, what happens to the lights?
  b When they are riding along, how can they make the lights brighter?

6 Arrange the following actions in the correct order to show how a dynamo is used to operate a set of bicycle lights:
- electrical energy turned into light in lamps
- current flows in coil
- move magnet into coil
- voltage induced in coil

Higher Questions

Current, voltage and resistance

By the end of these two pages you should know:

- that there is a relationship between current, voltage and resistance, represented by $V = I \times R$
- that changing the resistance of a circuit can change the current
- how current varies with voltage for fixed-value resistors.

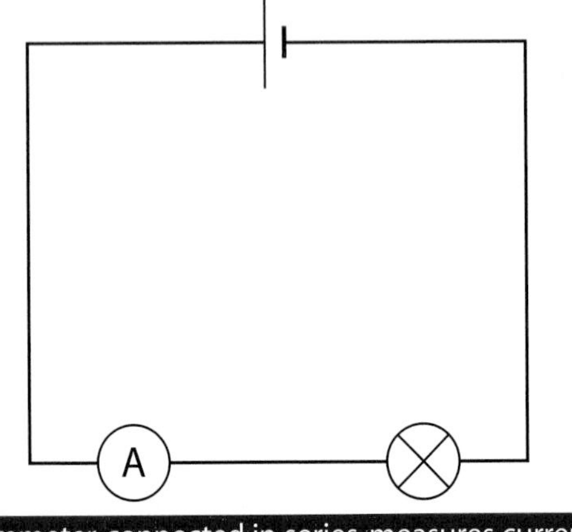

A An ammeter, connected in series, measures current.

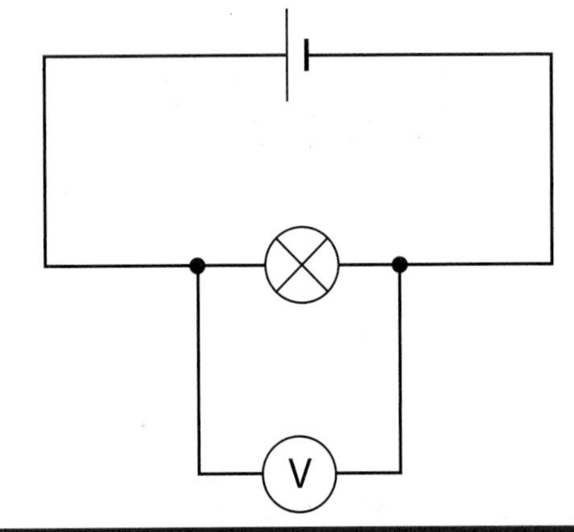

B A voltmeter, connected in parallel, measures voltage.

In any electrical circuit there are two very important quantities that can be measured. The voltage is measured in volts or millivolts with a voltmeter placed in parallel. The flow of electricity is called the current. This is measured in amps or milliamps and by connecting an ammeter in series in the circuit.

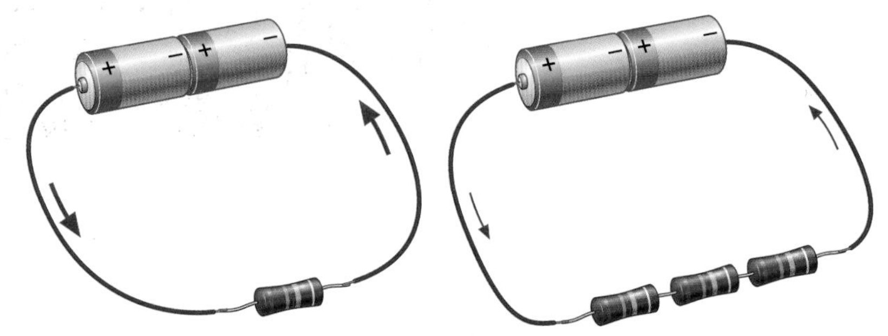

C The second circuit has three resistors. This means that the resistance is three times greater than in the first circuit and so less current will flow.

Resistance is used to describe how good a material is at conducting electricity. It is measured in ohms (Ω). If you add more components to a circuit then you will increase its resistance. This makes it harder for the current to flow and so for a fixed voltage the current will be less. Air has a very high resistance so does not normally conduct electricity. In a thunderstorm the voltage is so high that it causes the air to conduct and create lightning.

1 What units is resistance measured in?

2 What happens to the resistance of a circuit if you add more components?

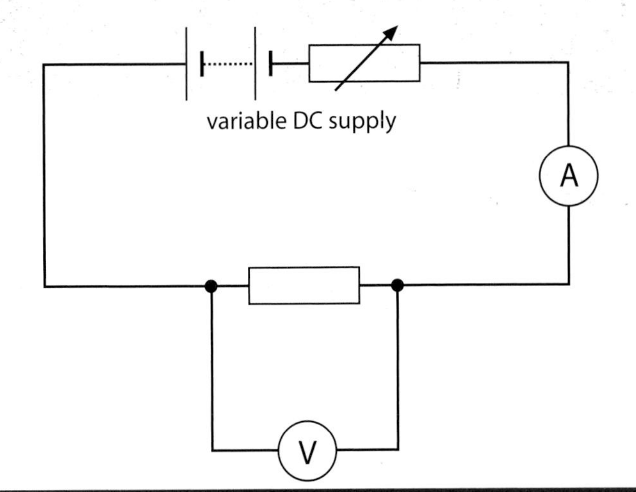

D This circuit can be used to measure the pattern between voltage and current for a resistor.

In a circuit there is a relationship between voltage and current. By carrying out the experiment shown above with fixed-value **resistors** a pattern can be found. If you increase the voltage then the current increases. Whatever fixed-value resistor you use, the line is always straight and goes through the origin (0,0). The bigger the resistance, the shallower the line, as shown in diagram E.

3 Sketch a graph showing what a graph of voltage and current for a fixed-value resistor would look like.

4 If the resistance of the resistor is increased, how would the graph look different?

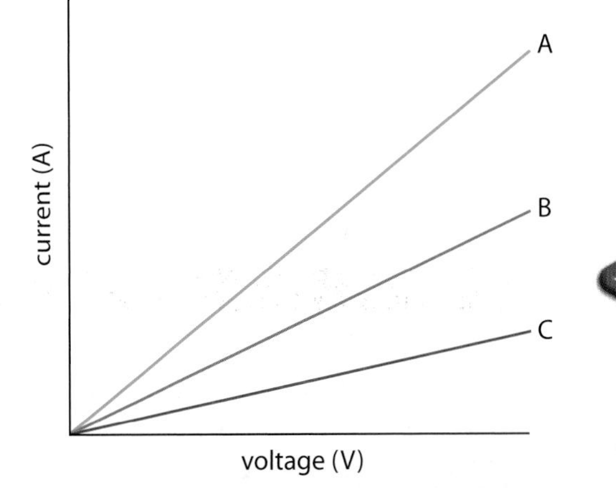

E The graph shows a straight line for each resistor, A–C. The higher the resistance the shallower the line. Thus we can see that A has the lowest resistance and C the highest.

The relationship between voltage, current and resistance can be represented by the equation below. V is the voltage (in volts), I is the current (in amps) and R is the resistance (in ohms).

$$\text{voltage (volts)} = \text{current (amps)} \times \text{resistance (ohms)}$$
$$V = I \times R$$

Example: What is the voltage in a circuit with a current of 3 A and a resistance of 4 Ω?

Solution: current (amps) × resistance (ohms) = voltage (volts)

$$3\text{ A} \times 4\,\Omega = 12\text{ V}$$

5 What is the equation that connects current, voltage and resistance?

6 What is the voltage in a circuit with a current of 8 A and a resistance of 2 Ω?

7 Describe how you would carry out an experiment to investigate the relationship between current and voltage for a resistor. State what measurements you would take, and what patterns you would expect.

Summary Exercise

Higher Questions

P

P

P1a.9.5

Resistance, lamps and computers

By the end of these two pages you should be able to:

- explain how current varies with voltage for a filament light bulb
- describe what happens to the resistance of a filament light bulb as the voltage increases
- describe how electrical sensors can be used to take readings that it would not be possible for a human to do.

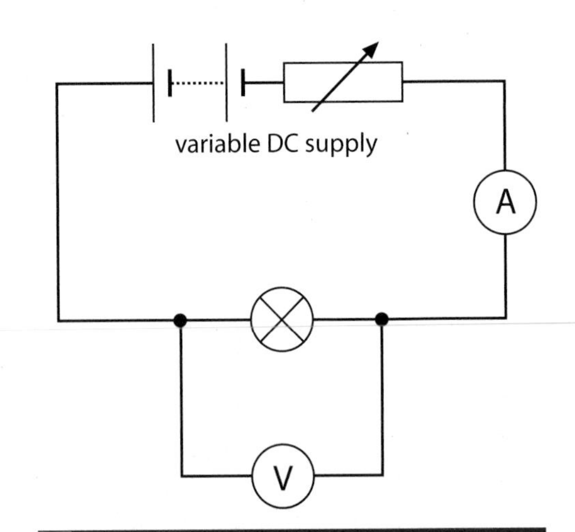

A Measuring the relationship between current and voltage for a filament lamp.

The relationship between voltage and current for a filament lamp is not the same as for a resistor. A graph and table of the results for the experiment illustrated in figure A are shown below.

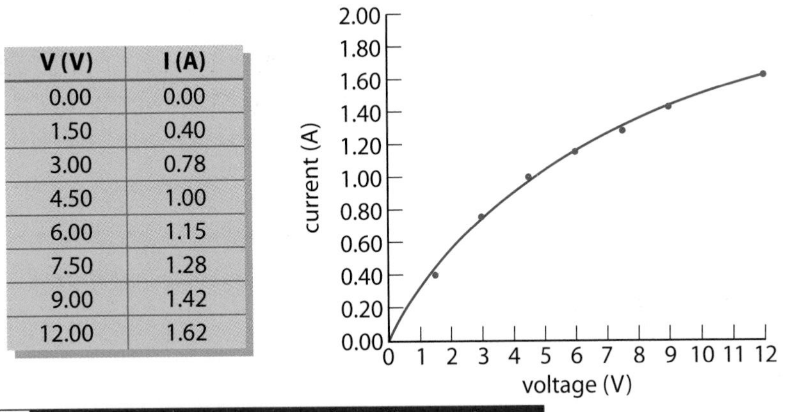

V (V)	I (A)
0.00	0.00
1.50	0.40
3.00	0.78
4.50	1.00
6.00	1.15
7.50	1.28
9.00	1.42
12.00	1.62

B Current and voltage results for a filament lamp.

The line of the graph for the lamp goes through the origin (0,0) but, unlike the straight line of the resistor, it is curved. The graph starts steep and gets shallower, showing that as the voltage increases the resistance increases. This is because as the wires in the lamp get hotter their resistance increases, making it harder for the current to flow.

1. What happens to the current through a filament light bulb when the resistance increases?
2. What is the shape of the line on a graph of current against voltage for a filament lamp?
3. What happens to the resistance of the filament lamp as it gets hotter?

If you have ever done an experiment then you will know that it is easy to forget to take a reading. It can also be easy to make a mistake when reading a value off a scale or display. In many modern situations computer sensors are used to take many readings in a fraction of a second and then use this information immediately to change or control something.

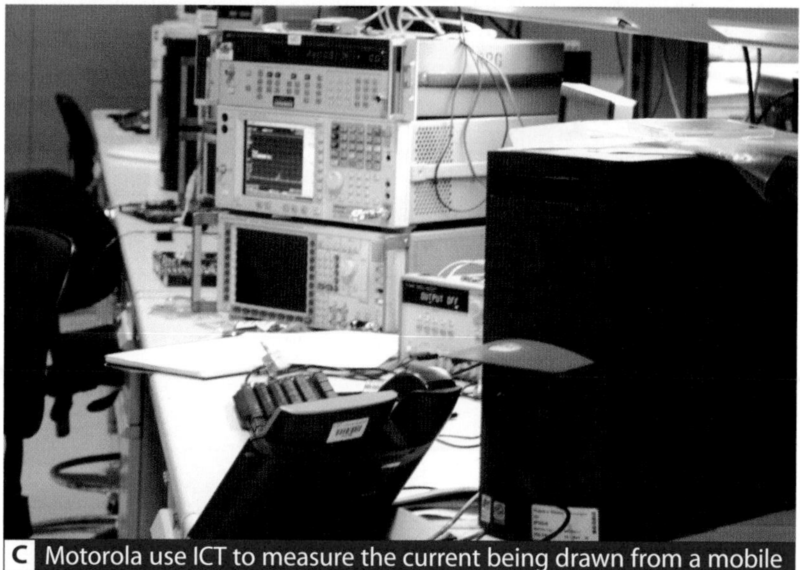

C Motorola use ICT to measure the current being drawn from a mobile phone.

When designing and testing modern electrical equipment such as mobile phones, the manufacturers use ICT to take many readings. This is to check that they operate as expected. By taking the readings automatically it is possible to collect many sets of readings and take many readings every second. This level of detail and accuracy would not be possible if the readings were taken by hand.

D Electrical and computer equipment is used to measure the mobile phone's audio response and antenna performance.

Other advantages of using ICT to measure and analyse data are that it makes it very easy to display the readings that have been taken instantly on a screen. Also, many thousands of readings can be recorded, and looked at or analysed at a later date.

4 What measurements are taken by the mobile phone company when testing their phones?

5 What are the advantages of using ICT to take these readings?

6 Suggest at least two disadvantages of taking experimental measurements yourself.

7 Write down two places where ICT is used to collect and display measurements.

8 Write a sensor report, of no more than 200 words, for your home. Write down all the things in and around your home that use sensors to measure anything and, if possible, a little bit about them. (*Hint*: The kitchen will probably be a good place to start.)

Cells, time and recharging

By the end of these two pages you should be able to:

- understand the term amp-hour (Ah) when describing cells
- work out how long a cell will last in a certain situation
- describe the advantages and disadvantages of different types of cell.

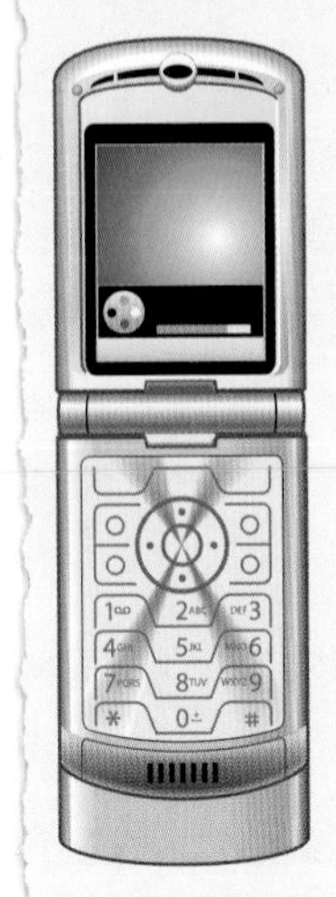

A How long a charged mobile phone battery will last depends on what it is used for.

Have you ever wondered:

How can I make the batteries in my MP3 player last longer?

Appliances use different amounts of current depending what they are doing. Mobile phones have a different standby time and talk time value for a full battery charge. In standby mode a phone uses the battery's energy much more slowly, so it will last longer.

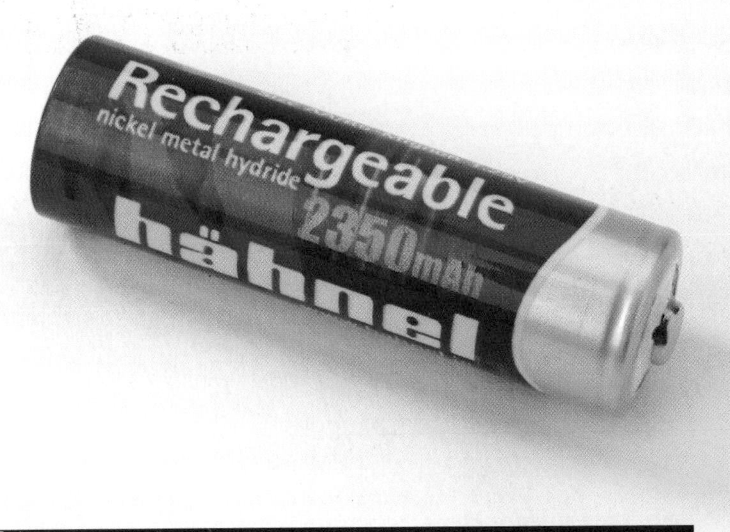

B The amp-hour value indicates how long before a cell runs down.

The **capacity** of a cell is given in amp-hours (Ah; or milliamp-hours, mAh). It is a measure of how much energy a cell can store. The time a cell can be used on one charge depends on how quickly you empty the energy – that is, the current that you use – and how long the appliance is turned on for.

To work out how long the cell will last you need to divide the value in amp-hours by the current the appliance uses.

Example: If a high-power lamp needs a current of 2 A, how long will it run on a 14 Ah cell?
Solution:

$$\text{Time in hours} = \frac{\text{battery rating in amp-hours}}{\text{current used in amps}} = \frac{14\ \text{Ah}}{2\ \text{A}} = 7\ \text{h}$$

1 What two things will affect how long it takes for a battery to run down?

2 Use the equation in the example on page 212 to copy and complete this table

How long will a 10 Ah battery last?	
Current (A)	**Time (h)**
10	1
5	
1	
0.1	

C Not all batteries and cells can be recharged.

Rechargeable batteries can be recharged hundreds of times, do not produce as much chemical pollution as dry-cell batteries and provide cheaper electricity in the long run. Some of the drawbacks of using them are that they cost more to buy than non-rechargeable batteries, you need a recharging unit and the amount of power that they can store is sometimes less than dry-cell batteries.

The amount of power you get out of a dry-cell battery slowly falls over time as you use them. The output from a rechargeable battery remains constant for most of its life but drops quickly to zero at the end.

3 How many times can you recharge a rechargeable battery?

4 Which type of battery provides a more constant supply of power when used?

Millions of batteries are bought in the UK each year but less than 5% of these are recycled. Batteries make up over 20,000 tonnes of landfill waste. However, over 90% of car batteries are recycled. The chemicals used in batteries can leak out and pollute soil and water, harming plant and animal life.

5 Which type of battery is nearly always recycled?

6 What are the main advantages and disadvantages of using rechargeable batteries?

7 Write a guide of about 200 words to help people choose which batteries or cells to use. You will need to include information about the types available, what the amp-hour rating means, how long they will last and any environmental issues.

Summary Exercise

Higher Questions

Controlling the flow of electricity

By the end of these two pages you should know:

- how the resistance of a thermistor changes with a change in temperature
- how the resistance of a light-dependent resistor changes with light intensity.

Many electrical components have sensors that detect changes in the environment and then adjust to what is happening. A smoke or fire alarm is a simple device that measures the presence of smoke or a rapid change in temperature and then sounds an alarm. All around your house are many electrical sensors that take measurements and control devices. These include the temperature sensors that help control the temperature of your fridge and of your central heating system. If you have an outside light that turns on automatically at night or when someone moves near it then this will contain both light and movement sensors. Some new cars even have distance sensors built into the bumpers so that the car gives off a signal if you reverse too close to something to stop you crashing.

There are two common electrical components that are used in many modern electrical devices. They behave like resistors but also act as sensors that measure things around them and are used to control a circuit. They are the **thermistor** and the **light-dependent resistor** (**LDR**).

A thermistor is a type of resistor that responds to changes in temperature. As the temperature of the thermistor increases then its resistance decreases. This will allow more current to flow. A very simple circuit with a thermistor is shown in figure A.

The flow chart below shows what happens to a thermistor when the temperature rises and how its change in resistance affects the current in a circuit that it is connected to.

If the temperature goes down then the resistance will increase and so the bulb will get dimmer.

1 What does a sensor do?

2 List at least three places where sensors are used. Say what each sensor is measuring.

3 Give examples of two electrical components that can be used as sensors.

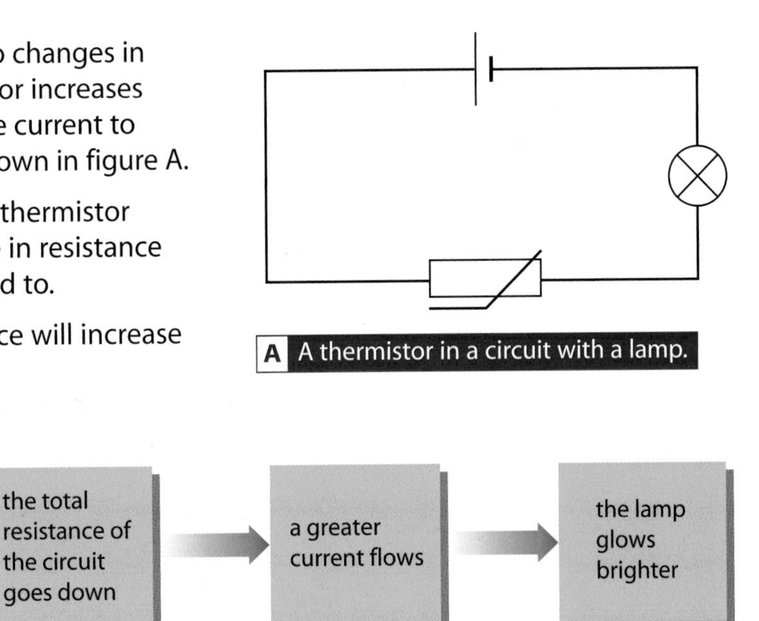

A A thermistor in a circuit with a lamp.

temperature rises → the resistance of the thermistor goes down → the total resistance of the circuit goes down → a greater current flows → the lamp glows brighter

B How a thermistor indicates a temperature change.

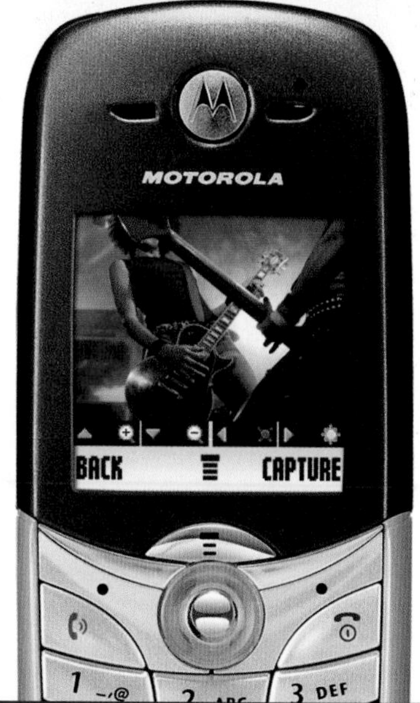

C Many camera phones use LDRs to measure light levels to improve your pictures.

An LDR is a type of resistor that responds to changes in light intensity. As the light intensity increases then its resistance decreases. This will allow more current to flow. A very simple circuit with an LDR is shown in figure D.

The flow chart below shows what happens to an LDR when the light intensity rises and how its change in resistance affects the current in a circuit that it is connected to.

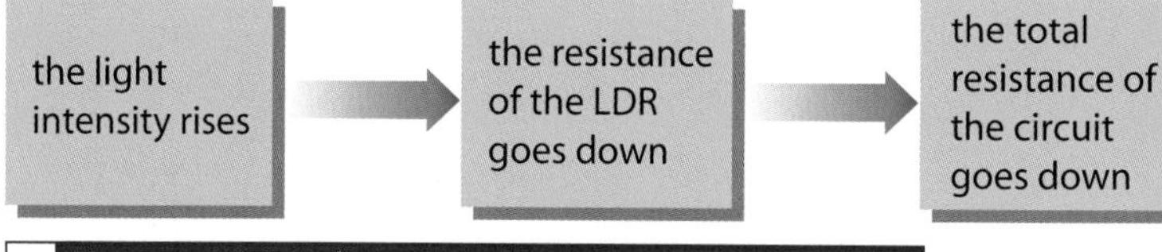

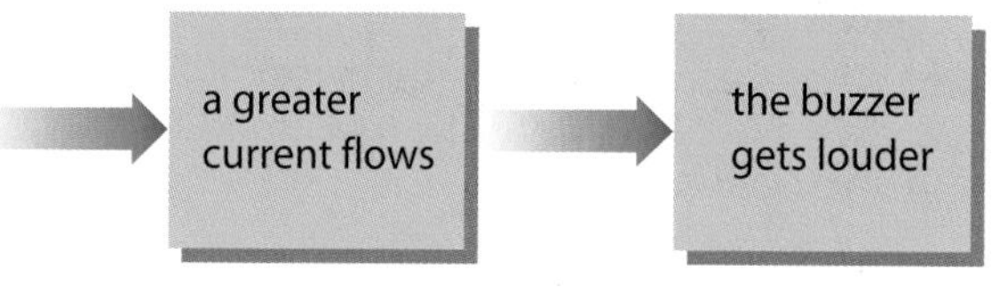

E How an LDR indicates a change in light intensity.

If the light intensity goes down then the resistance will increase and so the buzzer will get quieter.

6 When the light intensity changes, what property of an LDR changes?

7 When the light intensity decreases, what happens to an LDR?

The circuits above are very simple but the same ideas are used in more complicated equipment. The thermistor can be used anywhere where the temperature needs to be monitored, from keeping fridges cold to keeping swimming pools warm. The LDR can form part of many different pieces of equipment that need to monitor light intensity. These include digital cameras, sensors that are designed to turn car and street lights on or off automatically when it gets dark, and security alarms.

4 When the temperature changes, what property of a thermistor changes?

5 When the temperature increases, what happens to a thermistor?

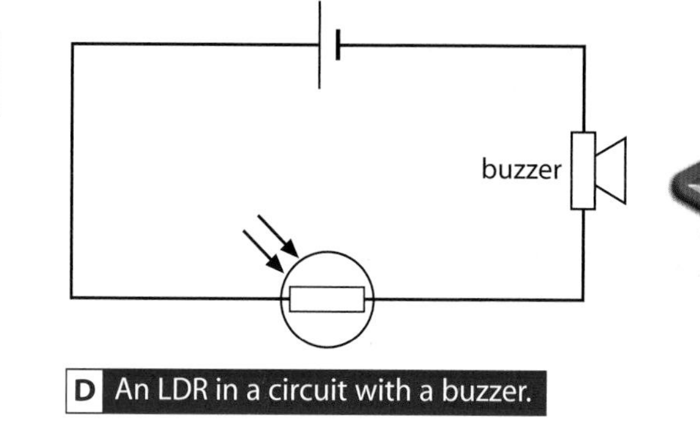

D An LDR in a circuit with a buzzer.

8 **a** List two places where thermistors might be used.
b List two places where LDRs might be used.

9 Describe how the LDR and thermistor work.

Higher Questions

Changing resistance

By the end of these two pages you should:

- be able to recognise and explain applications depending on resistance change
- understand how a digital camera's shutter depends on resistance change.

In many houses you will find light dimmer switches that are used to control the lighting levels. The main component inside a dimmer switch is a variable resistor. As you turn or slide the dimmer switch you change the resistance inside and this controls the brightness of the lights. Exactly the same idea is used in other places like the volume control on a stereo, the controller on a car's windscreen wipers that changes the speed they move and the switch that makes an electric fan spin at different speeds. In every case the change in resistance is used to control something.

A A light dimmer switch and stereo volume control use changing resistance.

1 What is the main component inside a dimmer switch?

2 List three places where a change in resistance is used to control something.

Have you ever wondered:

How does my digital camera take great pictures automatically?

B Digital cameras measure the light intensity to help take better pictures.

Most digital still and video cameras adjust to the lighting levels automatically so that the image is not too bright or too dark. This works by using a light-dependent resistor (LDR) as mentioned on pages 214 and 215. The camera has a built-in LDR and as the light intensity falling on it changes, its resistance changes. This change of resistance is measured and used to control the time the shutter is kept open. If the light reading is low then the shutter is kept open for longer so enough light gets in and a good picture can be taken. If the light reading is high the shutter is kept open for less time.

3 What component is used in a digital cameral to measure the light intensity?

4 If the light level is too high what happens to the time the camera shutter is kept open?

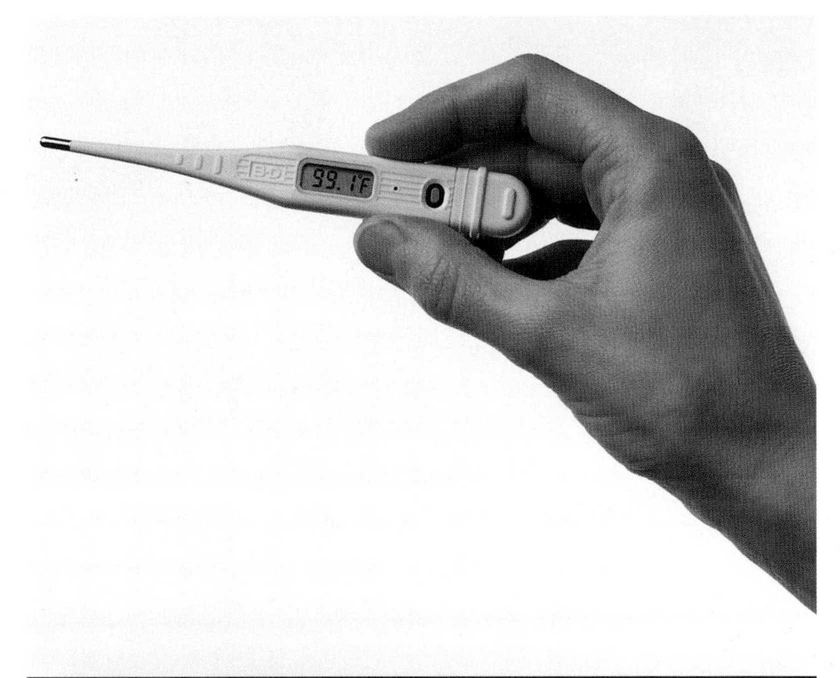

C Many digital thermometers use a thermistor to measure the temperature.

Thermistors are also used in many common applications. In this case the resistance of the thermistor changes with temperature. Some greenhouses have automatic ventilation systems that use a thermistor to measure the temperature and control the opening and closing of vents. If the temperature gets too high, the resistance of the thermistor drops and this is used to control the opening of the vents. Modern thermostats that measure the temperature of a room and control the heating system also use the change in resistance of a thermistor. Other applications include digital thermometers and industrial ovens.

5 Describe how a thermistor can be used to help in a greenhouse ventilation system.

6 List at least two other applications that use a thermistor to measure temperature to control something.

7 Write a short set of notes about situations where a change in resistance can be used to control a situation. Include at least one example that uses each of the following: a variable resistor, a thermistor, a light-dependent resistor.

Smaller and more powerful

By the end of these two pages you should:

- know about how the size of electrical components has changed over time
- understand how mobile phone technology has changed and advanced
- be able to suggest possible future applications of technology.

A B A laptop is thousands of times more powerful than some of the first computers, but a fraction of the size.

In the early days of modern computers in the 1960s, a computer that filled an entire room had less processing power than the average mobile phone made 40 years later.
If you bought a home computer in the 1980s then the amount of memory it had in total was about the same as is needed to send one mobile phone picture message today.

The key component in many electrical circuits is the transistor (a transistor is used as a switch and an amplifier in circuits). Over time, ways have been found to make them smaller and computers have become faster and more powerful. Transistors were first developed in the 1940s and at that time you could fit only one in your hand. If you buy a computer today the microchips that provide the main computing power are much smaller than your thumbnail, but may well have over 300 million transistors inside.

1 When were modern computers first developed?

2 What has happened to the size of the transistors used in computers since they were first developed?

3 How has this changed the speed and power of computers?

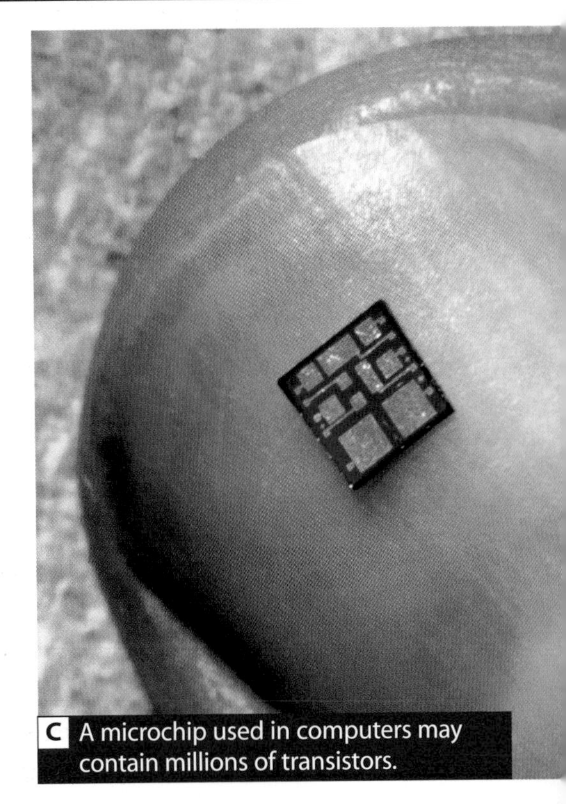

C A microchip used in computers may contain millions of transistors.

D As mobile phones have gained more processing power they have also got smaller.

The first mobile phones in the mid-1980s were massive and had a mass of several kilograms. The headset was the size of a house brick and all they could do was make phone calls! Now it is possible to have a phone that is five or ten times lighter, fits in the palm of your hand and also acts as a camera, games console, music player, diary and computer.

4 What has happened to the mass of mobile phones since they were first developed?

5 List at least four features that modern mobile phones have that were not available in the first ones.

Have you ever wondered:

Is it true my clothes will soon become wearable computers?

As the processing power of computers increases and they become smaller this creates opportunities for many possible applications in the future. Perhaps your kitchen cupboards will order the shopping for you. Your sofa might be able to check your health and make a doctor's appointment if you need one. Pretty soon we should be able to have fully functional computers that we can wear. Maybe after that we will be able to have them implanted inside our bodies.

6 Write down at least two possible future applications of new computer-based technology.

7 For any one of the applications above write down what the advantages would be of using it.

8 Write a short set of notes about what has happened to the processing speed of computers and the size of the components inside them over time; include information about current applications such as mobile phones as well as some suggestions as to possible future applications.

Higher Questions

V

P

P

?

Science leading to technology

H By the end of these two pages you should be able to:

- explore how a new technology develops as a result of scientific advances
- understand how the development of superconductors led to the Maglev train.

Have you ever wondered:

How can a train possibly go at 500 kilometres per hour?

A A Maglev train uses magnets and superconductors to allow it to travel at very high speeds.

You can now get a train at Shanghai in China that will take you 30 km in just over 7 minutes, reaching a top speed of 431 km/h and an average speed of 250 km/h. As well as travelling so fast, the train is levitating, so it hovers a few centimetres above the track. It is supported by very strong magnets, which reduces friction. This means that it can travel much faster than a conventional train.

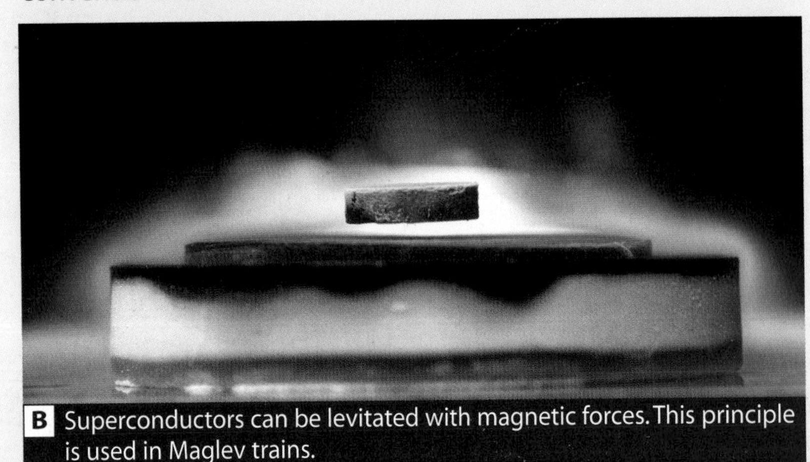

B Superconductors can be levitated with magnetic forces. This principle is used in Maglev trains.

The development of scientific ideas and discoveries has enabled us to advance new technologies, although these do not always seem obvious at first. At the start of the twentieth century scientists were trying to find things that conducted electricity really well. This led to the discovery of **superconductors**.

Superconductors are very important because they can be used to make incredibly strong **electromagnets**. These are more powerful than any other type of magnet because they are so good at conducting electricity.

As better superconductors were developed it was possible to make stronger and stronger electromagnets. This made possible the development of the first magnetically levitating train (shortened to Maglev) in the mid-1980s.

A Maglev train works by having superconducting electromagnets on both the train and the rails it sits above. As the train moves along, the electromagnets on the track are changed constantly by a computer so that the one in front always attracts and the ones behind always repel. This constantly pulls (and pushes) the train along the track.

C It is not just trains that can be levitated using magnetism.

You may have thought that humans being able to float was just science fiction, but it is not just trains that can be levitated. So far scientists in the Netherlands have managed to levitate a tomato, a strawberry, a frog and a grasshopper (all were unharmed). There is a chance that in your lifetime this technology may be developed to work on human beings.

1 What is the top speed of the Maglev train in Shanghai?

2 Why is there so little friction between the train and the track?

3 What is a superconductor?

4 What is Maglev an abbreviation of?

5 What is inside the train and track that moves the train along?

6 What things other than trains have scientists been able to levitate?

7 Describe how the discovery of superconductors has led to the development of Maglev trains. Make sure you explain what superconductors are and how they are used to make Maglev trains work.

Questions

Multiple choice questions

1 What do the terms AC and DC stand for?
- **A** Alternative connection and distribution connection.
- **B** Alternative current and direct current.
- **C** Alternating carrier and distribution carrier.
- **D** Alternating current and direct current.

2 What is another name for a normal non-rechargeable cell?
- **A** chemical cell
- **B** dry-cell
- **C** current generator
- **D** solenoid

3 Which of the following statements correctly describes electrical current?
- **A** It is a flow of negatively charged particles.
- **B** It is a flow of positively charged particles.
- **C** It is a flow of positively and negatively charged particles.
- **D** It is a flow of particles that have no charge.

4 What is the name of the equipment used to measure electrical current?
- **A** ampmeter
- **B** ammeter
- **C** voltmeter
- **D** ohmmeter

5 What of the following statements is true?
- **A** An ammeter can be connected in series or in parallel.
- **B** An ammeter must always be connected in parallel.
- **C** An ammeter must always be connected in series.
- **D** An ammeter must always be connected to a battery.

6 How many milliamps are in one amp?
- **A** 1,000,000
- **B** 100
- **C** 1000
- **D** 10

7 Which of these units is resistance measured in?
- **A** amperes
- **B** ohms
- **C** resistons
- **D** hertz

8 Which is the correct equation that connects current, voltage and resistance?
- **A** $V = I \times R$
- **B** $V = C \times R$
- **C** $V = I/R$
- **D** $V = R/I$

9 Which of the following statements about a filament light bulb is true?
- **A** As the voltage increases the resistance goes down.
- **B** As the voltage increases the resistance goes up.
- **C** As the voltage increases the resistance stays the same.
- **D** As the voltage increases the resistance goes up and then down.

10 Which is the correct term, sometimes written on a cell, that tells us how long it will last?
- **A** kilowatt-amps
- **B** units
- **C** volts
- **D** amp-hours

11 What does LDR stand for?
- **A** light-dependent rheostat
- **B** light-dependent reverser
- **C** light-displacement resistor
- **D** light-dependent resistor

12 What happens to a thermistor as it gets hotter?
- **A** Its resistance goes down.
- **B** It produces more current.
- **C** Its resistance goes up.
- **D** It produces more voltage.

13 Which of the following would increase the amount of current produced in a dynamo?
- **A** Spinning the dynamo faster.
- **B** Changing the wire connections in the dynamo.
- **C** Using a weaker magnet.
- **D** Using a wire coated in plastic.

14 A magnet is moved near a wire to generate electricity. Which of the following would change the direction of the current produced?
- **A** Moving the magnet faster.
- **B** Turning the magnet around.
- **C** Using a stronger magnet.
- **D** Using more coils of wire.

H **15** Which of the following statements is *not* true about Maglev trains?

A They use electromagnets to make the train move along.

B They can travel faster than conventional trains.

C The friction between the train and the track is very high.

D They do not touch the track they run on.

Short-answer questions

1 The mains electricity we receive is AC and the electricity we get from batteries and cells is DC.

a Write down what the abbreviations AC and DC stand for.

b Explain the difference between them.

2 Write a brief, newspaper-style report of around 200 words on how the size and processing power of computers have changed over time. Try and include some examples and applications of technology in your answer.

3 Name three different types of cell, explaining how they are different.

4 Write down three places in your home or school where you think there are light sensors.

5 Which of the following graphs shows the pattern between voltage and current for a fixed-value resistor?

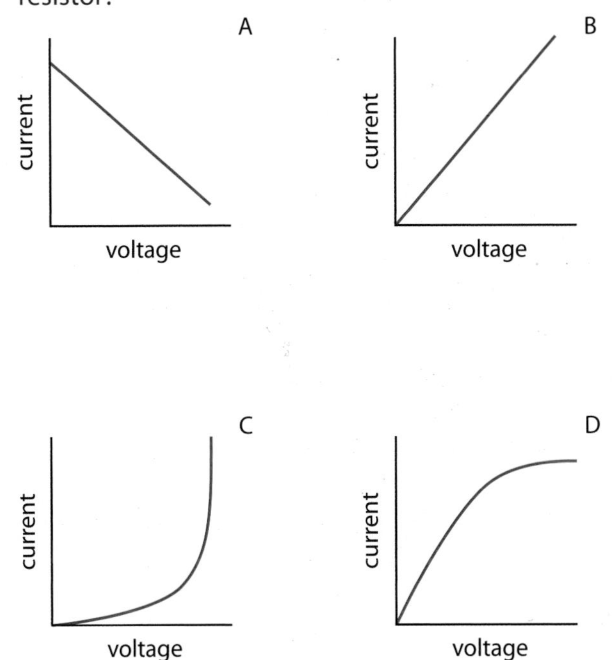

6 The table below shows some current, voltage and resistance readings for an experiment. Using the equation you have learned, calculate the missing numbers.

Voltage (V)	Current (A)	Resistance (Ω)
10	5	
	4	8
20		2

7 Brian is explaining to Robert that electricity goes from positive to negative but Robert says that it really goes the other way. David says that they are both right. Explain what David means. You will need to use the terms *conventional current* and *flow of electrons*.

D

P

Glossary

ammeter A device used to measure electrical current.
battery A collection of more than one cell connected together.
capacity A measure of the amount of electrical energy that a cell or battery can hold.
cell An electrical component that is used to create a voltage.
circuit A collection of electrical devices connected together to perform some function.
conventional current The description of the flow of electrical current that goes from the positive to the negative terminals around a circuit.
current The flow of electricity around a circuit. Measured in amperes, or amps (A), or milliamps (mA).
dry-cell The most common type of cell. The voltage is produced by a chemical reaction that normally takes place inside a metal cylinder.
dynamo A device used to generate electricity by rotating a magnet inside a coil of wire.
electromagnet A magnet that is created by the flow of current through a wire.
electromagnetic induction The process that happens when a magnet is moved near a wire that induces a voltage, making a current flow.
electron A very small negative particle that flows around an electrical circuit creating an electrical current.
light-dependent resistor (LDR) A resistor whose resistance changes with light intensity.
magnet Any object that has a magnetic field around it.
magnetic field The area around a magnetic object where other magnetic objects will feel a force.
rechargeable A type of cell or battery that it is possible to use many times. After its electrical energy has been used it can be 'refilled' with electrical energy and used again and again.
resistance A measure of how difficult it is for current to flow around a circuit. Measured in ohms (Ω).
resistor An electrical component that restricts the flow of electrical current. Fixed-value resistors do not change their resistance, but with variable resistors it is possible to vary the resistance.
series A type of circuit where all components are connected together in a single line or loop.
solar cell Something that converts solar power into electricity, also known as a photovoltaic cell.
superconductor A material that has an extremely low resistance and so conducts electricity incredibly well.
thermistor A resistor whose resistance changes with temperature.
voltage The difference in electrical energy between two points that makes a current flow, measured in volts (V), or millivolts (mV). It is sometimes called the potential difference.

You're in charge

A At Hockerton they generate their own electricity.

Hockerton is an ecological housing development for sustainable living. It is known as a zero-energy site. They generate their own electricity on site from renewable sources such as wind and solar power. They have many energy-saving measures.

Have you ever wondered:

What if all the electricity in the world went off and stayed off?

Imagine what would happen in a power cut. There would be no TV and no lights. In 2003, 97% of the UK's energy supply came from sources that will have run out or be less easily available in 100 years. Almost all vehicles on the UK roads run on petrol and diesel but these fuels won't last forever.

As we find other ways to get our energy our lives will change. In 10 years time you could be driving a car powered by hydrogen and living in a house heated by solar energy. We need to know how developments in technology will change the way we live.

In this topic you will learn that:

- the rate of transfer of electrical energy and its efficiency can be calculated
- a motor can be controlled using electricity
- it is important to consider the economical costs and environmental effects of energy use
- safety issues must be fully considered when working with electricity.

Copy and complete the table below.

My ideas about key words at the start of this topic.		
Key word	What I think it means	How sure am I? ***=certain **=pretty sure *=not sure
power		
efficiency		
earth wire		
insulation		
solar power		

V

?

D

Oil, coal and gas won't last forever

By the end of these two pages you should know:

- the different ways in which the UK could generate electricity
- the benefits and some of the costs of each method
- that the ways that we generate electricity can affect our lives
- how much these methods are used in the UK at the moment and what the future may hold.

Ninety-seven per cent of the **electricity** used in the UK comes from non-renewable fuels. Supplies of non-renewable fuels such as coal, oil and natural gas – fossil fuels – will never be replaced. We need other ways of generating electricity.

Have you ever wondered:

What kind of car will you be driving in 10 years' time?

A Many conventional power stations produce polluting gases.

B Renewable resources provide a future for energy supply.

Renewable resources do not have a limited lifespan. There are no fuel costs, only running and maintenance costs. In the UK, the renewable options include wind, wave, tidal and solar power. These sources are free and do not produce pollution such as carbon dioxide and sulphur dioxide gases.

Have you ever wondered:

Could your bedroom be powered by renewable energy?

1 What is the name for the energy sources such as wind, wave and tidal?

2 How much of the UK energy supply comes from sources like this?

Hockerton has two wind turbines to generate electricity. **Wind power** has a number of advantages. Very little maintenance is needed, and the **energy** is free as long as the wind blows.

One disadvantage of wind power is that wind strength can be unreliable and so a backup is needed. Other drawbacks include the cost of building the turbine and the fact that many turbines are needed to produce electricity on a large scale. Many proposed wind farms in the UK have been strongly opposed. Some people feel that they have a negative visual impact. There is also resistance to such dramatic change.

In 2005 there were over 1200 wind turbines in the UK and we have the potential for over 40% of all of Europe's wind energy. The future will almost certainly see more wind turbines. A site with 18 wind farms could power 1 million homes.

3 How much of Europe's wind energy is available in the UK?

4 Write down two advantages and two disadvantages of wind power as an energy source.

C A wind turbine at Hockerton.

D The solar cells at the Hockerton site.

Solar power can be used to heat water using passive **solar panels**. It can also be used to generate electricity directly using **photovoltaic cells** or **solar cells**. Solar panels heat the water without a boiler being needed at all and even work on cloudy days. Solar cells can meet the electricity needs of a normal house. This saves money. The use of solar power is becoming more widespread. Hockerton uses solar cells and solar panels to save money and reduce polluting emissions.

5 Write down the names of two devices that use solar power.

6 Write down two advantages of using either of these devices.

7 Write a newspaper article of between 150 and 200 words about the advantages and disadvantages of using solar and wind power in the future.

Higher Questions

The National Grid

By the end of these two pages you should:

- know how domestic electricity is distributed around the country
- H understand the issues around using new technology to distribute electricity.

Have you ever wondered:

Why don't many people in rural Africa have electricity at the flick of a switch?

At Hockerton the wind turbines are very near the houses and so it is easy and cheap to connect the houses to the power supply. If houses are very far away from a source of electricity it can be very expensive to connect them up to the mains. This is why small-scale electricity production can be a very good thing in remote areas.

A Hockerton is unusual as most people in the UK get their electricity from the National Grid.

B The National Grid distributes electricity across the whole of the UK.

All across the UK is a network of pylons and cables that distributes electricity to our houses. There are not power stations everywhere in the country and so the National Grid is needed so that we can all have the electricity that we need.

1 What is the name for the network of cables and pylons that carries electricity around the whole country?

2 What is the advantage at Hockerton of having the wind turbines near the houses?

C The National Grid is a network of pylons and electricity cables across the country.

H In the UK, most of the electricity we use comes from fossil fuel or nuclear power stations. It is known that the supplies of non-renewable fuels will not last forever and so things will have to change over time.

One alternative to a National Grid would be to use small-scale, renewable power generators such as wind turbines connected directly to local homes. This would have a positive environmental effect because it would not need a large distribution network. Drawbacks would include the high cost of changeover and that the reliability of output from renewable supplies cannot always be guaranteed. Small-scale power may not be as practical for larger towns and cities.

To rip down the existing National Grid and start again would be very costly and disruptive. It is more likely that the National Grid will start to be supplied by more renewable sources than non-renewable ones. This already happens on a small scale and some people who run wind turbines sell any excess electricity back to the National Grid.

P

3 Will the National Grid go on forever being supplied by power stations running on fossil fuels?

4 What alternatives are there to a National Grid for electricity supply?

5 What are the drawbacks of ripping down the existing National Grid and replacing it with something else?

D Solar power can be used in rural areas.

Some countries do not have a distribution network like the National Grid. When deciding to build one there are many things to be considered. Building costs for a grid would be high. Construction would have an adverse environmental effect and there are potential safety issues if people are not used to large-scale electricity distribution.

There would also be benefits of a system like this. A reliable electricity supply would help support local services and industry and aid further development. This could transform a community.

P

6 What types of places would benefit best from small-scale local electricity production?

7 Why is it expensive to connect some houses to the National Grid?

Summary Exercise

Higher Questions

Motors

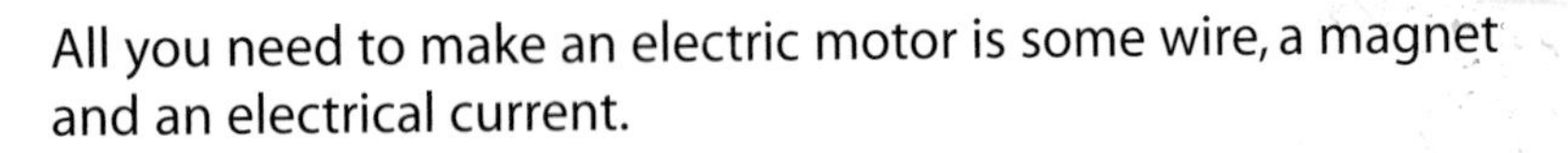

By the end of these two pages you should be able to:

- explain how a simple electric motor works.

Michael Faraday built the first electric **motor** in 1821 and changed the world. Motors are everywhere, in fridges, computers, portable music players and vibrating mobile phones. Cars engines may run on petrol but they use an electric motor to get started. Without electric motors our lives would be very different.

All you need to make an electric motor is some wire, a magnet and an electrical current.

1 What three things are needed for an electric motor?

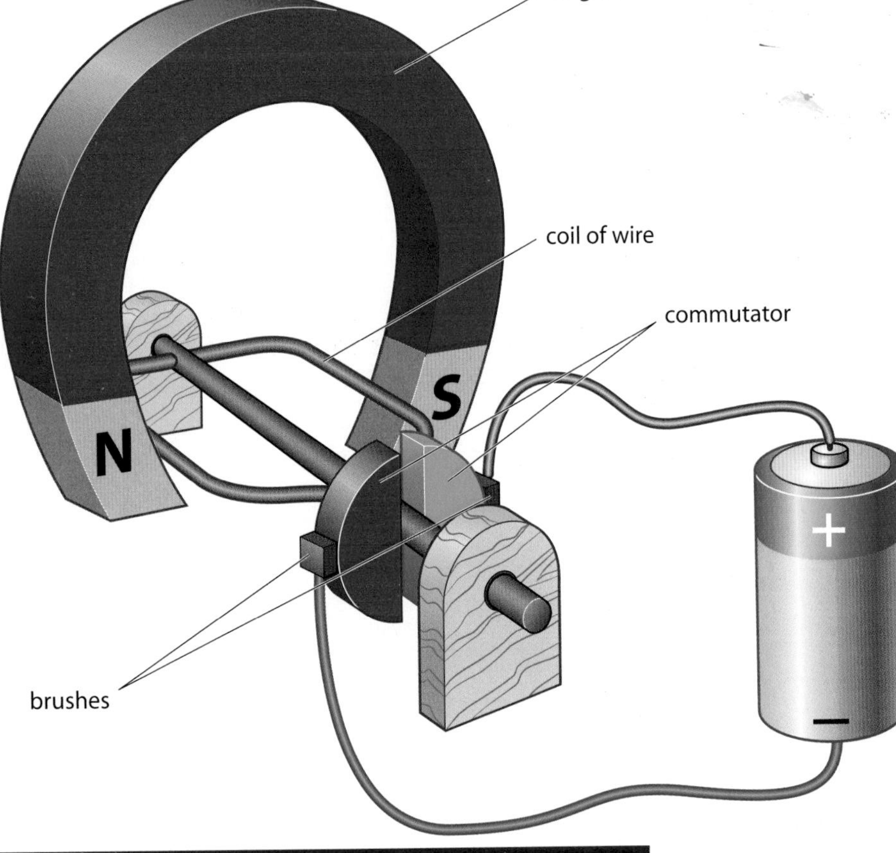

A An electric motor turns electrical energy into movement.

When a current is passed through a wire it produces a **magnetic field** around it. If this wire is near a permanent magnet then the wire will move. (If you were able to hold the wire still, the magnet would move instead.) This is called the motor effect.

2 What happens around a wire when current passes through it?

3 What happens when a current is passed through a wire that is near a magnet?

4 What is the name for this effect?

The force on a single wire carrying a current can be small, but the force can be made larger and make the wire move more.

If you increase the size of the current passing through the wire then this will make the wire have a stronger magnetic field. This means that it will be affected more by the permanent magnet and so will move more.

If you increase the strength of the field of the permanent magnet then this will have a greater effect on the wire and so it will move more.

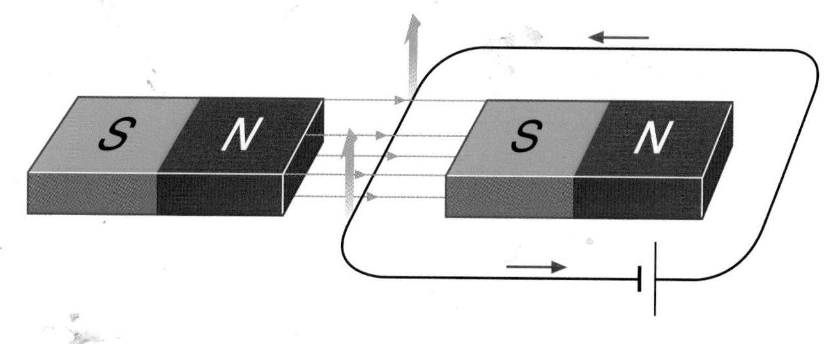

B If a wire carrying a current is near a magnetic field then it will move.

The direction of the force on the wire can also be changed by

- reversing the direction of the current through the wire (by swapping the electrical connections around).
- reversing the direction of the permanent magnetic field (by turning the magnet around).

5 What are the ways you can increase the force on a wire?

The principles described above are used to make electric motors. Every electric motor will contain magnets and wires. They all pass currents through the wires near the magnets and this makes them move. This list summarises how a motor works.

- Inside a magnetic field a coil of copper wire is wrapped around a core (magnet) that can spin on its axis.
- The wire is free to spin without getting tangled up.
- The current always flows in the same direction in the complete circuit but the current is actually continually reversing in the wires in the motor themselves because of the spinning motion of the motor. This means that the motor always spins in the same direction.
- A current passed through the coil makes the coil magnetic. It interacts with the permanent magnet and spins.

P

6 List as many different pieces of electrical equipment in your house that have electric motors as you can.

7 If you put an electric motor near some steel paperclips, they stick to the motor. Why?

8 Describe how an electric motor works.

Summary Exercise

Higher Questions

Wires, fuses and safety

By the end of these two pages you should:

- know the different parts of a domestic plug
- understand the purposes of the earth wire and fuse, and how they work
- H know why a residual current circuit breaker (RCCB) is useful.

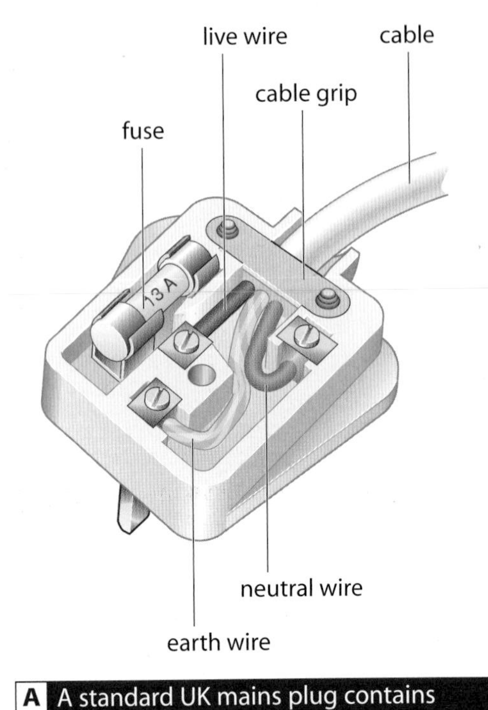

A A standard UK mains plug contains three wires and a fuse.

Have you ever wondered:

How many devices can you safely plug into one wall socket?

Most domestic electrical **appliances** are connected to the mains electricity supply using a cable and three-pin plug. The UK mains is normally 230 V (V means volts) but does vary a little. The colours, names and jobs of the wires in the plug are shown in the table below.

Colour of wire	Name of wire	Job of wire
brown	live	carries current into the appliance
blue	neutral	carries current away from the appliance
yellow/green stripes	earth	prevents electrocution

All the wires must be connected firmly, with no bare wires showing. The outer casing of the cable must be secured in the cable grip.

1 What are the names of the three wires in a domestic three-pin plug?

2 What is the voltage of the UK mains supply?

Have you ever wondered:

Will a 230 V electric shock kill you?

The **fuse** and **earth wire** are two ways to protect the user from electrical hazards. The fuse is a thin piece of wire with a low melting point inside a ceramic or glass and metal casing. If the current flowing into the appliance through the fuse gets too high then the fuse wire gets hot and melts. This breaks the circuit and so the current no longer flows, protecting the appliance and cables from overheating and catching fire.

All appliances with a metal outer casing must be connected to earth with an earth wire. This yellow and green wire could save your life. Plugs for appliances without a metal casing don't have an earth wire because they could never give you an electric shock.

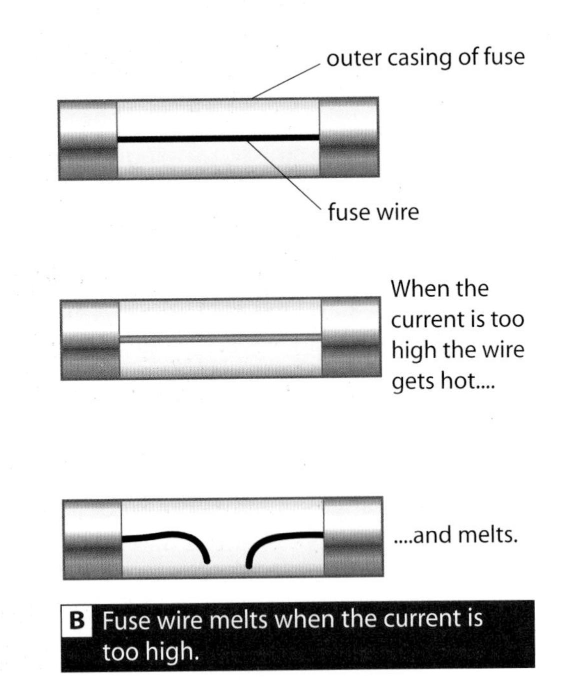

B Fuse wire melts when the current is too high.

If a fault in an appliance makes the live wire touch the metal casing, the whole appliance becomes 'live'. The metal casing is attached to the earth wire, so current goes in through the live wire and out through the earth wire. The current increases because the earth wire offers less **resistance** than the appliance. The fuse wire heats up and melts. The circuit becomes safe.

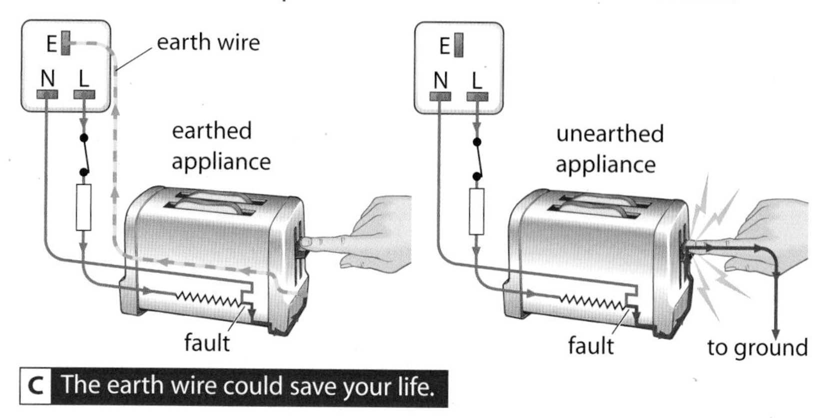

C The earth wire could save your life.

3 What is inside a fuse?

4 Do all appliances need to be earthed?

H Many modern houses have safety devices known as **residual current circuit breakers (RCCB)**. Their job is different from a fuse; they measure the current flowing into and out of a circuit. Normally these currents should be the same but if there is a fault then the RCCB will shut down the circuit. RCCBs are easy to reset.

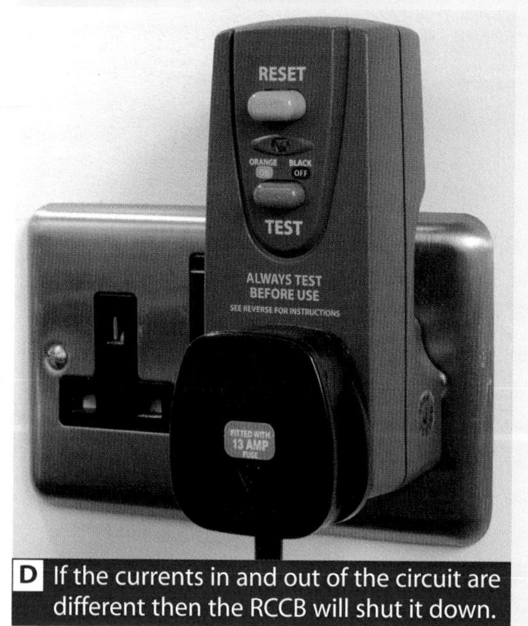

D If the currents in and out of the circuit are different then the RCCB will shut it down.

5 What does the fuse protect the appliance and cable from?

6 Describe the way the following two domestic electrical safety features work: the fuse and the earth wire.

Higher Questions

Power

By the end of these two pages you should:

- understand the idea of electrical power
- know and be able to use the formula power = current × voltage.

A One of the residents at Hockerton.

Imagine that the Hockerton Housing Project wanted to build an additional building and had to decide which construction company to employ. They could devise a test to see which company had the better crane. The test could find out how many bricks the cranes could lift and how quickly they could lift them.

Crane name	Total weight lifted (N)	Time taken (minutes)	Rate of lifting (N per minute)
Frippatron	2000	100	20
Enobulk	2800	200	14

In both cases the weights were lifted the same distance but the time it took in each case was different. By looking at how quickly the cranes lifted the bricks we can see that the Frippatron is the 'better' crane as it can work faster.

Power is not just about how much **work** something does, it is also about how quickly it does it. When you measure how quickly something happens it is called a **rate**.

Electric motors convert electrical energy into kinetic energy (the energy of movement). The rate at which motors convert this electrical energy is called the **power** of the motor. A powerful motor will spin fast because it is converting electrical energy into movement very quickly. Power is measured in units know as watts (W) or kilowatts (kW). 1 kW is equal to 1000 W.

B C More powerful electric motors use energy more quickly and so spin faster.

1 What does the word rate mean?

2 What units is electrical power measured in?

3 A motor converts electrical energy very slowly into kinetic (movement) energy. Is this motor high power or low power?

In any electrical appliance the flow of current (in amps) and the voltage (in volts) will affect what happens and how quickly it happens. The power of an appliance can be worked out as long as you know the **current** and **voltage**. The equation that is used is shown below.

power (watts) = current (amps) × voltage (volts)

Example: What is the power of a hair dryer that uses 3 A at a voltage of 230 V?
Solution: current (amps) × voltage (volts) = power (watts)
3 A × 230 V = 690 W.

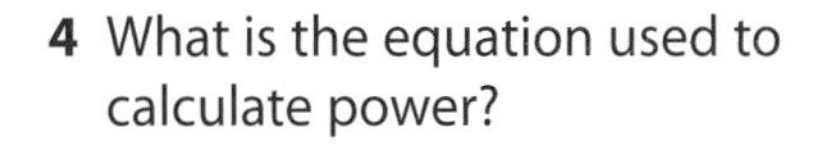

4 What is the equation used to calculate power?

5 What is the power of a fridge that has a current of 2 A and a voltage of 230 V?

Mains appliances that have a high power rating will have a large current. This means they need thicker wires so they don't overheat and melt or catch fire. The wire that goes to an electric cooker (over 2000 W) is much thicker than the wire that goes to a table lamp (60 W).

6 If an appliance has thick wires what does this tell you about the current it uses?

7 What might happen if a current of 10 A was passed through a wire designed for 2 A?

8 Two hair dryers run on 230 V mains, one is labelled 3 A and the other is labelled 4 A. Which has the greater power? Show any calculations and explain your answer.

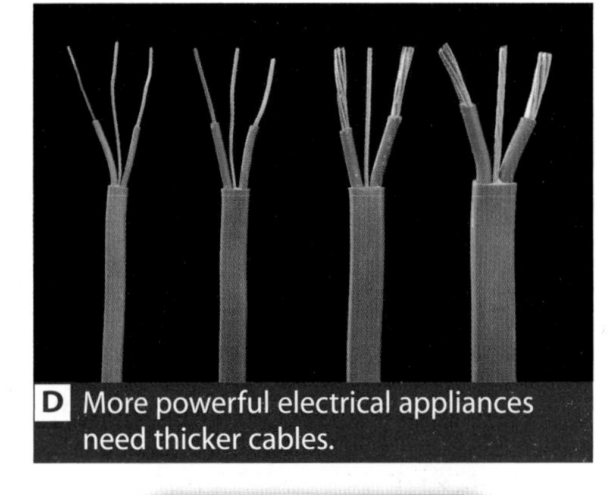

D More powerful electrical appliances need thicker cables.

Summary Exercise

Higher Questions

Efficiency

By the end of these two pages you should be able to:

- understand the term efficiency
- calculate efficiency for simple situations.

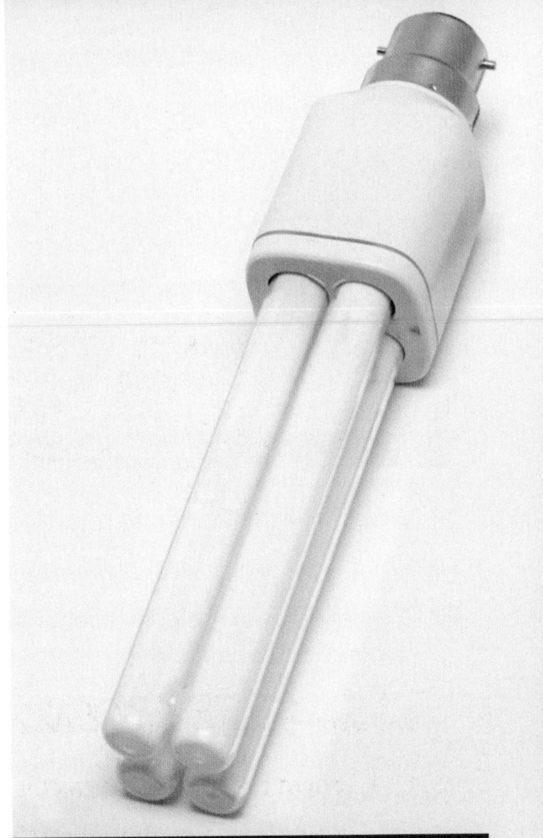

A At Hockerton they use efficient, low-energy light bulbs.

The power of an appliance indicates how quickly it converts energy or how quickly work is done. The amount of energy something uses is measured in joules (J). Not all the energy we get out from something is in a form that is useful to us. A light bulb converts electrical energy into heat and light but it is only the light that we want and is useful. Normal filament light bulbs get very hot because they create so much heat. Energy-saving light bulbs are much cooler because more of the energy going in comes out as light and less as heat.

1 In each of the following appliances, what is the useful energy output?
   a A torch.
   b A car.
   c An electric oven.

2 Why are energy-saving light bulbs much cooler than normal filament ones?

Efficiency tells us how good something is at converting energy in one form into useful energy in another form. Electrical appliances convert electrical energy into another form of energy that is wanted. If something is *efficient* then most of the energy goes into the form that is wanted. Efficiency is measured as a percentage and is calculated using this equation:

$$\text{efficiency} = \frac{\text{useful output}}{\text{total input}} \times 100\%$$

For example: What is the efficiency of a food processor that uses 500 J of electrical energy to do 200 J of cake mixing?
Solution:

$$\text{efficiency} = \frac{\text{useful output}}{\text{total input}} \times 100\% = \frac{200}{500} \times 100 = 40\%.$$

Some energy is always transferred into an unwanted form. It is not possible to have something that is 100% efficient. Also, it is never possible to have an efficiency of over 100%, otherwise you would get more energy out than you put in!

B This food mixer is only 30% efficient.

3 Write down the formula to calculate efficiency.

4 Is it possible to have something that is 110% efficient?

When you buy a new appliance like a washing machine or a fridge you will normally find an energy efficiency sticker on it. Appliances are rated from A (most efficient) to G (least efficient). Although they do not give detailed calculations, they do offer an excellent guide to how much energy they waste and so how much they cost to run. The most efficient appliances may cost a little more but they can save you money in the long run.

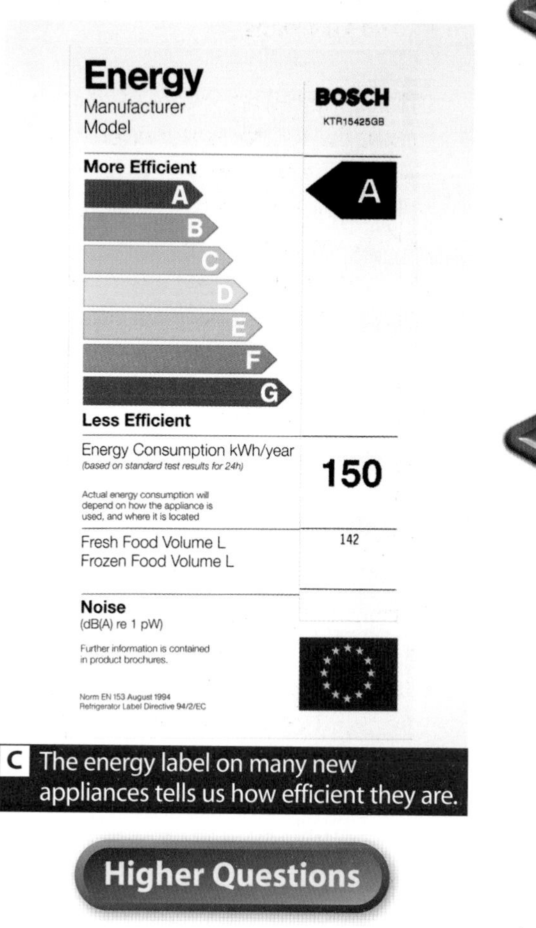

C The energy label on many new appliances tells us how efficient they are.

5 As well as the cost to buy an appliance, what else should you look for to help you to decide which one to buy?

6 An 11 W energy-saving light bulb gives out the same amount of light as a 60 W filament light bulb. Why does this make the energy-saving light bulb better?

7 If a cooker used 200 J of electrical energy to generate 80 J of heat, what is its efficiency?

8 Some people confuse the terms power and efficiency. Write a definition of each of them including the units that they are measured in.

Higher Questions

Electricity, units and money

By the end of these two pages you should:

- know how domestic electricity is measured
- be able to calculate how much it costs to run a particular electrical appliance
- understand how you can monitor your use of electricity.

Have you ever wondered:

Could you increase your allowance by saving electricity?

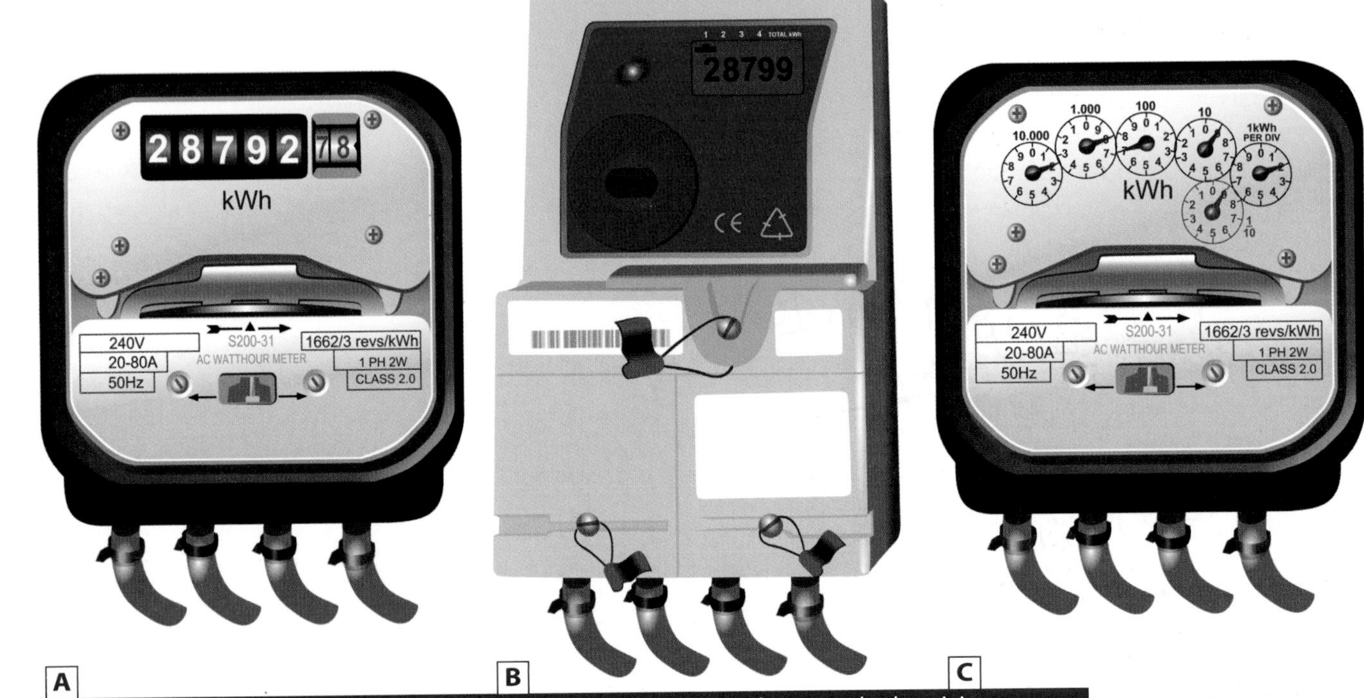

There are many different types of electricity meters, all measuring how much electricity you use.

Electricity meters tell you how many kilowatt-hours of electricity have been used. They do not tell you how much money you have spent. A kilowatt-hour of electricity (sometimes called a unit) usually costs less than 10 p and contains a fixed amount of energy. Appliances with a large power rating use energy more quickly and so cost more to run.

To calculate the number of units of electricity that something uses you need to know:

- the power rating of the appliance (in kilowatts)
- the time that it is on for (in hours).

energy transferred (kWh or units) = power (kW) × time (h)

Many appliances have their power rating listed on them in watts. It is important to convert this into kilowatts. To do this you just divide the power in watts by 1000.

1 What is the other name for a kilowatt-hour of electricity?

2 Express the following power values in kilowatts.
a 1500 W
b 80 W

3 What is the energy transfer in kWh when a 2 kW fire is kept on for 3 h?

To work out the total cost of electrical energy used, you use the equation below.

$$\text{total cost} = \text{power (kW)} \times \text{time (h)} \times \text{cost of 1 kWh}$$

This can be written as

$$\text{total cost} = \text{total number of units (kWh)} \times \text{cost of 1 kWh}$$

Example: If a household uses 3 kW for 20 h, with each kWh costing 6 p, how much does it cost?

Solution: total cost = power × time × cost of 1 kWh

$$360\text{ p} = 3 \times 20 \times 6$$

4 A household uses 5 kW for 10 h. How many kWh do they use?

5 If 1 kWh of electricity costs 6 p, how much does the electricity in question 4 cost?

Normal electricity meters don't tell you how much money you are spending. In some European countries and parts of the USA they use smart meters, designed to make it much easier to work out how much you are spending. They can even be checked remotely.

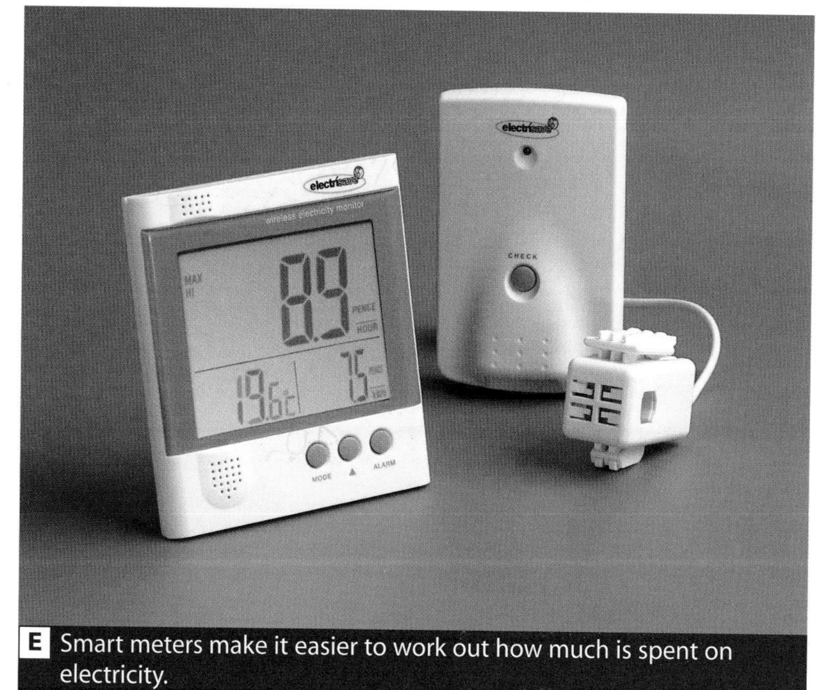

E Smart meters make it easier to work out how much is spent on electricity.

The hope is that if people can see how much it actually costs to use electrical appliances they will think about their energy usage and use less. It is estimated that if the UK reduced its energy usage by 5% per year then household bills would drop by over £35 a year. Across the country this would be more than £40 million a year.

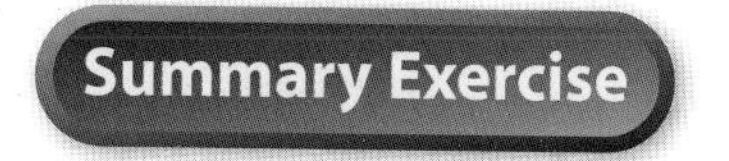

D Electricity bills list how many kilowatt-hours you have used and how much each one costs.

6 What is a smart meter?

7 Which will cost more to run, a 6 kW fire for 2 h or a 2 kW motor for 6 h?

8 Calculate how much it will cost to run each of the following appliances when the cost of 1 kWh is 6 p.

a A 2 kW fire for 3 h.

b A 60 W light bulb for 24 h.

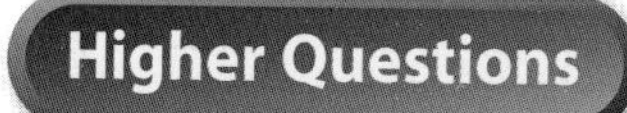

Energy-efficiency and saving money

By the end of these two pages you should:

- know the ways in which a house can be insulated
- understand the term cost-effective
- be able to work out how cost-effective an energy-efficient measure is
- know ways to test different energy-efficiency measures.

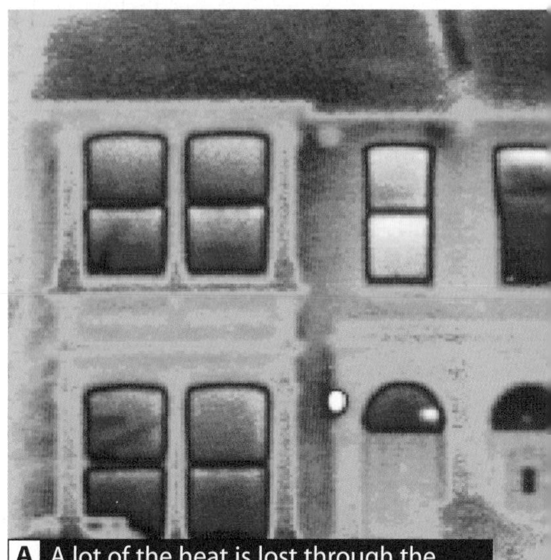

A A lot of the heat is lost through the windows, which is why they are lighter in colour.

The average UK household spends over £600 a year on gas and electricity. A large part of this is spent on heating. Much of this heat can escape from the house into the air outside your home. It is possible to save up to £200 per year with different energy-saving measures. Some energy-saving measures are very expensive and some may cost only a few pounds. This can make it hard to know which ones to choose.

1 What is the average energy bill for a UK household?

2 What can be done to make this smaller?

To help decide which energy-saving measure to choose you need to know two things:
- the cost of the measure
- the amount of money it will save each year.

From this information we can calculate the pay-back time. This is the time it will take to save enough money to cover the cost of the energy-saving measure.

$$\text{pay-back time} = \frac{\text{cost of energy-saving measure}}{\text{amount of money saved each year}}$$

If something has a short pay-back time then we say it is cost-effective.

Example: It costs £200 to fit loft **insulation** in a house and once fitted it will save £40 each year in bills. What is the pay-back time?

Solution: $\text{pay-back time} = \frac{\text{cost}}{\text{saving}} = \frac{200}{40} = 5 \text{ years.}$

3 How do you calculate pay-back time?

4 What does the term cost-effective mean?

B At Hockerton, the grass roof provides insulation.

The table shows some of the places in the home where heat is lost and ways in which this heat loss can be reduced.

Place where heat is lost	Percentage of household heat lost this way	Way to reduce it
roof	25	loft insulation
doors	15	draft excluders
windows	10	double glazing, thick curtains
walls	35	cavity wall insulation, silver coating behind radiators
floor	15	sealed floorboards, fitted carpet

5 Which part of the home allows the biggest loss of heat?

6 What two things can be done to reduce heat loss through windows?

It is not possible to completely stop all the heat loss from a house but it can usually be reduced a great deal. This will save money.

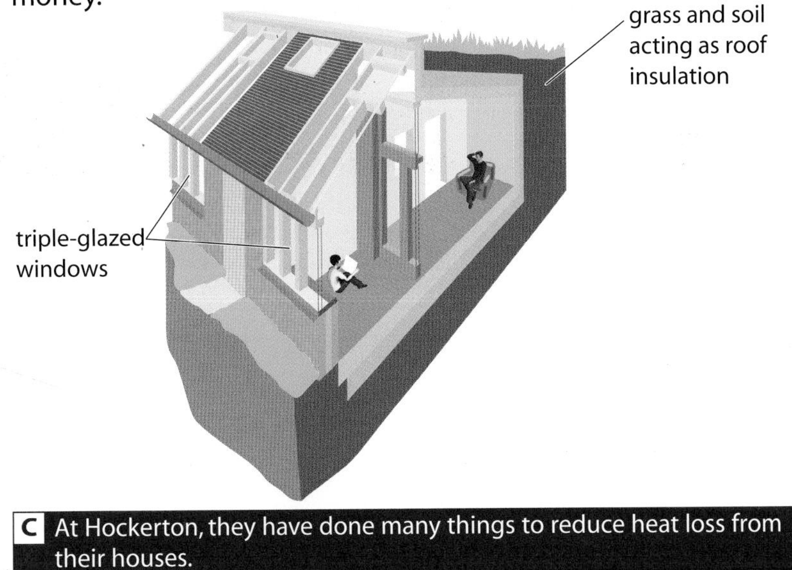

C At Hockerton, they have done many things to reduce heat loss from their houses.

7 Many of the windows at Hockerton are triple glazed. How does this help to reduce the heating bills?

8 The table contains some information about energy-saving measures in a house. Copy the table and fill in the missing values and write down which is the most cost-effective energy-saving measure.

Energy-saving measure	Cost to buy (£)	Money saved each year (£)	Pay-back time (years)
loft insulation	200	40	
draught excluders	20	10	
double glazing	2000	50	
fitted carpet	400	20	

Summary Exercise

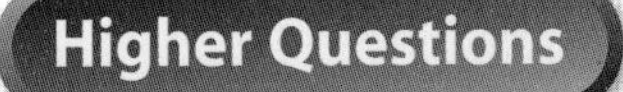

Solar cells

By the end of these two pages you should:

- know the advantages and disadvantages of using solar cells on a domestic level
- understand how efficient solar cells can be
- know why solar cells are not widely used in the UK.

Solar power can be used in many different ways but it has two common practical uses. Solar panels use solar power to heat water. Solar cells, also known as photovoltaic cells, turn solar energy into electrical energy.

In both cases the energy supply is free and once the cells or panels have been bought they require very little maintenance. On a small scale, solar cells can be used for appliances that don't use much electrical power like radios, calculators, watches and even satellites. On a larger scale they can be used to power a whole house.

1 Name three things that can be powered by solar cells.

2 What is another name for a solar cell that generates electricity?

B The solar panels at Hockerton use a photovoltaic cell to power a pump.

Solar cells can provide a source of free electricity at the point of use. Many homes have a large roof area and every day large amounts of wasted solar energy falls on them. If you install around 30 m^2 of solar cells then you could power a normal household in the UK. There are sometimes grants and discounts available to pay for the cells as the government want people to use renewable energy sources. In some cases solar cells generate more electricity than is needed. The excess electricity is sold back to the National Grid for use by other people.

A At Hockerton they use solar power as an energy source.

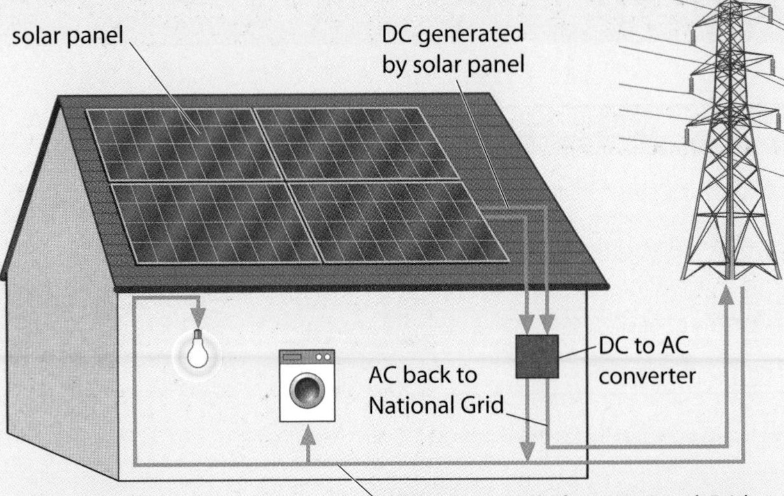

C Solar cells can provide electricity for a house and even let you sell some back to the National Grid.

One of the drawbacks of using solar cells is that they are still not very efficient. In the best cells only 20% of the solar power is turned into electricity. This means that over 80% of the possible energy is lost and not converted.

As well as the low efficiency there are other reasons why solar cells are not used more widely in the UK. Many people will continue using conventional electrical supplies because it is relatively cheap and very easy to do so. Installing solar cells can be expensive at first and requires many years of use to save enough money to cover the cost of installation. The area needed to create enough electricity for a domestic house can be quite large. Unless you have a way of storing the electricity, such as rechargeable batteries, then nothing will work at night.

D Solar cells can convert up to 20% of the Sun's energy falling on them into electricity.

Summary Exercise

3 Why does the government sometimes help with the cost of installing a set of solar cells?

4 If you install solar cells and generate more electricity than you need, what can you do with the extra?

5 How efficient are the best solar cells?

6 How can solar cells be used to provide electricity at night?

7 Write a short report of around 200 words, listing all the key features of solar cells. Explain how they are very useful but also why they are not used more widely.

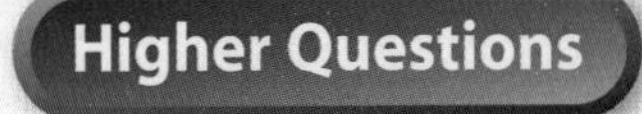

Higher Questions

Technology, electricity and medicine

H *By the end of these two pages you should:*

- know about how electricity has been and is being used in medicine
- be able to describe how our scientific ideas on the use of electricity have changed
- know some of the possible future uses of electrical technology in medicine.

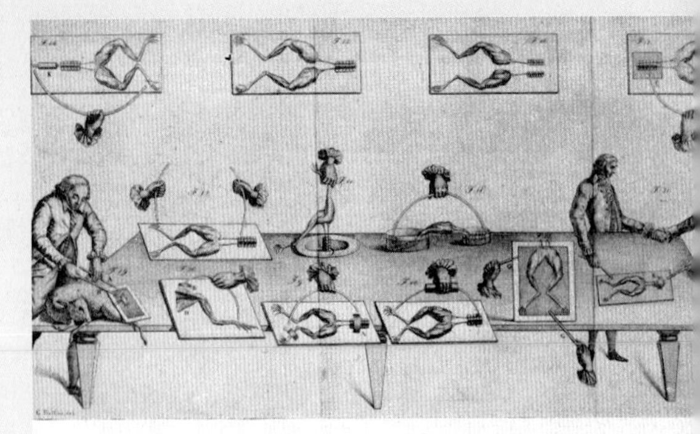

Expériences de Galvani sur les grenouilles. (Extrait des Œuvres de Galvani.) (Collection de M. E. Sartiaux.)

A A flow of electric current will make a dead frog's leg move.

In the 1780s Luigi Galvani discovered that if you pass an electric current through a dead frog's legs they will move. This told him that electrical currents are used to control muscles in animals. Before then people thought that the nerves were like pipes. Now we know that they are a network of electrical conductors.

Ever since then there have been many ideas about electricity in medicine. Many of the historical uses of electricity in medicine have had some useful effects. However, some could be painful, not very reliable and sometimes dangerous. Electricity could not be seen and was hard to measure, so people had to imagine what might be going on. As we have learned more we have been able to develop better and more successful treatments. Some common ones are listed here.

Use in medicine	What it does
transcutaneous electrical nerve stimulation (TENS)	uses electrical signals to help reduce pain, often used during childbirth
conduction testing	electrical signals are sent along nerves to test how well they are working
electronic muscle stimulation (EMS)	electrical signals are used to stimulate and move the muscles of paralysed people in physiotherapy – this is exactly the same idea as Galvani's frog's legs
pacemaker	implanted in the chest to provide a regular pulse of electricity to regulate the heartbeat

P

1. How was it discovered that muscles are controlled by electrical signals?
2. How is electricity used to help exercise paralysed patients?
3. Write down two other ways in which electricity is used in modern medicine.

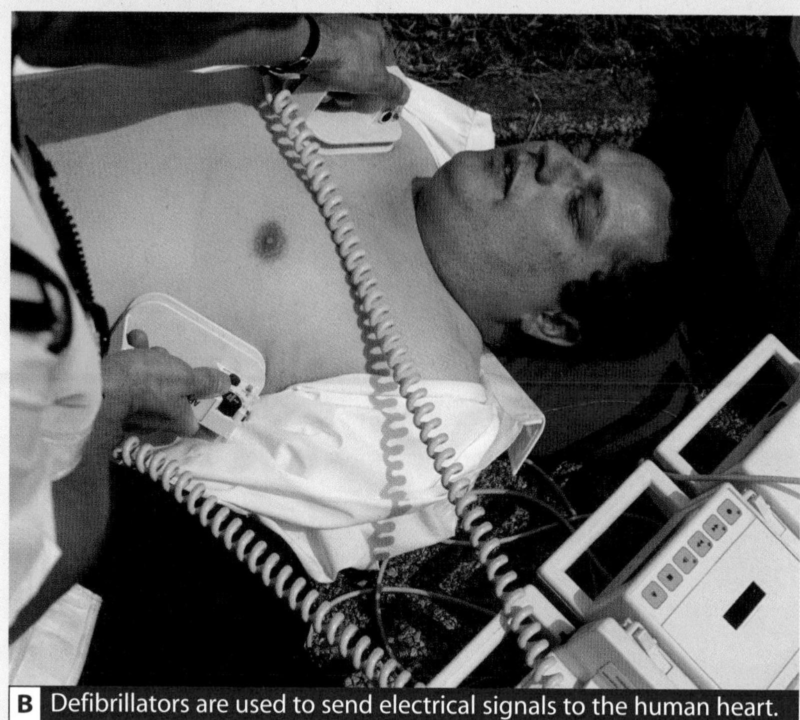

B Defibrillators are used to send electrical signals to the human heart.

Like all muscles, the human heart is controlled and regulated by electrical signals sent to it. Below are two pieces of equipment that are used in diagnosing and treating heart problems.

- Defibrillators are used to treat a heart that is not beating correctly. They give the heart an electric shock to make it return to beating normally.
- The electrocardiogram (known as ECG) helps doctors to diagnose heart conditions. Electrodes are placed across the body and measure the small electrical currents generated during heart activity.

As electrical components get smaller and smaller the possible uses of electricity in medicine grow. Some of the applications being developed and used around the world include implants that deliver drugs when and where needed, defibrillators implanted inside the body, and sensors that monitor brain activity of patients in a coma.

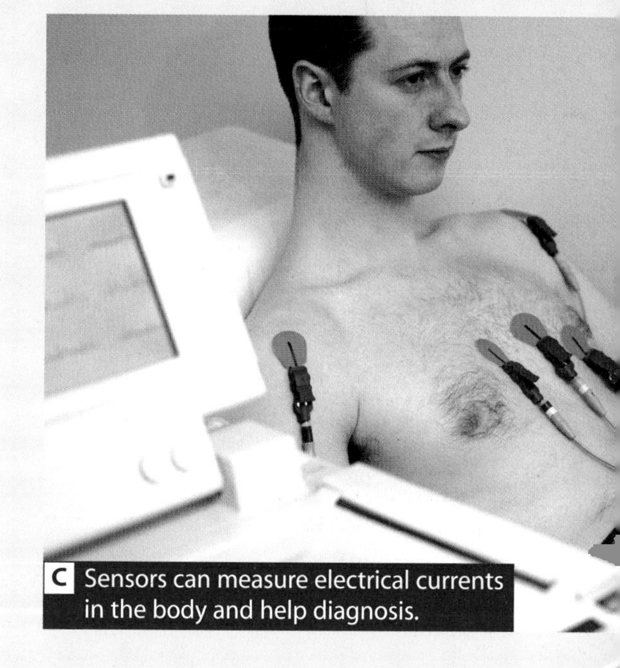

C Sensors can measure electrical currents in the body and help diagnosis.

4 What device uses the signals that the heart produces to diagnose heart conditions?

5 What is the name for the sensors attached to the body that measure these signals?

6 Write a letter to Luigi Galvani explaining what people used to think about electricity and medicine many years ago and how these ideas have changed. Explain how electricity is used in medicine today.

Higher Questions

Questions

D

D

Multiple choice questions

1 Which of the following is not a renewable source of energy?
A solar power
B wind power
C nuclear power
D tidal power

2 Here is a list of statements about using wind power in the UK. Which of them is untrue?
A The climate and geography of the UK provides many possible sites.
B The energy supply is free.
C It does not produce pollutant gases.
D The energy supply will run out eventually.

3 What is the name for the effect that makes a wire carrying a current move when it is near a magnet?
A The generator effect.
B The motor effect.
C The current effect.
D The voltage effect.

4 A coil of wire carrying a current is placed near a magnet and it moves. Which of the following would not make it move more?
A Using a larger current.
B Passing the current in the other direction.
C Using a stronger magnet.
D Using more coils of wire.

5 What is the name for the network of cables that connects houses to the mains electricity supply?
A Electrical Grid.
B National Mains.
C National Network.
D National Grid.

6 Which of the following statements about the distribution system from question 5 is **not** true?
A The network is already built.
B It is difficult to connect remote houses to the system.
C It is only powered by wind turbines.
D It can provide electricity all day and night.

7 Which of these statements about a fuse is false?
A It is made of wire.
B It is a safety device.
C It will melt if the current gets too low.
D It is connected to the live wire in a plug.

8 What is the name for the units we buy electricity in?
A kilowatt-hour
B power
C watt
D joule

9 What is the colour of the earth wire in a plug?
A yellow and green
B red
C blue
D blue and yellow

10 Which is the correct equation for electrical power?
A power=current×voltage
B $\text{power}=\frac{\text{current}}{\text{voltage}}$
C power=current×time
B $\text{power}=\frac{\text{voltage}}{\text{time}}$

11 What is the equation for efficiency?
A $\text{efficiency}=\frac{\text{total input}}{\text{useful output}}\times 100\%$
B $\text{efficiency}=\frac{\text{useful output}}{\text{total input}}\times 100\%$
C efficiency=useful output×total input×100%
D efficiency=total input−useful output×100%

12 What is 1500 W in kW?
A 1.5 kW
B 15 kW
C 0.15 kW
D 150,000 kW

13 A car engine uses 500 J of chemical energy in the petrol to generate 200 J of movement energy. What it its efficiency?
A 60%
B 250%
C 40%
D 100%

14 Which of the following statements about energy-saving light bulbs is false?
A They cost less to run than filament light bulbs.
B They do not get as hot as filament light bulbs.
C They have a lower efficiency than filament light bulbs.
D They are better at converting electricity into light than filament light bulbs.

15 Which of the following would not reduce heat loss in a domestic home?
- **A** Sealing the floor boards.
- **B** Installing loft insulation.
- **C** Installing cavity wall insulation.
- **D** Replacing double glazing with single glazing.

Short-answer questions

1 In an electric motor, electrical energy is turned into movement.
- **a** Describe the key features of an electric motor.
- **b** Explain the different ways in which a motor could be made to spin faster.
- **c** Write a brief description of what modern life might be like if there were no electric motors.

2 You have been asked to write a report about heat loss in your school.
- **a** Write down some different places in your school where you think heat loss might be happening.
- **b** For each of these ways write down how this heat loss could be reduced.
- **c** For two of these methods of heat-loss reduction describe how they work.
- **d** Write a *polite* letter to your head teacher explaining the problems and why they should do something about it.

3 Each year millions of pounds are spent on domestic fuel bills. It is important to know about how we pay for electricity and how we can measure what we use. This can help to save money. For this question use 6 p as the cost of one unit of electricity.
- **a** The table below gives an idea of how much it costs to run different appliances. Copy and complete the table.

Appliance	Power rating (kW)	Time used in a day (h)	Total units (kWh)	Cost (p)
cooker	2.5	1		
kettle	1.0	2		
fridge	0.1	24		

- **b** The readings on an electricity meter are used to calculate bills. The readings below are from two different dates. How much did it cost to use the electricity between these two dates? Show your working.
  Reading last month: 8560 kWh
  Reading today: 8940 kWh

When buying new appliances, knowing the power rating can give you an idea as to how quickly they use energy.
- **c** The label on a fridge has the following information on it: 230 V, 0.5 A. What is its power rating?
- **d** How much would it cost to run this fridge for a year? (1 year=8760 hours)

Glossary

appliance A household device that uses electricity to do something.
current The flow of electricity around a circuit. Measured in amps (A) or milliamps (mA).
earth wire The green and yellow wire in a plug that protects the user from an electric shock.
efficiency A measure of how good something is at turning energy from one form into another useful form.
electricity The energy involved when charged particles flow from one point to another.
energy Whenever something happens, energy is involved. It can exist in many different forms but is always measured in joules (J).
fuse A device in a plug that protects the equipment and cable from overheating and fire if the current becomes too large.
insulation Materials that can help reduce heat loss.
magnetic field The area around a magnetic object where other magnetic objects will feel a force.
motor Something that turns one form of energy into movement.
photovoltaic cell Another name for a solar cell.
power A measure of how quickly something converts energy from one form to another. Measured in watts (W) or kilowatts (kW).
rate A measure of how quickly something happens.
renewable An energy supply that does not run out or is easily replenished, such as solar or wave power.
residual current circuit breaker (RCCB) A device that protects from electric shock by shutting a circuit down.
resistance A measure of how difficult it is for current to flow around a circuit. Measured in ohms (Ω).
solar cell Something that converts solar power into electricity.
solar panel Something that harnesses solar power for heating water.
solar power The energy that comes to us from the Sun.
voltage The difference in electrical energy between two points that makes a current flow. Measured in volts (V) or millivolts (mV). It is sometimes called the potential difference.
wind power The energy that can be harnessed from moving air currents using wind turbines.
work The energy transferred by a force on a moving object.

Now you see it, now you don't

A Waves play an important part in making air travel safe.

A million passengers travel through Heathrow airport each day. Security is a top priority. Operations manager Arthur Mount needs to ensure safety at the airport.

Wave energy is used to help make airports and aeroplanes safe. Arthur's team uses radio waves to communicate. Radio waves also help aeroplanes to take off safely. The pilot communicates with on-board electronic equipment and air traffic control. Aeroplanes are tracked using radio waves.

Waves are also used in security. Bags are scanned using X-rays, CCTV cameras record activities and security guards use mobile phones. All these security measures use energy in the form of waves.

Waves are used during aeroplane maintenance. Infrared is used to check that different parts of the aeroplane are at the correct working temperature.

In this topic you will learn that:

- different types of waves have similar properties
- waves carry energy that enables them to penetrate materials
- the reflection and absorption of waves can be used for a variety of scanning applications
- wave energy can be a risk to health.

Which type of wave energy do you think could be used for each application?

visible light waves, radio waves, X-rays, gamma-rays

- Scanning the contents of a suitcase for metal objects.
- Tracking and guiding a plane in to land.
- Scanning an aeroplane wing to check for cracks.
- Scanning a person's iris to check their identity.

V

D

Wave basics

By the end of these two pages you should:

- be able to explain the terms amplitude, speed of a wave, wavelength and frequency
- understand the similarities and differences between transverse and longitudinal waves.

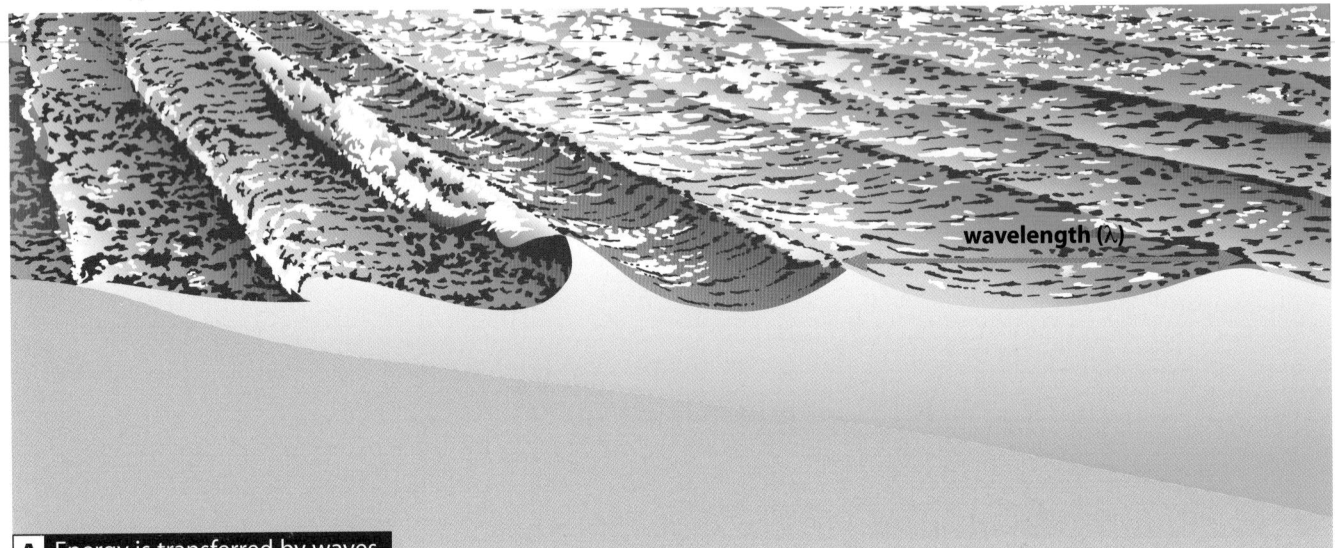

A Energy is transferred by waves.

Sea **waves** are made up of water particles moving up and down from their normal positions. If you follow the movement of a water particle it keeps returning to its starting point. This movement of a particle from a starting point and back again is called an **oscillation**. Water waves are an example of a **transverse** wave. The wave moves forwards at right angles to the 'up and down' movement of the water particles.

Amplitude means the height of a wave. It is the maximum distance the water particles move. The **speed of a wave** at sea is the distance it moves along in a second. The **wavelength** is the distance between two crests (or two troughs). It is the distance between two points on the wave that have moved a maximum distance from the middle of the wave. The symbol for wavelength is λ, or lambda (pronounced *lam-da*). The **frequency** of a wave is the number of oscillations of a particle in 1 second. It is measured in hertz (pronounced *hurts*), which can be shortened to Hz.

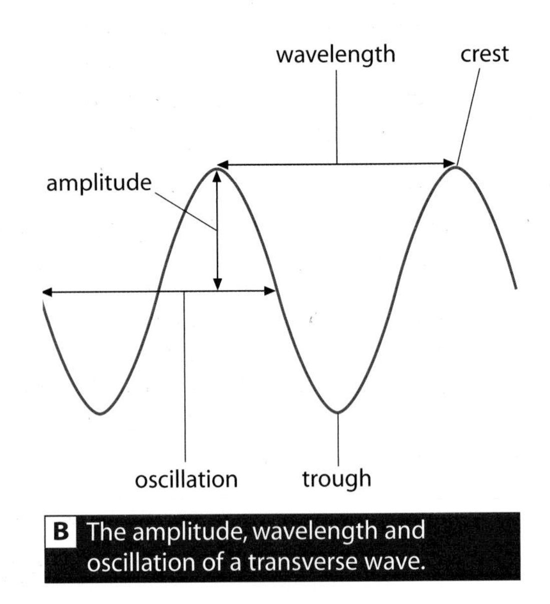

B The amplitude, wavelength and oscillation of a transverse wave.

1 Fill in the missing word.
  a The maximum distance a particle moves in a wave is the… .
  b The movement of a particle back to its starting point is called an… .
  c The distance between two peaks on a wave is the… .

Particles move backwards and forwards in both **longitudinal** and transverse waves. These are called oscillations. In transverse waves the particles move *at right angles* to the direction of the wave. However, in longitudinal waves the particles move *parallel* to the wave.

2 Sound waves travel through air. Air is moved backwards and forwards parallel to the direction of the sound. Explain whether sound is a longitudinal or a transverse wave.

particles in a longitudinal wave move as if someone has pushed a slinky spring from one end

wavelength

C In a longitudinal wave the particles move parallel to the direction of the wave.

D Airports use waves of different frequencies to communicate with aeroplanes.

3 Complete the following sentences:
 a The number of waves per second is the… .
 b The distance between the peak of one wave and the next is the… .

4 A student writes an answer to the question, What is a transverse wave?
'A transverse wave is one in which the particles move <u>parallel</u> to the direction of movement of the wave.'
A mistake has been underlined. Replace this word with a correct phrase.

5 What is meant by a longitudinal wave?

6 The wavelength of microwaves can be about 3 cm. Explain what is meant by wavelength.

7 A singer discovers she can break a wine glass by singing a note with a precise *frequency* or *wavelength*. She has to sing the note with a large *amplitude*. Sound is an example of a *longitudinal* wave. Write a definition for each word that is shown in italics.

Higher Questions

Reflecting waves

By the end of these two pages you should be able to:

- understand that waves can be reflected from different surfaces
- explain how scanning by reflection can be used to scan a foetus during pregnancy
- use the equation speed=distance/time to calculate the distance to a reflecting surface if you know the time the reflected wave takes to return
- discuss the advantages and disadvantages of scanning by reflection.

Have you ever wondered:

How do we know the Moon is 380,000 km away?

A wave's speed can be calculated using this equation:

$$\text{speed} = \frac{\text{distance covered by the wave}}{\text{time taken}}$$

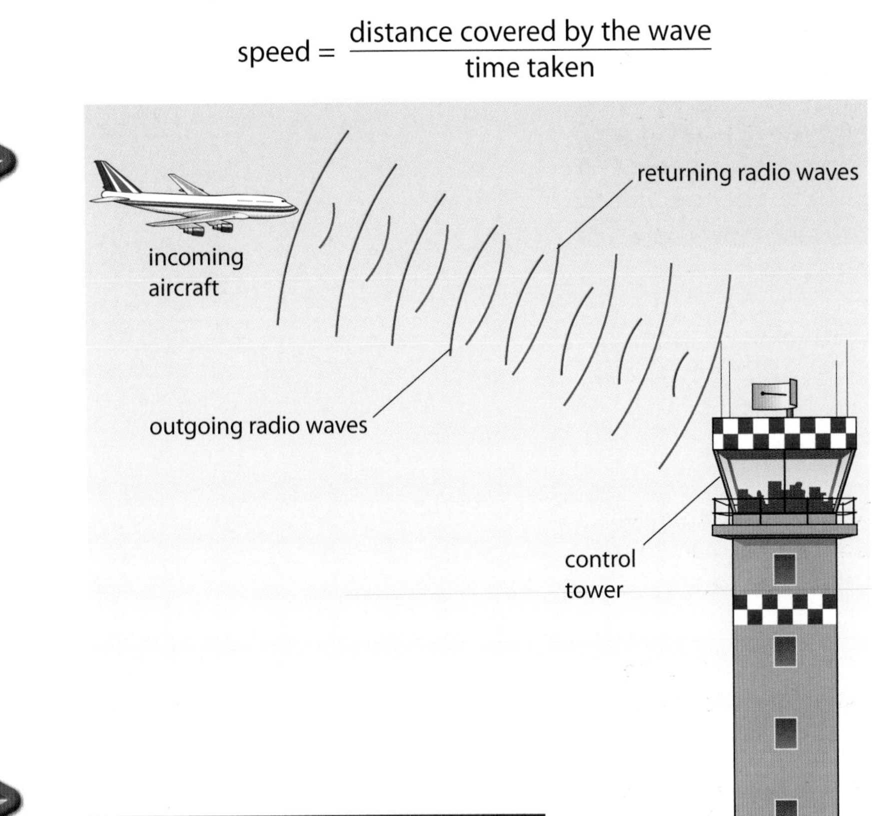

A Radar uses radio waves to find aircraft.

Radio wave transmitters at an airport send out signals which reflect off aircraft. The returning signal tells air traffic control exactly where each aircraft is. The distance to the aircraft is calculated by multiplying the speed of the radio wave by half the time it takes the wave to return to the control tower. Air traffic control stop aircraft flying too close to each other.

1 A sound wave generated by a jet aeroplane travels 4200 m from the aeroplane to a person on the ground. It takes 13 s. Calculate the speed of the sound wave.

2 A radio wave is sent by a control tower towards an aeroplane. The reflected wave returns to the control tower 0.000033 s later. Calculate the distance from the control tower to the aeroplane. The speed of the radio wave is 300,000,000 m/s, which can be entered on a calculator as 3×10^8.

Have you ever wondered:

How do you see an unborn baby?

Ultrasound is high-frequency sound which we cannot hear. It can go through the softer tissues of the human body. It is reflected whenever it travels from one type of material to another.

B Ultrasound produces a scan of a foetus.

Scanning is building up a picture in sections or pieces. During an ultrasound scan of an unborn baby the head shows up clearly because it has greater **density** than the surrounding tissue. This causes a better **reflection**. An advantage of ultrasound scans is that the wave's energy is reflected rather than absorbed by the foetus, so it does not harm the baby. A disadvantage is that the wave is scattered in all directions when it is reflected.

3 Ultrasound is a high-frequency sound wave. Is it a transverse or a longitudinal wave?

4 Shout loudly at a cliff face and you will hear an echo a short time later. Walk away from the cliff face, shout again and you will again hear the echo. Would the time between you shouting and hearing the echo increase or decrease?

5 A student shouts at a large wall. He times the returning echo to be 1.0 s from when he shouted. The distance to the wall is 150 m. Calculate the speed of sound.

6 Fishermen send ultrasonic waves into the sea to find shoals of fish. Explain how this method can find a shoal and tell the fishermen how deep to trail their nets.

7 Explain how ultrasound is used to measure the width of a baby's head. Use at least one diagram in your answer.

Summary Exercise

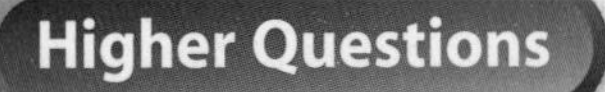

Earthquake waves

By the end of these two pages you should know:

- how differences in the density of materials will cause waves to be reflected or refracted
- H how earthquake waves can be used to decide which types of material are inside the Earth
- why scientists find it difficult to predict earthquakes and tsunami waves.

Material A	Material B	Density A (kg/m^3)	Density B (kg/m^3)	Percentage reflection
air	skin	1.2	1060	99%
skin	water	1060	1000	2%
bone	water	2000	1000	47%
water	bone	1000	2000	47%

A Densities of materials and the percentage of an ultrasound signal that is reflected as it travels from A to B.

1 Different materials reflect ultrasound by different amounts. Use the table to say what happens when ultrasound travels between materials:
a of similar density.
b of very different density.

Any wave energy that is not reflected will continue to travel through the new material. The new material will usually cause the speed of the wave to change. This is called **refraction** and the wave usually changes direction.

Earthquake waves, known as **seismic waves**, can cause a frightening amount of damage. They can also tell you about what lies beneath the Earth's surface.

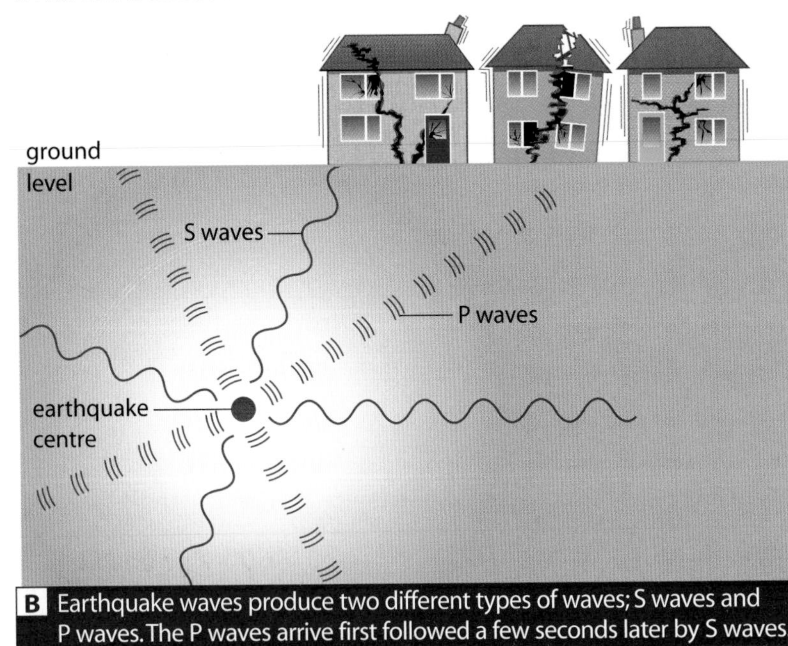

B Earthquake waves produce two different types of waves; S waves and P waves. The P waves arrive first followed a few seconds later by S waves.

A town near the source of an earthquake (the **epicentre**) will experience two types of tremor. **P waves** are longitudinal waves and **S waves** are transverse waves.

H The waves leave the epicentre and travel at different speeds. The speed depends on the density of the material that the waves travel through. S waves cannot travel through liquids.

Depth (km)	Velocity of wave (km/s)	
	P wave	S wave
0	7.00	3.00
40	8.00	4.50
1000	11.00	6.50
2900	13.50	7.00
3000	7.00	0.00
4999	10.75	0.00
5000	11.00	0.00
6000	12.00	0.00

C Speeds of P and S waves travelling towards the centre of the Earth.

2 The speed of S waves below 3000 km beneath the Earth's surface is zero.
- **a** What can you say about the Earth at this depth?
- **b** Can S waves travel any deeper than 3000 km? Explain your answer.

3 What patterns do you see as the P waves travel deeper?

An earthquake underneath the sea bed can cause a tsunami. Huge waves are created by vibrations from the earthquake that pass energy into the sea water above. Although this is a rare event, a tsunami occurred in December 2004. It affected countries around the Indian Ocean, such as Indonesia. Huge waves from the sea destroyed villages along the coastline. The waves travelled at fast speeds across the ocean.

Scientists use movements in the Earth to rate the strength of an earthquake. It is not possible to predict where or when the next earthquake will occur. Certain areas are more likely to have earthquakes because large sections of the Earth's crust are moving in different directions. Although small earthquakes can be a warning of a major earthquake, we cannot predict when the earthquake will occur.

Have you ever wondered:

Why do scientists believe there could be an even more catastrophic tsunami than the last one?

4 Suggest what happens to ultrasound as it passes from fat into muscle during a scan of a human body. Muscle has a density of about 1100 kg/m³ and fat has a density of about 900 kg/m³.

5 Some watery jelly is placed on the end of the ultrasound probe. The probe, which sends out and receives ultrasound, is placed on the skin. The jelly spreads out between the probe and the skin. Suggest why the jelly might help the ultrasound work better.

Summary Exercise

The wave equation

By the end of these two pages you should be able to:

- use the relationship speed = frequency × wavelength
- describe how similarities between waves can be shown in the electromagnetic spectrum
- recall that electromagnetic waves all travel at the same speed in a vacuum.

Have you ever wondered:

Why does helium make your voice go high?

A loudspeaker makes a sound of frequency 200 Hz. The wavelength of the sound is 1.5 m.
The speed of the wave can be calculated by the wave equation:

$$\text{speed} = \text{frequency} \times \text{wavelength}$$

So, the speed of the sound waves described above is calculated as the frequency (200 Hz) multiplied by the wavelength (1.5 m), which is equal to 300 m/s.

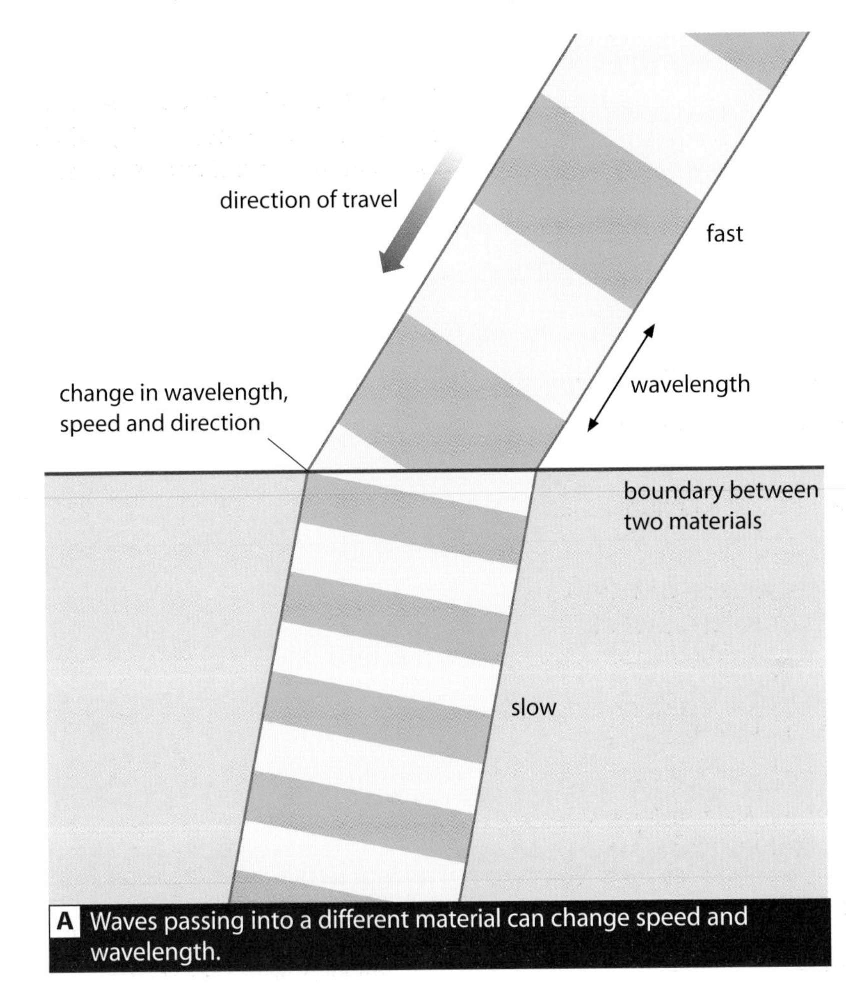

A Waves passing into a different material can change speed and wavelength.

1 The frequency of an ultrasound wave is 500 kHz. Note that 1 kHz means 1 kilohertz and is equal to 1000 Hz. The wavelength is 0.0006 m. Calculate the speed of the ultrasound.

When waves meet a boundary between two different materials of different density some of the wave will be reflected. The amount of reflection will depend on how different the densities of the two materials are. The rest of the wave will continue with a different speed but the same frequency.

2 A sound wave slows down as it moves into a new material. The frequency remains the same. Use the wave equation to predict whether the wavelength will be decreased or increased.

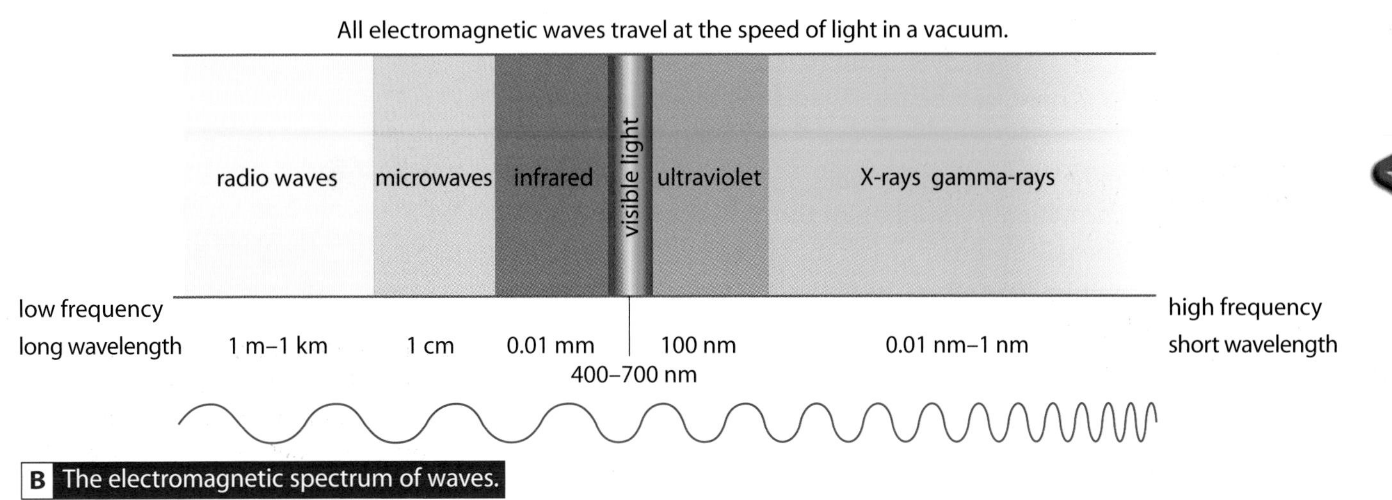

B The electromagnetic spectrum of waves.

The **electromagnetic spectrum** of waves includes radio waves, **microwaves**, **infrared**, visible light, **ultraviolet** light, **X-rays** and **gamma-rays**. Waves on the left of figure B, such as radio waves, have low frequencies and long wavelengths. As you move towards the right the frequency increases and the wavelength gets shorter. X-rays and gamma-rays, on the right, have high frequencies and short wavelengths.

All electromagnetic waves travel with the same speed in a **vacuum**. This speed is called the speed of light. They are all transverse waves.

3 The wavelength of a radio wave is 693 m. The frequency is 432,900 Hz. Calculate the speed of this wave.

4 Name the family of waves which all travel at the speed of light.

5 Which region of the electromagnetic spectrum lies between infrared and ultraviolet?

6 Can you name one property that is shared by X-rays and radio waves?

7 The wavelength of waves used to track the position of aircraft is 0.25 m. The frequency is 1200 Mhz. 1 MHz means 1,000,000 Hz or 1×10^6 Hz.

a Calculate the speed of these waves in m/s.

b Using diagram B, which family do these waves belong to? Explain your answer.

8 Draw a diagram showing each region of the electromagnetic spectrum. Indicate which region has the highest frequency.

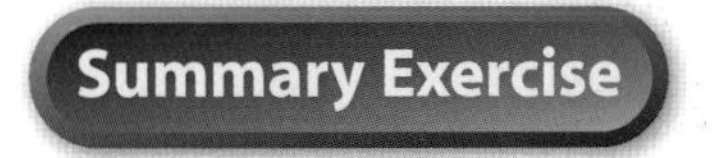

Summary Exercise

Higher Questions

Using electromagnetic waves

By the end of these two pages you should know:

- the differences between waves within the electromagnetic spectrum
- how visible light can be used for iris recognition
- the advantages and disadvantages of iris recognition
- how infrared sensors can be used to measure temperature.

Radio waves can get through brick walls, so you can listen to the radio at home. When a wave passes through something it **penetrates** it. There are limits to what materials radio waves can penetrate. Car radios fade in tunnels. Different parts of the electromagnetic spectrum can penetrate different materials.

Mobile phones use microwaves. Microwaves will also penetrate through materials. Your mobile phone works indoors. Microwaves are easier to direct than radio waves. You can use your mobile phone as long as there are no large objects like hills between you and the receiver.

1 State two differences between radio waves and microwaves. Use figure B on p257 to help you.

2 State two similarities between radio waves and microwaves.

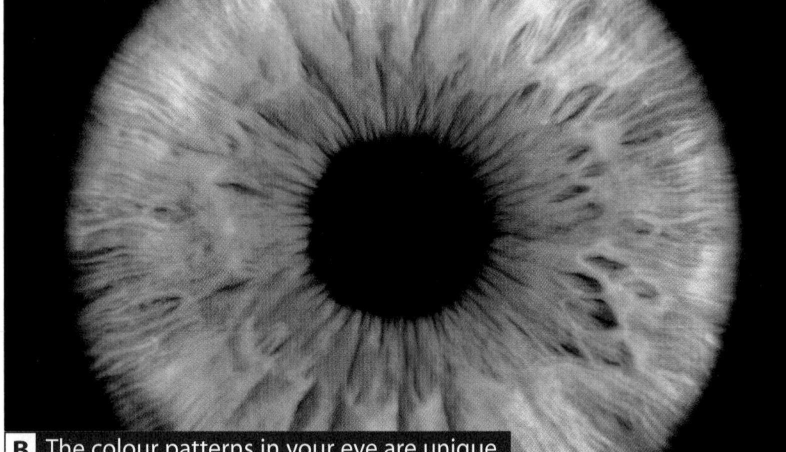

B The colour patterns in your eye are unique.

If you travel through an airport you need your passport for identification. Many airports now scan your **iris**. This is called **iris recognition**.

Iris scanning is like fingerprinting – we all have a unique iris pattern. A camera collects the reflected light from your iris. The pattern is checked on computer. But everyone will need their iris scanned beforehand. Some people think the system could be fooled by changing someone's computer record or even scanning a photograph of someone else's iris.

A Radio waves are used for communication in airports.

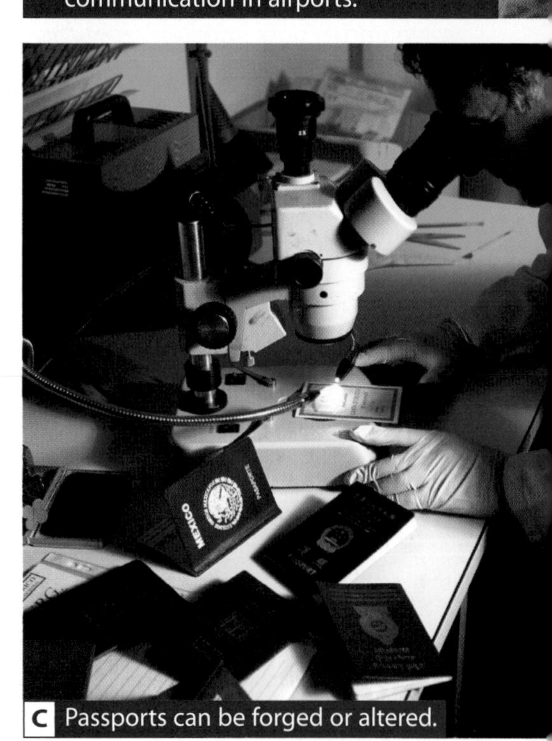

C Passports can be forged or altered.

3 Look at your iris in a mirror. Draw the pattern of colours.

Have you ever wondered:

Do night vision goggles you see in the movies really work?

Objects emit **radiation** due to their temperature. You may see the **emission** of visible light if objects are very hot. Hot objects actually emit a range of different types of electromagnetic radiation. A light bulb emits visible light because the filament is very hot.

Hot objects emit infrared radiation. Infrared radiation carries most of the heat from the Sun to Earth. The surface temperature of an object can be measured by detecting the amount and type of infrared radiation. This is called infrared thermography.

6 In the film *Predator*, an alien can only detect infrared light. The hero makes himself invisible to the alien by covering himself with wet mud. Why might this make the hero invisible to the alien?

F An aircraft wing in infrared light showing cracks that have let in water, which has frozen.

This wing is made of new material which is light and strong. But if water gets into the wing it could freeze and cause damage. An infrared detector will show water in the wing if it is inspected soon after landing. The water will have frozen because of the cold temperature at high altitude. Ice will emit less infrared than the rest of the wing.

4 List three things that emit visible light when they are hot.

5 Give two examples of electromagnetic radiation which have a lower frequency than visible light.

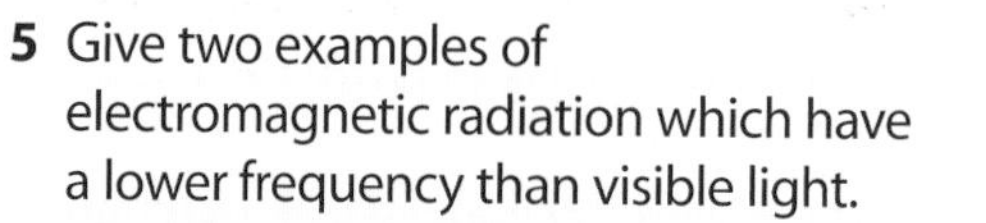

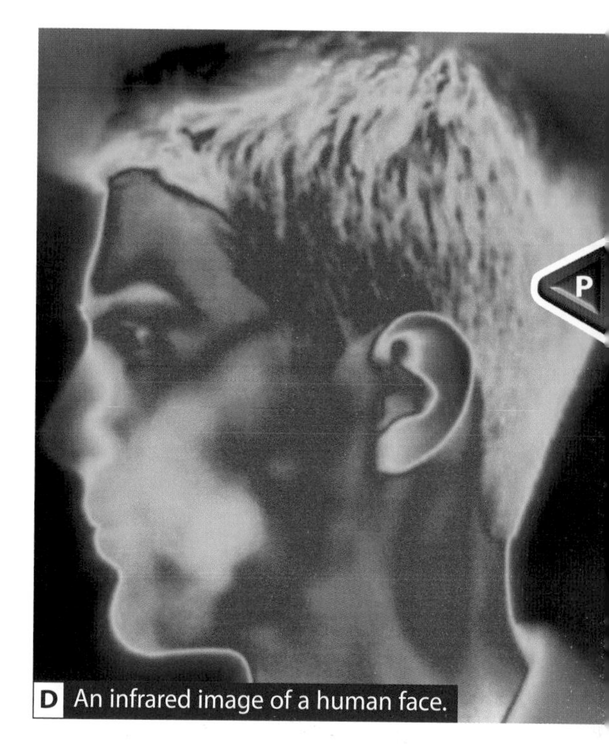

D An infrared image of a human face.

7 State one use of microwave radiation.

8 Gamma-rays are used to detect cracks in aircraft wings. Describe two differences between gamma-rays and visible light. Use figure B on p257.

Higher Questions

More uses of electromagnetic waves

By the end of these two pages you should know:

- how X-rays can be used to see inside luggage and to detect bone fractures
- how microwaves are used to monitor rain
- how ultraviolet can be used to detect forged banknotes.

Have you ever wondered:

How do X-rays work?

A X-rays are used to inspect luggage.

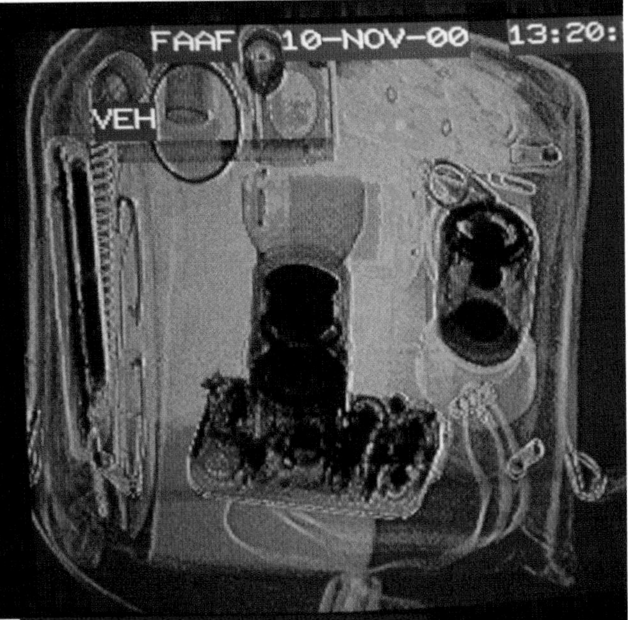

B An X-ray of objects inside a suitcase.

X-rays penetrate the softer materials in luggage. Denser objects block or absorb the X-rays. A film is placed on the opposite side of the luggage from the source of the X-rays. The film goes white in the presence of X-rays and remains dark elsewhere. This shows up the outline of dense objects in the suitcase.

1. List two places, other than airports, where bags are checked for undesirable objects.
2. Why use X-rays to inspect luggage rather than asking people to open their suitcases?

Broken bones can be detected using X-ray **absorption**. Absorption means that some of the wave energy is transferred to the material. If there is a break or fracture in the bone then X-rays penetrate through the break.

3. Explain why X-rays can be used to image bones.

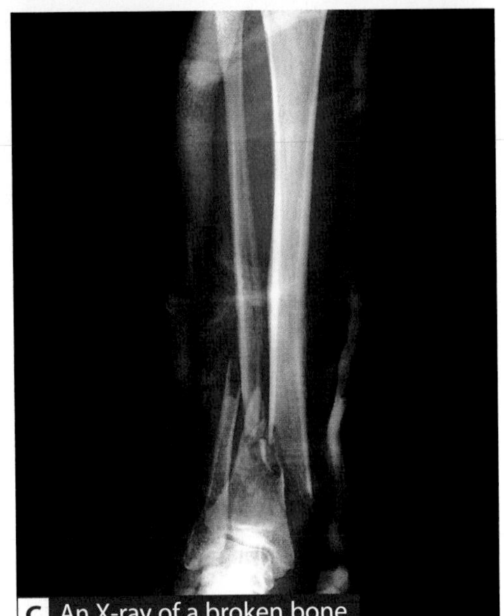

C An X-ray of a broken bone.

Have you ever wondered:

How can microwaves be used to forecast the weather?

Meteorologists need information about the current climate to predict the weather. In heavy rain your mobile phone will have poor reception. Mobile phones use microwave radiation. Microwaves are absorbed by water.

D A microwave image showing rain clouds.

To make microwave images of rain clouds, emitters on the Earth send microwaves to satellites in orbit. Detectors on the satellite will measure any decrease in the strength or amplitude of the microwaves due to rainfall or cloud. The decrease in strength is colour coded and shows areas of rain cloud on a computer.

Have you ever wondered:

How can forged banknotes be detected?

A white shirt will look blue in a dark nightclub. Chemicals in the shirt absorb ultraviolet light and then emit blue light. This is called fluorescence. The paper used for banknotes should not fluoresce – emit blue light – when radiated with ultraviolet light. The chemicals in ordinary white paper will fluoresce. **Fluorescent** inks are used to add invisible symbols to banknotes. These symbols show up under ultraviolet light.

4 Describe the appearance of a forged banknote when held under an ultraviolet light.

5 Name the region of the electromagnetic spectrum between gamma-rays and ultraviolet light.

6 What is meant by absorption?

7 Amy decides to investigate the absorption of microwaves using a microwave oven. She places an empty plastic mug in the oven and switches it on for 5 minutes. After 5 minutes the mug has not been heated at all. She fills the mug with water and repeats her experiment. This time the mug and its contents are hot. Which is better at absorbing microwaves, plastic or water? Explain your answer.

8 List the three types of electromagnetic wave studied in these two pages in order of decreasing wavelength. Explain the role of absorption in the use of each wave.

Higher Questions

Microwaves and mobile phones

By the end of these two pages you should be able to:

- describe the effects of microwaves on body tissue
- discuss whether microwave radiation from mobile phones or masts could pose health risks
- discuss how the media report scientific issues such as mobile phone risks.

Have you ever wondered:

Is too much exposure to mobile phone radiation dangerous?

A Microwave ovens use microwave radiation to cook food quickly.

During the Second World War, soldiers realised they could warm themselves by standing close to microwave transmitters used for radar. They also noticed that dead birds had been cooked by the microwave radiation. This observation led to the invention of microwave ovens.

Many households have a microwave oven. They use microwaves to heat food very quickly. Microwave radiation is absorbed by water and fat. Food with a high water content is particularly suitable for microwave cooking. At first people thought that they might absorb escaping microwaves, but microwave ovens have a metal grid in the front glass door to prevent microwaves escaping.

1 Explain why people might absorb microwaves.

2 Look at figure B on p257. What two areas of the electromagnetic spectrum are on either side of the microwave region?

Mobile phones also make use of microwaves to communicate. When you make a call, microwaves are sent out from your phone. Your head will receive a dose of microwave energy when you hold your phone next to your ear. Some of the energy will be absorbed and it will heat some of the tissue in your head. Some scientists suggested that long-term exposure to microwaves might lead to some health problems. There might be a higher risk of certain types of brain tumour. Newspapers reported these theories with headlines such as:

Mobile phone warning causes confusion

Parents campaign to halt mobile phone mast in school grounds

Using a mobile phone might buzz brain into action

B Typical headlines about possible mobile phone risks.

We could not be sure about the long-term risks until at least 10 years after the introduction of mobile phones.

3 Why does this need to be studied for such a long time?

The latest findings suggest that 10 years of regular mobile phone use shows no significant increase in the risk of brain cancer. Scientists do not yet know whether there are longer-term risks. Mobile phones can interfere with electronic communication systems. This is why you must switch them off in airports and hospitals.

4 If you were asked to study the risk of mobile phones on health, what measurements would you take?

5 Do you think the newspaper headlines are right?

6 Explain why some people have suggested that 'hands-free' mobile phone sets reduce the risk of brain tumours.

7 You must switch off your phone in airports and on aircraft. A significant amount of communication both between people and between equipment is carried out using microwaves. Why do you think you are asked to switch off your mobile phone?

8 Summarise the mobile phone debate as a newspaper article. Write one paragraph arguing that mobile phones are safe and another suggesting that there is a risk.

ATTENTION

USE OF MOBILE PHONE, COMPACT DISC PLAYER AND ANY PERSONAL ELECTRICAL DEVICE IS PROHIBITED ON BOARD

C Mobile phone frequencies might interfere with aircraft electronics.

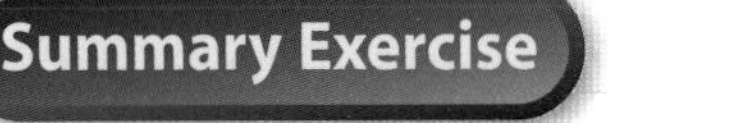

Higher Questions

Harmful electromagnetic waves

By the end of these two pages you should be able to:

- describe the harmful effects of too much exposure to infrared, X-rays and gamma-rays
- discuss the characteristics of ultraviolet light and relate them to the dangers of over-exposure
- discuss why too much sunbathing can harm the body
- H explain the damage caused by each type of wave in terms of increasing frequency.

Have you ever wondered:

Why does your skin burn quicker in the midday sun?

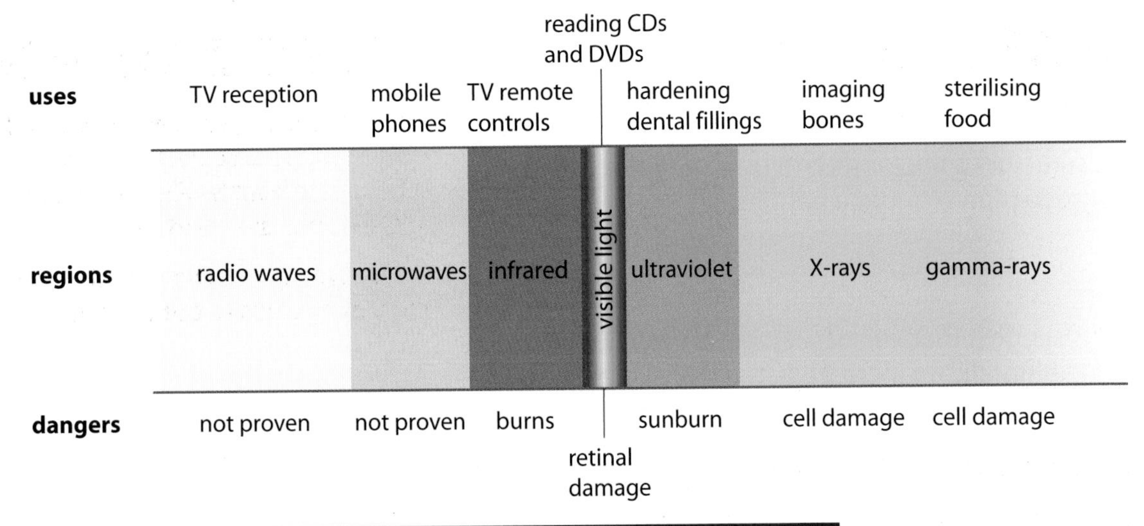

A Uses and dangers of different regions of the electromagnetic spectrum.

The dangers of radiation increase as the frequency increases. Infrared radiation carries a considerable amount of the Sun's energy to Earth. The strength of this radiation does not burn our skin. If the strength (or amplitude) of infrared radiation becomes very high (such as when there is a fire) then it can cause skin burns. Firemen wear protective clothing to reduce this risk.

X-rays are used for medical purposes, such as imaging bone, but even short exposure to X-rays causes some tissue damage. X-rays can knock out electrons from atoms. If the atoms are in a human cell then X-rays can destroy the cell. X-rays also damage DNA in the cell. Damage to DNA causes **mutation** which can lead to malfunction of cells and possibly cancer.

When you have an X-ray in a hospital the radiographer (the person taking the X-ray) will leave the room while the X-ray machine is switched on.

1 Why do radiographers leave the room when X-rays are being taken?

X-rays and gamma-rays have similar properties. They can both be used positively to treat certain cancers by destroying cancer cells. Background radiation is constantly all around us and consists mainly of gamma-rays. These come from many sources such as the rocks on which our houses are built. Certain atoms in the rock are unstable and emit gamma-rays. Scientists know how much of this radiation we can be exposed to without having serious health problems.

Description	Absorbed dose	Risks
background radiation	0.36	average health expectancy
nuclear power worker	0.66	no additional risk
minor nuclear accident	1.00	0.08% increased chance of cancer
major nuclear accident	100	significant increase in chance of cancer; direct illness resulting from exposure

B Absorbed doses of gamma-rays and health risks.

UV A		UV B		UV C	
400 nm	315 nm	280 nm		200 nm	wavelength
7.5×10^4 Hz				15×10^4 Hz	frequency

Type of ultraviolet (UV) radiation	Wavelength	Effect of ozone layer	Danger
UV A	longest	no effect	milder effects compared with UV B
UV B	medium	partially absorbed by ozone layer	can cause severe sunburn and cell damage
UV C	shortest	completely absorbed	very dangerous – will cause severe cell damage

C Some dangers of different types of ultraviolet radiation.

The ultraviolet region of the electromagnetic spectrum can be divided into three different bands, depending on wavelength or frequency. The ozone layer (part of the atmosphere) absorbs each of these bands of light produced by the Sun. Ozone is a form of oxygen and is easily damaged by certain pollutants such as CFCs.

Ultraviolet radiation can **ionise** atoms. It can knock electrons out of atoms in a similar way to X-rays and gamma-rays. In cells this can lead to destruction or mutation and ultimately to cancer.

2 Waste chemicals, such as CFCs in old fridges, can reduce the ozone layer. What risk to health will a reduction in the ozone layer lead to?

3 Which type of ultraviolet radiation is most dangerous to our health?

4 Name one type of wave which might harm us.

5 Why would you expect X-rays to be potentially more damaging to human tissue than radio waves?

6 Some forms of granite rock produce a radioactive gas called radon. This can lead to increased background radiation levels in some buildings. Suggest how this problem might be reduced.

7 Why will a dentist only take an X-ray of your jaw once every 2 years unless you need emergency treatment?

8 Copy and complete this table:

Electromagnetic radiation	Risk
UV A	premature ageing of skin
X-rays	
	can cause skin burns

Higher Questions

Digital information

By the end of these two pages you should be able to:

- describe the advantages of sending information in the form of a digital signal compared with analogue
- describe how optical fibres use total internal reflection of light waves to transfer information over long distances.

Have you ever wondered:

Why is the picture better on a digital TV?

A A needle picks up vibrations in the grooves of an LP.

Vinyl records store music in **analogue** form. Analogue means that a continuously varying signal is produced when the record is playing. This is amplified and sent to a loudspeaker. The advantage of analogue is that it uses relatively cheap and simple equipment.

About 150 years ago a very simple system was invented to communicate across large distances. It used an electric switch in one location connected to a buzzer at another location. The switch could be tapped quickly – *dot* – or held for a moment – *dash*. All letters in the alphabet were represented as different combinations of dots and dashes. The letter S was dot dot dot. Messages could be sent between towns down electric wires. A telegrapher would code and decode messages. This was called Morse code.

1 Give two disadvantages of recording music on vinyl records.

2 Give two disadvantages of Morse code as a system of communication.

Morse code is a **digital** system of coding information. It uses two codes, *dot* or *dash*. Modern digital communication systems use two states – *on* (1) or *off* (0) – or bits. Digital signals are processed directly by computers. The signal quality is better than analogue. Interference can still affect a digital signal but the decoding system only has to interpret 0s and 1s. Digital information can be sent using electric wires but it can also be sent using anything that can be switched on and off, or turned low and high.

An effective way of sending a great deal of digital information very fast is to turn the ons and offs into a flash of light (on) and no light (off). Light can be trapped in an **optical fibre** and sent hundreds of kilometres from one town to another. This is how most TV and telephone systems work.

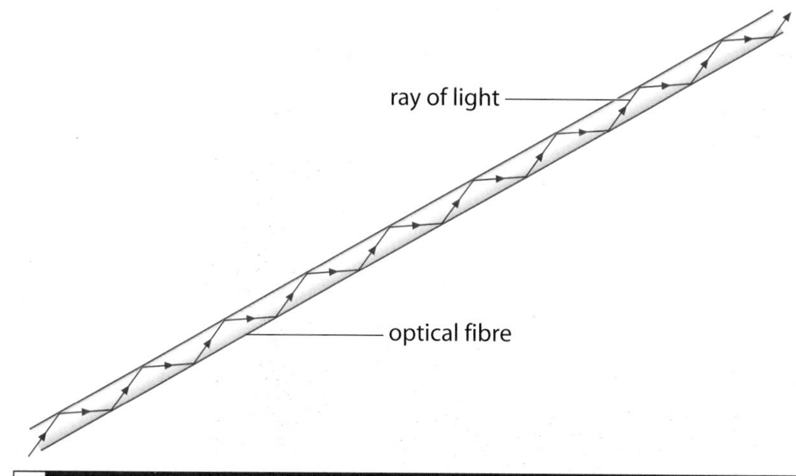

B Light is totally internally reflected off the inside edge of a glass fibre.

Light rays can be made to reflect off the inside edges of an optical fibre. This is called **total internal reflection**. You may have seen Christmas trees made from optical fibres carrying light to each end. Light is then transmitted along the fibre. Very little light is absorbed by the fibre so it can travel several kilometres before it needs **amplification**.

4 In Morse Code the letter O is three dashes and S is three dots.
  a Translate this Morse Code symbol: • • • – – – • • •
  b Is Morse Code digital or analogue? Explain your answer.

5 Explain what is meant by an analogue signal.

6 Give two advantages of digital coding compared with analogue coding.

7 **a** Explain how an optical fibre works.
  b Explain why the quality of music downloaded using optical fibre communication is high.

3 A neighbour lives just out of earshot but within sight. You both own torches. What simple digital system would allow you to send simple messages?

The digital revolution

By the end of these two pages you should be able to:

- discuss how digital signals have changed the way music can be made, distributed and listened to
- H discuss the benefits and drawbacks to society of a technology that is based on the properties of waves.

Have you ever wondered:

Why do CDs store music using a digital format?

A Music storage has changed over the past 30 years.

The way we store music has changed over the last century. Notice how the size has reduced. The cost has also changed in real terms. The cost of a record in the 1970s was the equivalent of working a Saturday at a supermarket. The equivalent cost of a CD is less than half of that.

Methods of reducing the amount of memory required have resulted in different digital formats. MP3 is an example and requires about 4 MB per track. Music files can be downloaded using the Internet. You can pay to download music tracks from some websites.

Recording companies now sell individual tracks for downloading rather than a full album. This means you can pick the tracks you like and only pay for those. Electronic synthesizers can reproduce the sounds of real musical instruments. They do this by generating and mixing sound waves together so that the result sounds almost exactly the same as a particular instrument. Synthesizers use digital code to recreate the effect of an analogue signal.

1 Divide each method of music storage shown in figure A into digital or analogue.

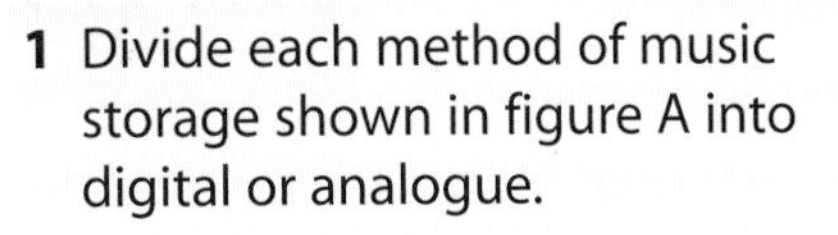

2 A compact disc has a typical memory capacity of about 700 megabytes (MB). Each audio music track requires about 35 MB. How many tracks would you expect to find on a compact disc?

3 A digital music player is small enough to fit into your pocket and has a memory capacity of 20 GB. Calculate the number of MP3 tracks it can store. 1 GB is equal to 1000 MB.

B

Our leisure has changed because of waves.

C

In the Victorian age the only waves being produced were the sound waves of our voices or from a piano. People and families made their own entertainment and talked about things together. We are now surrounded by electronic devices that can process information and entertain us. We spend more time doing our own activities, without other people. Games that once needed another person to play against can be programmed. A computer can take the place of a human opponent.

4 Write a list of all the activities people do now compared to 100 years ago because of waves. You can start with 'listen to the radio'.

5 Cassette tapes stored music in analogue form. State two disadvantages of cassettes compared with CDs.

6 Digital signals can be processed using computers. Describe one use of the computer for audio purposes.

7 A typical phone call exchanges information at a rate of 64,000 bits per second. The latest technology allows about 2×10^9 bits per second to be transmitted along an optical fibre. Calculate the number of different phone calls that can be carried by just one fibre.

8 Explain how the digital revolution has changed the way music is made, heard and sold.

Summary Exercise

Higher Questions

Questions

Multiple choice questions

1 Ultrasound can be used for scanning
 A the iris.
 B forged banknotes.
 C bone fractures.
 D a foetus during pregnancy.

2 Microwaves can be used to monitor rain because
 A rain absorbs microwaves.
 B microwaves can penetrate rain.
 C rain reflects microwaves.
 D rain emits microwaves.

3 Our bodies are usually warmer than their surroundings. This means our bodies will emit
 A X-rays.
 B infrared radiation.
 C ultrasound.
 D ultraviolet radiation.

4 Which list shows types of radiation in order of increasing frequency?
 A ultraviolet light, radiowaves, visible light, X-rays
 B radiowaves, infrared radiation, visible light, X-rays
 C X-rays, visible light, microwaves, radiowaves
 D visible light, X-rays, gamma-rays, radiowaves

5 Which of the following properties do all the waves in the electromagnetic spectrum have in common?
 A They are all longitudinal waves.
 B They all have the same wavelength.
 C They all travel at the same speed in any material.
 D They are all transverse waves.

6 Information can be sent as bits. This format is known as
 A frequency.
 B digital.
 C analogue.
 D transverse.

7 The diagram shows a wave. Which letter correctly shows the wave's amplitude?

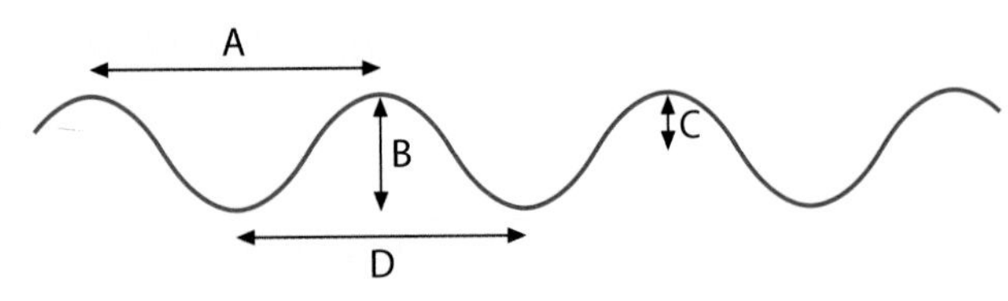

8 The frequency of a dipper in a ripple tank is 5 Hz. The wavelength of the water waves generated is 2 cm. The speed of the waves is
 A 0.4 cm/s.
 B 2.5 cm/s.
 C 10 cm/s.
 D 5 cm/s.

9 Which of the following is the correct definition of transverse waves?
 A The oscillations are parallel to the direction of the wave.
 B The direction of the wave is at right angles to the wave.
 C The oscillations of the wave are at right angles to the wave direction.
 D The direction is parallel to the oscillations.

10 Which of the following properties of seismic S waves is incorrect?
 A They are created at an epicentre.
 B They cannot travel through liquids.
 C Their speed depends on the density of the material in which they are travelling.
 D They are longitudinal.

11 The speed of light in a vacuum is 3×10^8 m/s. The distance between the Sun and Earth is about 1.5×10^{11} m. The time taken to travel from the Sun to Earth is
 A 500 seconds.
 B 0.002 seconds.
 C 36 hours.
 D 23 minutes.

12 Which of the following statements is correct?
 A Infrared radiation can cause mutation of cells in the skin.
 B Infrared can cause skin burns.
 C Radiowaves can cause mutation of cells in the body.
 D Visible light can cause destruction of cells in the skin.

13 Waves can change direction at a boundary between one material and another. This is called
 A transverse.
 B reflection.
 C refraction.
 D absorption.

14 A DVD has a memory capacity of about 36.8 Gbit (1 Gbit = 1,000,000 kb). The playing time on a DVD recorder is 120 minutes. The processing rate of the DVD player is about

A 306,666 kb per second.
B 0.31 kb per second.
C 4416 kb per second.
D 5111 kb per second.

15 Tropical rainforests contain some trees that are over 500 years old. Cells taken from the tops of the trees have been shown to contain different DNA to those at the bottom. Which of the following types of radiation could be responsible?

A ultrasound.
B visible light.
C radiowaves.
D ultraviolet light.

Short-answer questions

1 The record number of dominoes that has been set up and then toppled by knocking the first one into the second and so on is about 303,000. Ten seconds after the first domino was toppled the furthest toppled domino was 8 m from the start.

a Explain whether this an example of a transverse or a longitudinal wave.
b Calculate the speed of the 'wave'.
c If the separation between each pair of dominoes was 2 cm and all 303,000 dominoes had been set up in one long row, calculate the time taken for all the dominoes to topple.
d In practice the time taken for all 303,000 dominoes to topple was significantly less than this. Comment on how this could be achieved.

2 Terahertz radiation (or T waves) is a sub-section of the electromagnetic spectrum lying between microwaves and infrared. Terahertz radiation is absorbed by water. If it is not absorbed it will penetrate up to several millimetres of human tissue before reflecting. This has made it an excellent tool for detecting early forms of cancer tissue close to the surface of the body. Cancer tissue has a higher water content compared with ordinary tissue.

a State one property that all waves in the electromagnetic spectrum share.
b Three regions of the electromagnetic spectrum are mentioned in the paragraph. State another region that has a smaller wavelength than those mentioned.
c How does the penetrating power of T waves compare with visible light?
d Explain how T waves can detect cancerous tissue.

3 In old-fashioned cowboy films outlaws would sometimes plot a hold-up of a train to steal loot. To see if the train was coming they would put their ear to the rail of the track and listen.

a What does this suggest about the way sound travels through the metal rail compared with the air?
b The train is 5 km away. How long would it take the sound to travel along the rail? The speed of sound in the rail is 5000 m/s.
c How long would the sound take to travel through the air? The speed of sound in air is 300 m/s.
d What will happen to the difference between these two times as the train gets nearer to the outlaws?

4 A digital camera has the following specification: 3.2 million pixels, 128 MB memory card, 0.8 MB per picture.

a Explain what is meant by 'digital'.
b Calculate the maximum number of pictures that can be stored in the memory.
c A full memory card of pictures is going to be downloaded into a computer at a rate of 1.2 Mbits per second. Calculate how long this will take. 1 MB = 8 Mbits.
d The pixel number describes how many tiny segments make up the image. Give one advantage and one disadvantage of a high pixel number.

Glossary

absorption When some of the energy of a wave is transferred to the material in which it is travelling.

amplification The increase in size of a signal.

amplitude The maximum distance of particles in a wave from their normal positions.

analogue A signal sent in the form of a wave.

density The mass of a unit volume of material.

digital A signal which has two levels or states.

electromagnetic spectrum A group of waves that all travel at the same speed in a vacuum and are all transverse.

emission The production and sending out of waves.

epicentre Centre or source of an earthquake.

fluorescent A substance which absorbs ultraviolet light and emits visible light.

frequency The number of oscillations of a particle in 1 second.

gamma-rays Part of the electromagnetic spectrum. They are radiation with the highest frequency.

infrared Part of the electromagnetic spectrum usually associated with heat.

ionise To knock electrons out of an atom. Ultraviolet light, X-rays and gamma-rays are forms of ionising radiation.

iris The coloured part of a person's eye; the black hole in the centre is known as the pupil.

iris recognition A way of identifying a person by the coloured patterns of their iris.

longitudinal A wave that oscillates parallel to the direction of travel.

microwave Part of the electromagnetic spectrum.

mutation When the DNA of cells is altered.

optical fibre A thin glass or Perspex rod which carries light.

oscillation The complete movement of a particle from its starting point and back.

penetrate When waves pass into a material.

P wave A longitudinal wave produced by an earthquake.

radiation Energy transferred from a point.

reflection When the wave bounces off a boundary between materials.

refraction The change of speed and direction of a wave when it enters a new material.

scanning Building up an image of something in stages.

seismic wave A wave produced by an earthquake.

speed of a wave Calculated using distance covered divided by time taken.

S wave A transverse wave produced by an earthquake.

total internal reflection Light reflecting off the inside edge of a material rather than refracting outwards.

transverse A wave that oscillates at right angles to the direction of travel.

ultrasound High-frequency sound waves which we cannot hear.

ultraviolet Part of the electromagnetic spectrum.

vacuum A volume of space in which there is no matter.

wave Describes the way energy can be transferred by oscillations.

wavelength The distance between two points in a wave which have moved a maximum distance from their normal positions.

X-rays Ionising radiation of high frequency and part of the electromagnetic spectrum.

Space and its mysteries

A President Kennedy inspired the USA to land a person on the Moon.

At the beginning of the 1960s, President Kennedy inspired the USA to achieve the nearly impossible. The following is an extract from some of the speeches he made: 'I believe that this nation should commit itself to achieving the goal, before this decade is out, of landing a man on the Moon and returning him safely to the Earth. We choose to go to the Moon in this decade, not because it is easy, but because it is hard.'

The USA now intends to place an astronaut on Mars. If you are planning a trip to another planet you need to know how long it might take, what the journey will be like and what the place will be like when you get there.

In this topic you will learn:

- that planets in our solar system have different characteristics
- about the formation and evolution of the universe and its stars
- the requirements for travelling in space and taking holidays on different planets
- how we explore the universe and the benefits this can bring.

State whether you think each of the following statements is true or false.

- The distance to the Moon is about eight times the distance around the Earth.
- The Apollo space trips to the Moon and back took about seven days.
- There is no atmosphere on Mars.
- The centre of the universe is the Sun.
- The Sun has always been there.

A weighty problem

By the end of these two pages you should be able to:

- explain the role of gravity on Earth
- explain the difference between mass and weight
- use the equation: weight = mass x acceleration of free-fall
- use the unit of gravitational field strength, newton per kilogram.

A Astronauts weigh less on the Moon than they do on Earth.

Gravity is the force of attraction from the Earth that holds you on the ground. This force is called your **weight**. Your weight is six times greater on Earth than it would be on the **Moon**. Gravity affects our lives on Earth. Gravity forces the water out of a tap and it makes raindrops fall to the ground. To walk and run we have to overcome the force of gravity. Gravity traps our **atmosphere** and holds it close to Earth.

1 List four activities that would be difficult without gravity.

In the seventeenth century, Sir Isaac Newton realised that the force of gravity acted between objects with **mass**. We all attract each other but the force is very small. The force between an object and the Earth is large because the Earth is so large. Your mass remains exactly the same wherever you are. It is a measure of how much there is of you. Mass is measured in kilograms (or kg). Weight is a force; it is measured in newtons (or N).

B On the Moon, objects have the same mass but much less weight.

2 A person's mass is 60 kg on Earth. How much is their mass on the Moon?

3 A person's weight is 600 N on Earth. How much is their weight on the Moon?

C David Scott dropping a hammer and a feather on the Moon.

If a feather and a hammer are dropped at the same time and from the same height, they will hit the surface of the Moon at the same time. Each object falls with the same **acceleration** because of gravity. This acceleration is called acceleration due to free-fall. On the Moon it is about 1.6 m/s^2. This means that the speed of a falling object increases by 1.6 m/s each second. The acceleration due to free-fall is also referred to as the **gravitational field** strength. The units can be expressed as m/s^2 or N/kg.

The Earth pulls with more gravity than the Moon because it is much larger than the Moon.

4 The acceleration of an object as it falls to the surface of Earth is 10 m/s^2.
  a Calculate its speed 3 s after it was dropped.
  b State the gravitational field strength at the surface of the Earth.

You can calculate the weight of an object with this equation.

weight = mass × acceleration due to gravity, or $W = mg$

The mass of a person is 60 kg. The acceleration due to gravity on the Moon is 1.6 m/s^2. The weight of this person on the Moon can be calculated using weight (W) = mass (m) × acceleration due to gravity (g): $W = 60 \times 1.6 = 96$ N. Remember: weight is measured in newtons (N).

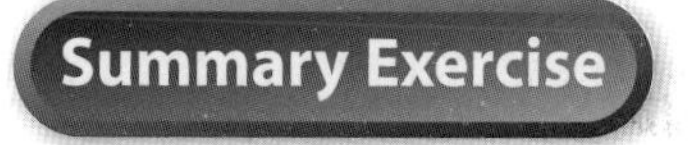

5 A pupil writes: 'The mass of an object is a force measured in kg.' Underline and explain one incorrect word.

6 Calculate the weight of an astronaut of mass 90 kg on the surface of Venus. Assume that the gravitational field strength on the surface of Venus is 9 N/kg.

7 Explain the difference between mass and weight. You should include the terms gravity, force and gravitational field strength in your answer.

Higher Questions

P1b.12.2

Our place in the universe

By the end of these two pages you should be able to:

- describe the solar system as part of the Milky Way galaxy and discuss how this is related to other galaxies and the universe
- explain the role of gravity in astronomy
- compare the relative sizes of Earth, our Moon, the planets, the Sun, galaxies and the universe, and the distances between them.

The **planets** move in **orbits** around the **Sun**. Planets do not produce their own light. They reflect light from the Sun. There are also **asteroids** and **comets**. This is known as the **solar system**. Our Sun belongs to a huge group of **stars** known as the Milky Way **galaxy**. Our galaxy is spiral-shaped, but we see it edge on. The galaxy is spinning around a centre. The Milky Way is one of millions of galaxies which make up the **universe**. Galaxies can be different shapes compared to the Milky Way.

1 Within the universe many things spin or orbit. Starting with the Earth, which spins on its axis, list any other examples of spinning or orbiting.

The Moon has an almost circular orbit of about 400,000 km around the Earth. 400,000 km is 400,000,000 m and can be simply written in standard form: 4×10^8 m. The last astronauts on the Moon were Eugene Cernan and Harrison Schmitt in 1972. They spent 3 days on the Moon and about 5 days getting there and back. Gravity keeps the Moon in orbit around the Earth. Gravity keeps the solar system orbiting the centre of the Milky Way.

2 The distance to Venus from Earth at their closest is about 40,000,000,000 m. Write this in standard form.

A Our place in the universe.

The nearest planet to Earth is Mars. The planets are all moving in orbits around the Sun. The further a planet is from the Sun the longer it takes to complete its orbit. The Earth takes 1 year. Pluto is about 6×10^{9} km away from the Sun. It takes Pluto about 250 years to follow an elliptical orbit around the Sun.

Object	Description	Distance	Time taken to orbit
Earth	our home planet	150,000,000 km from the Sun	365 days to orbit the Sun
Moon	about 100 times less mass than the Earth	385,000 km from the Earth	27 days to orbit the Earth
Venus	slightly less mass than the Earth	108,000,000 km from the Sun	225 days to orbit the Sun
Mars	about 10 times smaller than the Earth	228,000,000 km from the Sun	687 days to orbit the Sun
Saturn	about 100 times more massive than the Earth	1,400,000,000 km from the Sun	10,750 days to orbit the Sun
Pluto	about 500 times smaller than the Earth	5,870,000,000 km from the Sun	90,600 days to orbit the Sun
Proxima Centauri	nearest star to our Sun	39,900,000,000,000 km from the Sun	–
centre of Milky Way galaxy	possibly a black hole	35,000 light years from the Sun	–
edge of observable universe	the light from objects that are the furthest we can see with our telescopes	14,000,000,000 light years from the Sun	–

B Useful space data.

When astronomical distances become very large it is easier to measure them in light years. This is the distance that light travels in 1 year. It is equal to 9.5×10^{12} km. To calculate the distance to Pluto from the Sun in light years, divide the distance to Pluto by the distance that light travels in a year: distance in light years $= 6\times10^{9} / 9.5\times10^{12} = 0.00063$ light years. It takes light 0.00063 years to travel from the Sun to Pluto.

3 The Moon is very much smaller than a star. Why does the Moon appear much brighter and larger than the next-nearest star to someone on Earth?

4 How many stars are there in the solar system?

5 It took about 2.5 days to reach the Moon from Earth, a distance of 400,000 km. The distance to Pluto is about 6×10^{9} km. How long will it take, in days, if you travelled there at the same speed?

6 A student suggests that the Moon is a planet. Explain why this is incorrect.

7 Put the following objects in order of increasing size: Milky Way galaxy, universe, Pluto, the Sun, Earth, the Moon.

Comets and asteroids

By the end of these two pages you should be able to:

- explain the role of gravity in the solar system
- describe how the orbit of a comet differs from that of a planet or an asteroid
- discuss the risks of a global catastrophe such as a comet hitting the Earth.

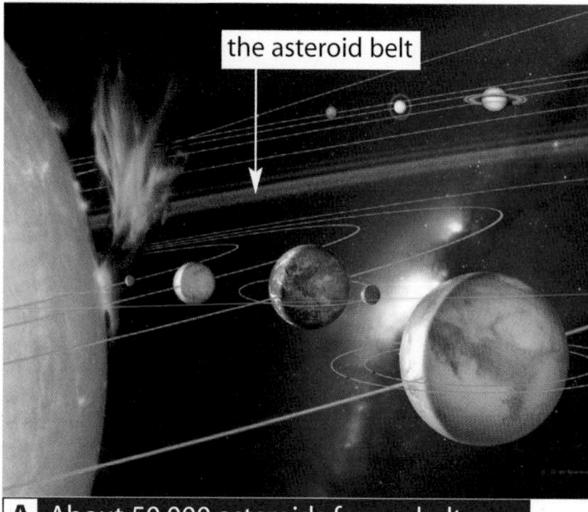

A About 50,000 asteroids form a belt between Mars and Jupiter.

Have you ever wondered:

The risk of dying from an asteroid impact is the same as being in an air crash. How can this be?

The gravity holding us on the ground is the same as that keeping the Moon in orbit around the Earth. Gravity attracts all the objects in the solar system towards the Sun and stops them flying away. Understanding the gravitational forces between planets and the Sun helps us to predict their motion.

1 Write down the numbers 1 to 9 and name each planet starting with the nearest one to the Sun.

2 Suggest what force holds a TV-broadcasting satellite in orbit around the Earth.

3 Explain why space travel beyond Mars has a very high risk of failure.

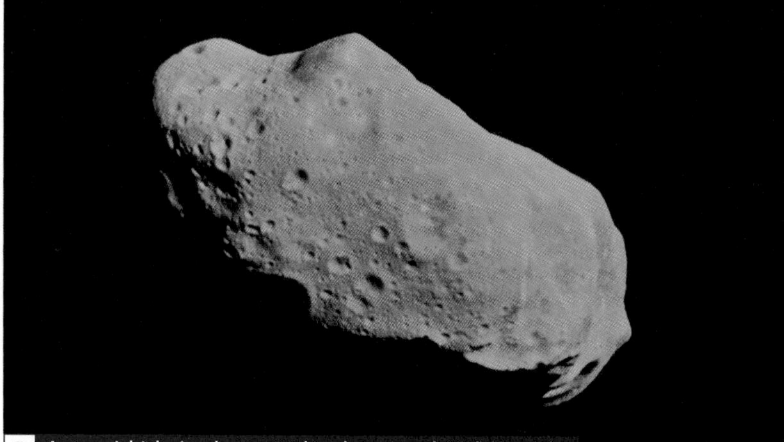

B Asteroid Ida is about 50 km long and 20 km wide.

An asteroid is a lump of solid rock tens or hundreds of kilometres across. Asteroids are left over from when the solid planets – the four closest to the Sun – were formed. The dinosaurs were probably wiped out by a large asteroid 65 million years ago. The impact would have thrown debris into the atmosphere and blocked the heat from the Sun. Those animals not killed by the impact would have died from cold.

Astronomers estimate that a medium-sized asteroid will collide with Earth every two or three hundred years. A collision in Russia in 1908 destroyed uninhabited forests over 2000 km^2. A large asteroid called 1950 DA has a 1 in 300 chance of hitting Earth in 2880. A collision with this asteroid would be devastating and would cause major climate change.

A comet is large ball of ice and dust. Comets are often several kilometres across. Planets have elliptical orbits around the Sun but comets have eccentric elliptical orbits. Sometimes a comet's orbit is very eccentric – or 'squashed'. Halley's comet follows a very eccentric orbit of about 75 years.

D A very eccentric orbit is more elliptical.

4 State the difference between asteroids and comets.

5 What is meant by a very eccentric orbit?

If a comet passes close to a planet it can change course. This makes it difficult to predict its precise course. Comets can become visible as they pass near to us. A 'tail' of debris stretches out behind the main comet.

In 1994 comet Shoemaker–Levy was tugged out of its orbit around the Sun by Jupiter. It caused an enormous explosion which would have been completely devastating had it happened on Earth. Because of its enormous size Jupiter attracts dangerous objects, but the Moon also shows evidence of many collisions.

C Comet Hale–Bopp.

6 Complete the sentence: Solar system objects are kept in their orbits around the Sun by … .

7 Suggest why the impact of a large asteroid or comet would cause a very long winter.

8 'Asteroid 1998 DA might collide with Earth in 2015. Its orbit cannot be predicted accurately but it may come within 800,000 km of Earth. This medium-sized asteroid would devastate the immediate area.' Comment on whether the public should worry.

Space travel

By the end of these two pages you should be able to:

- describe what the conditions are for astronauts in space
- discuss how scientists plan to overcome the problems of long space flight
- H describe how spacecraft are designed to cope with conditions in space.

A An astronaut 'floating' in space.

The USA wishes to send an astronaut to Mars. But **interplanetary** space is very hostile. The Earth's gravity will decline as the spaceship moves away from Earth into interplanetary space. The astronaut will float freely. In space there is no air to breathe. The **temperature** can vary widely between freezing cold in the absence of sunlight to very hot in the full glare of the Sun. Spacecraft have to provide astronauts with a comfortable environment.

1 State the value of gravitational field strength in the absence of gravity.

2 In terms of forces, what will happen to astronauts as their spaceship approaches Mars?

When there is very little gravity the muscles and bones of astronauts become weaker. They are not needed to work as hard as on Earth. Astronauts returning after long space flights cannot stand up until their bodies have readjusted properly. The heart also weakens gradually in space, as it doesn't work as hard. Astronauts must exercise on a space flight to reduce these effects.

B After a long time in space, an astronaut is weak.

Astronauts are at risk from cosmic **radiation**. This is usually high-speed protons. These carry a lot of energy and damage human tissue. Continuing damage over a long time can affect an astronaut's health. Cosmic radiation also damages sensitive equipment on spacecraft.

3 Why do you think our bones deteriorate in the absence of gravity?

H

C The international space station.

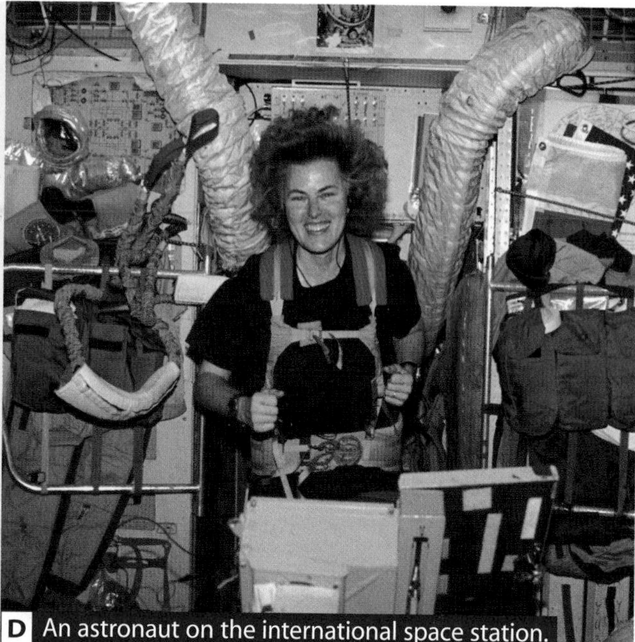

D An astronaut on the international space station.

Astronauts spend months on the space station getting used to life in space. The space station has a temperature-controlled environment. Reflective, silvered metals on the outside prevent the space station from getting too hot. The astronauts' spacesuits protect them from cosmic radiation. The walls of the space station are made of Kevlar™, a high-strength material, which can take the impact of cosmic radiation and even small bits of rock. The space station provides oxygen to breathe. It collects the carbon dioxide you breathe out and vents it into space.

4 List three things which a spacecraft must provide for the survival of astronauts.

In interplanetary space you are practically weightless because there is very little gravity. In outer space you would feel true **weightlessness** as there is no gravity. Your muscles and bones deteriorate if you spend several months in space unless you exercise. We 'feel' weight because the ground pushes upwards on our feet with the same force as our weight pushes down. The space station rotates to give a sense of 'weight'. The astronauts have the 'feeling' of weight because their feet are in contact with the outside walls.

5 What is meant by cosmic radiation and why is it a problem for astronauts?

6 Explain whether it is hot or cold in space.

7 Explain why astronauts need to exercise if they are in space for any length of time.

8 You have been selected for a 1 year trip to Mars. List the problems associated with a space journey of that length and suggest how each problem might be overcome.

Summary Exercise

Higher Questions

Launching into space

By the end of these two pages you should be able to:

- explain how spacecraft might be powered in terms of action and reaction
- recognise how force = mass × acceleration can be used to predict how an object behaves.

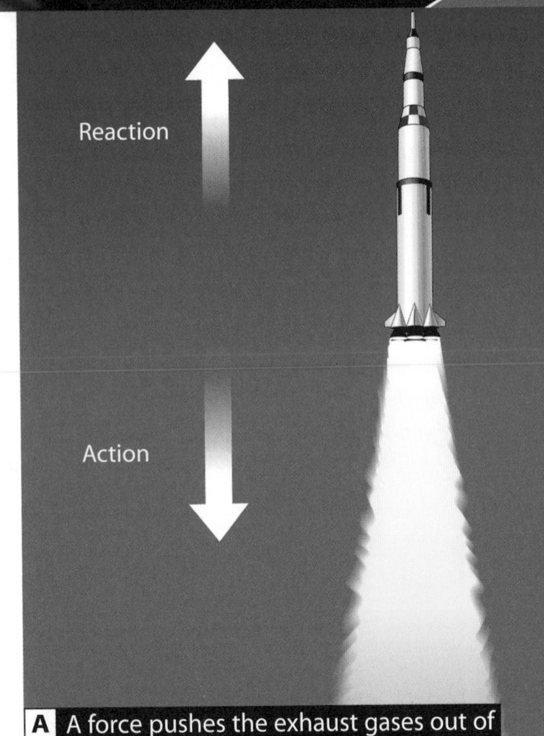

A A force pushes the exhaust gases out of the rocket. An equal and opposite force pushes the rocket upwards.

A rocket usually throws out exhaust gases. The rocket then moves in the opposite direction. The force pushing the exhaust gases out is usually provided by igniting them.

Forces usually come in pairs known as **action** and **reaction**. The force pushing the exhaust gases out is paired with an equal and opposite force pushing the rocket in the opposite direction.

1 Look at the two diagrams below. Copy each diagram and show the two forces with appropriate labelled arrows.

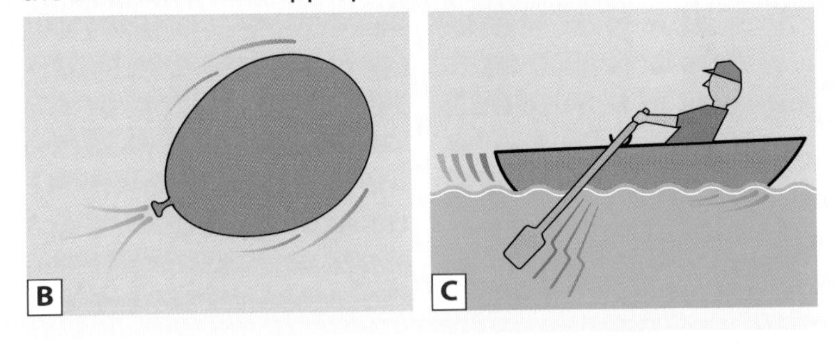

D A shuttle requires enormous forces to launch into space.

The forces involved in a shuttle launch produce an acceleration upwards. The acceleration shoots the shuttle to a speed of about 5000 km/h some 2 minutes after the launch. The formula below describes the link between the overall force and the acceleration of the shuttle. It can be used to predict the speed of the shuttle at various times after the launch. Force is measured in newtons (N).

$$\text{force} = \text{mass} \times \text{acceleration}$$

2 A car of mass 800 kg accelerates at 1.4 m/s^2. Calculate the force producing this acceleration.

3 The acceleration of the space shuttle is initially about 3 g (g is the acceleration due to gravity=10 m/s^2). Calculate the overall force if the mass of the shuttle is about 2,000,000 kg.

E A hammer thrower holds the hammer in orbit and then releases the cord, which allows the hammer to 'escape'.

When the space shuttle is orbiting the Earth, its direction of motion is continuously changing. The change in direction means that its **velocity** is also changing. A change in velocity is an acceleration. The acceleration must be caused by a force. The shuttle is kept orbiting in space around the Earth by the force of gravity acting on the shuttle's weight. Without this force the shuttle would fly off at a tangent into space.

4 State what force keeps the shuttle in orbit around the Earth.

5 Explain the force which acts on the Moon to keep it in an orbit around the Earth.

6 The tides are caused by the gravitational attraction of the Moon. Explain how the forces between the Earth and the Moon are an example of forces coming in equal and opposite pairs.

7 Describe the forces acting on a space rocket during its launch and explain how its acceleration can be predicted.

Higher Questions

Observations of the universe

By the end of these two pages you should be able to:

- describe ways of discovering information about the universe other than travelling there
- show an understanding of how data-logging and remote sensing can provide information about the universe.

Have you ever wondered:

Is it worth £25 billion to put astronauts on Mars, when we could just send robots?

A Two Viking spacecraft landed on the surface of Mars in 1976. This remarkable photograph showed the red landscape of Mars close up for the first time.

B NASA landed a Rover space probe called Spirit on the surface of Mars in January 2004.

We can find out a lot about our solar system by sending space probes. Space probes are like mini laboratories packed with measuring instruments, including cameras. NASA launched two spacecraft in 1977 called Voyager 1 and 2. Their mission was to photograph and transmit information while orbiting the outer planets in our solar system. The Voyager missions are still going on. Both spacecraft are now moving out of the solar system, having travelled about 13 billion km.

In 1976 two spacecraft known as Viking were launched to photograph and then land on the surface of Mars. In 2003 NASA launched two Rover spacecraft, Spirit and Opportunity. Their mission was to land on the surface of Mars. They were to collect rock and soil samples and test them. This was to collect evidence for the presence of water on Mars' surface in the past. By doing this NASA hopes to see if life ever developed on Mars. It also hopes to prepare for human exploration of the planet.

1 List two advantages of sending robotic probes rather than manned spacecraft.

2 What was the NASA Rover mission able to do which the Voyager missions could not?

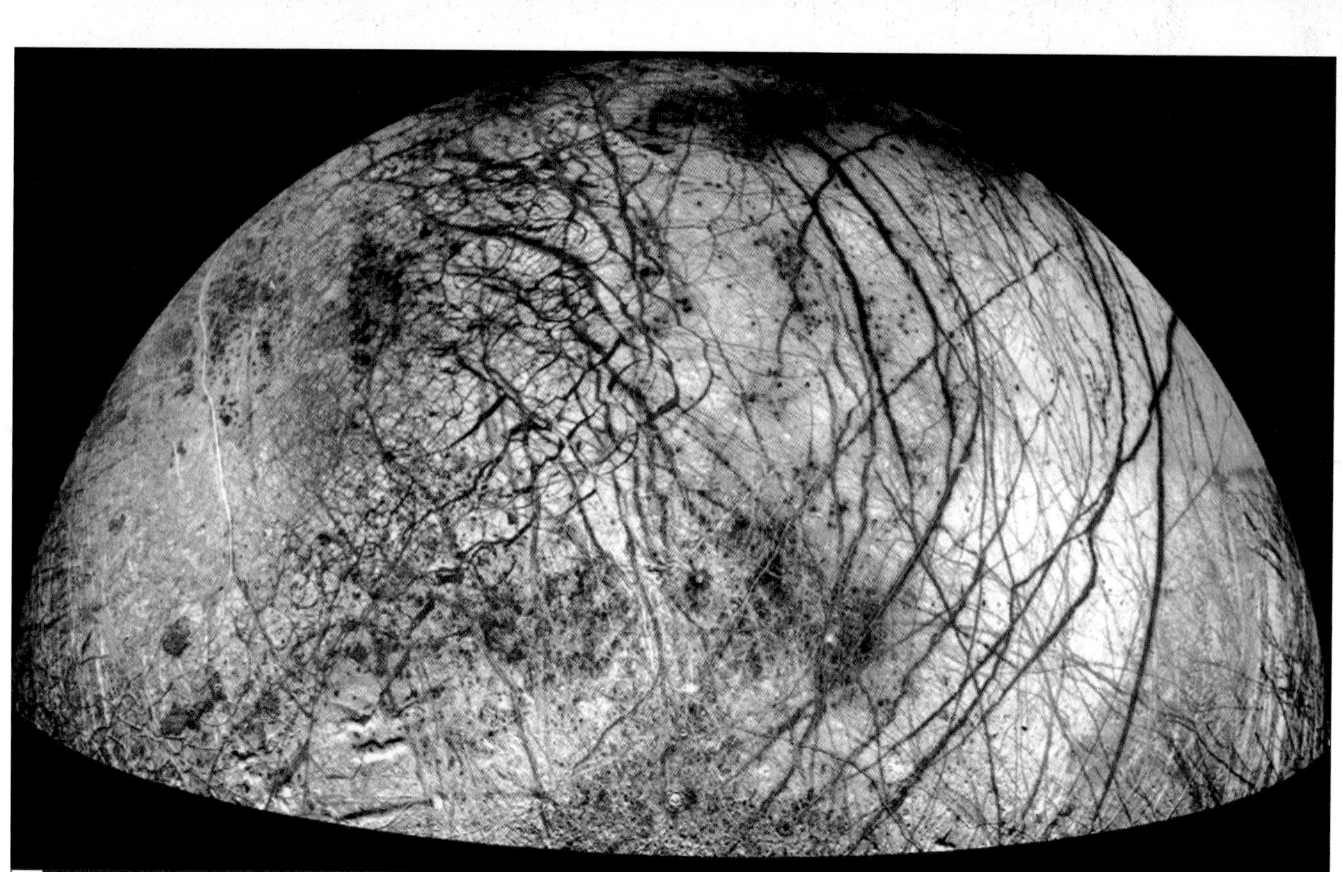

C One of Jupiter's moons, Europa. This shows a thin layer of ice, probably covering an ocean. Scientists believe this ocean has a chance of containing some primitive life.

For several years two spacecraft, Galileo and Cassini, sent back pictures from Jupiter. They produced some startling pictures of Jupiter's moons. Space probes have different types of detector. They can detect visible light just like an ordinary camera. They can also detect other types of radiation such as infrared and X-rays. Some astronomers believe that these moons could offer conditions for basic life forms to develop. NASA communicates with its space probes using radio waves. Radio waves travel at the speed of light.

3 Name one of NASA's space missions.

4 Name two types of electromagnetic radiation that space probes are equipped to detect.

5 Explain why there will be a delay between NASA scientists sending instructions to a space probe and the tasks being carried out.

Measurements can be collected by sensors. For instance, temperatures can be measured by collecting infrared radiation. The measurements provide data. The data are logged in a computer and transmitted back to Earth.

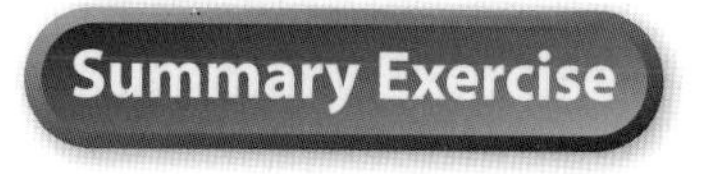

6 Why are scientists interested in whether the surface of Mars once contained water?

7 Infrared detectors have collected information about the temperature of the universe. The temperature of most of space is very cold at about −270°C. Suggest what could cause local areas of space to have much higher temperatures than this.

8 Describe the ways we can find out about space without sending astronauts.

Higher Questions

Searching for life

By the end of these two pages you should be able to:

- describe the Search for Extraterrestrial Intelligence (SETI) mission
- recognise that some scientific questions remain unanswered
- H use scientific evidence to argue for or against the idea that intelligent life exists elsewhere in the galaxy
- propose ways of finding intelligent life
- describe how the existence of life is determined by a planet's position and the age of the star it orbits.

Have you ever wondered:

The universe is full of planets where intelligent life could start, so where is everybody?

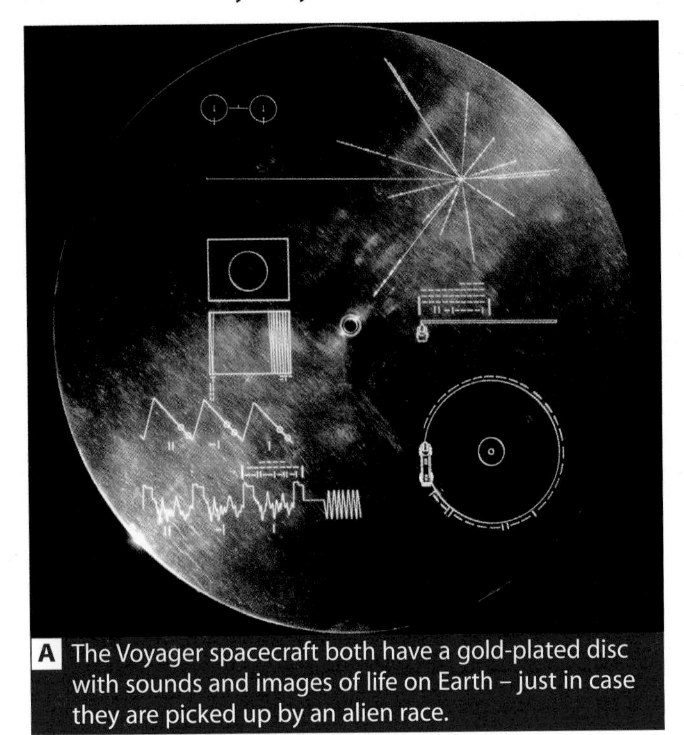

A The Voyager spacecraft both have a gold-plated disc with sounds and images of life on Earth – just in case they are picked up by an alien race.

'Is there anybody out there?' We have been sending radio signals into space for 100 years. If there is another intelligent life form on a planet within 100 light years they might have received our signals. They might have tried to reply. Signals from another civilisation might take a long time and be very weak when they get here. **SETI**, the Search for **Extraterrestrial** Intelligence, is a NASA-sponsored organisation. It listens for radio signals from space with some sort of 'message' attached to them. Signals will travel much faster and further than spacecraft. We are most likely to detect other intelligent life forms from their signals rather by than seeing their spacecraft.

1 Why is there no point, with our current technology, in sending a manned spacecraft to search for alien civilisations?

2 Explain why alien civilisations would need to be within about 100 light years to have received our signals.

The fact that we have not seen any signs of extraterrestrial life does not mean it does not exist. There are billions of stars in the universe. Some must have similar planets to Earth. Alien civilisations may be too far away, they may not have detected our signals or we may not have detected their replies. We do not yet know whether there is anyone else out there.

H

B The Earth has just the right conditions for life to survive.

There are an estimated 100,000,000,000 stars in the Milky Way. There may be other planets orbiting stars elsewhere that could support life. Earth is the right distance away from the Sun to have temperatures that can support life. The Earth is solid and has an atmosphere. The atmosphere traps liquids, particularly water, on the surface. Water is vital for life. Clouds are made up of liquid water and ice crystals because they condense from water vapour in the air. Our Sun has been in its current state for about 5 billion years. Life on Earth has probably been able to develop and evolve to its current state over a period of between 4 and 5 billion years.

3 Write a short paragraph arguing for or against the existence of extraterrestrial life.

4 Describe one sound and one image which you would try to transmit to summarise life on Earth.

5 The original SETI programme listened for radio waves with a regular pattern. Why?

6 Explain why the signals coming from intelligent life elsewhere in the universe are likely to be very weak.

7 List two reasons why life has not developed on
a the Moon.
b Mars.

Life of a star

By the end of these two pages you should be able to:

- describe the life cycle of a star
- H explain the role of gravity in astronomy, including the idea of black holes.

Our Sun is a typical star. Some stars are much brighter and some are much older. The coolest part of the Sun is its surface, with a temperature of about 6000°C. The Sun is about 5 billion years (5,000,000,000 years) old. The Sun's diameter is about 100 times greater than the Earth's.

A Different stages in the life of an average star.

The Sun was created by gravity slowly drawing in atoms of hydrogen and helium. A cloud of helium and hydrogen gas is called a **nebula**. As more atoms collect, the temperature increases. This collection of hot matter is called a protostar. When the temperature reaches about 10^7 °C hydrogen atoms collide violently, combining to form helium. This fusion produces a vast amount of heat energy. This is the first stage of **stellar** evolution. Our Sun is at this stage.

When stars have used up most of their hydrogen atoms they fuse helium atoms into carbon. The star throws out unused hydrogen. Its surface cools and appears red. This will happen to our Sun about 5 billion years from now. This stage of the star's life is called a **red giant**. Red giants are very large.

When most stars have run out of helium they contract into a ball of hot carbon surrounded by a halo of unused hydrogen and helium gas. The hot carbon is called a **white dwarf**. It gradually cools and fades from sight. The envelope of unused gas is called a planetary nebula.

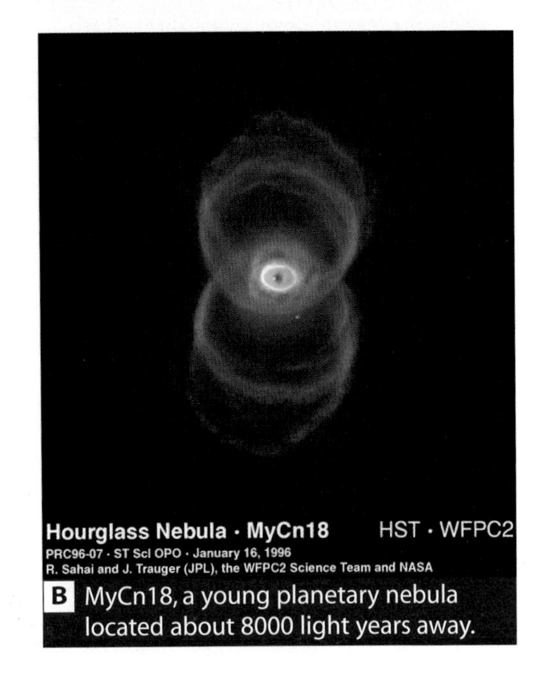

B MyCn18, a young planetary nebula located about 8000 light years away.

1 What causes the hydrogen and helium atoms to collect together to form a protostar?

2 What is the typical surface temperature of an average star?

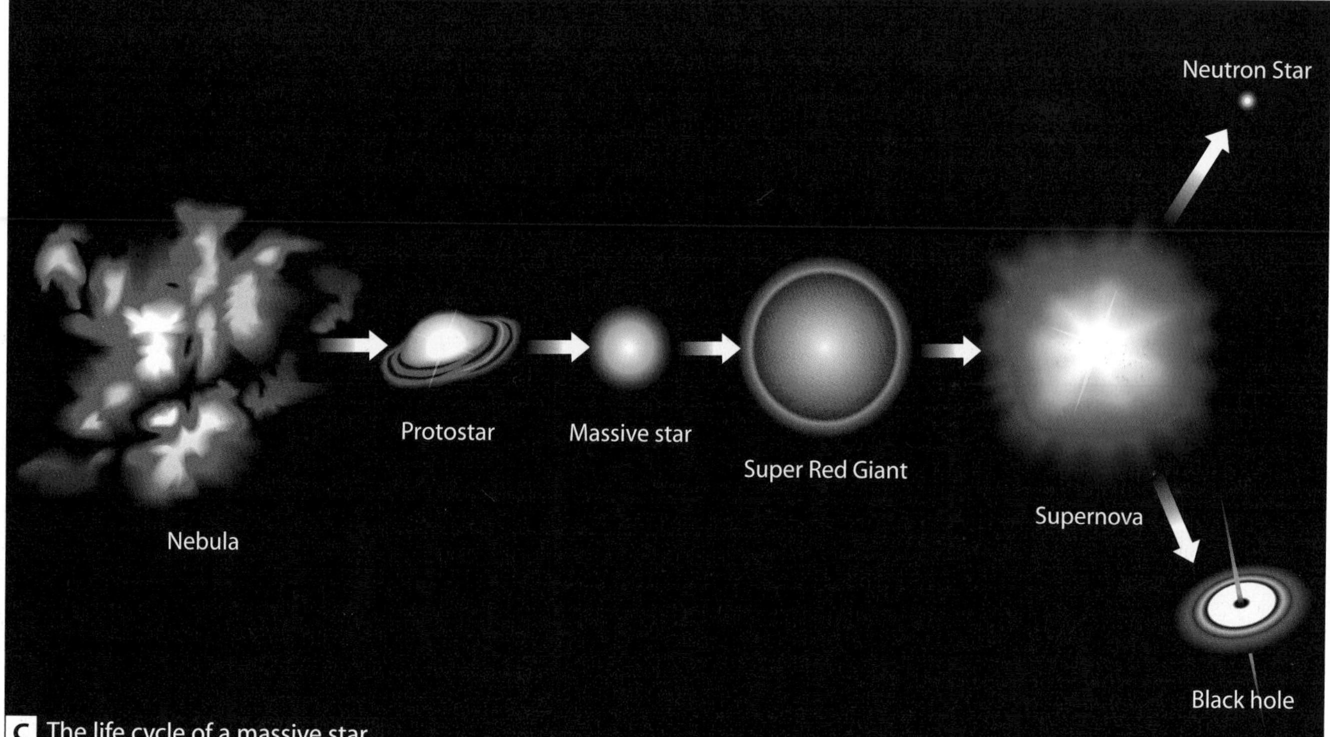

C The life cycle of a massive star.

Massive stars do not become white dwarves. They carry on fusing carbon atoms to form even heavier atoms. This needs much higher temperatures than the fusion of hydrogen into helium. These temperatures are only available if the star is very big. When these stars finally stop fusing atoms together enormous gravitational forces rapidly pull the star inwards on itself. There is a massive explosion called a **supernova**. Much of the star's contents are thrown outwards into space. In the heat and pressure of the explosion some heavier elements are formed.

Have you ever wondered:

How do we know black holes exist when they're completely black?

H Gravity causes a massive star to explode in a supernova. Sometimes this produces a neutron star. One cubic centimetre (the size of a sugar cube) of this material would have a mass of 100 million tonnes. Alternatively, what is left behind is so dense it has a gravitational field that can trap light. This is called a **black hole**.

3 How is a protostar different to a star?

4 What is a planetary nebula?

5 What is a supernova?

6 Explain why red giants are usually quite easy to see compared with ordinary stars.

7 On Earth there are plenty of heavier atoms such as iron. Suggest what must have happened in this region of space at some point in the past.

8 Sketch a diagram showing the various stages in the life of an average star like our Sun.

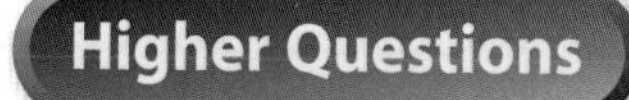

P1b.12.9

The universe: past, present and future

By the end of these two pages you should be able to:

- describe the origin, current state and fate of the universe
- describe the different scientific theories about the universe
- H explain the supporting evidence for these theories.

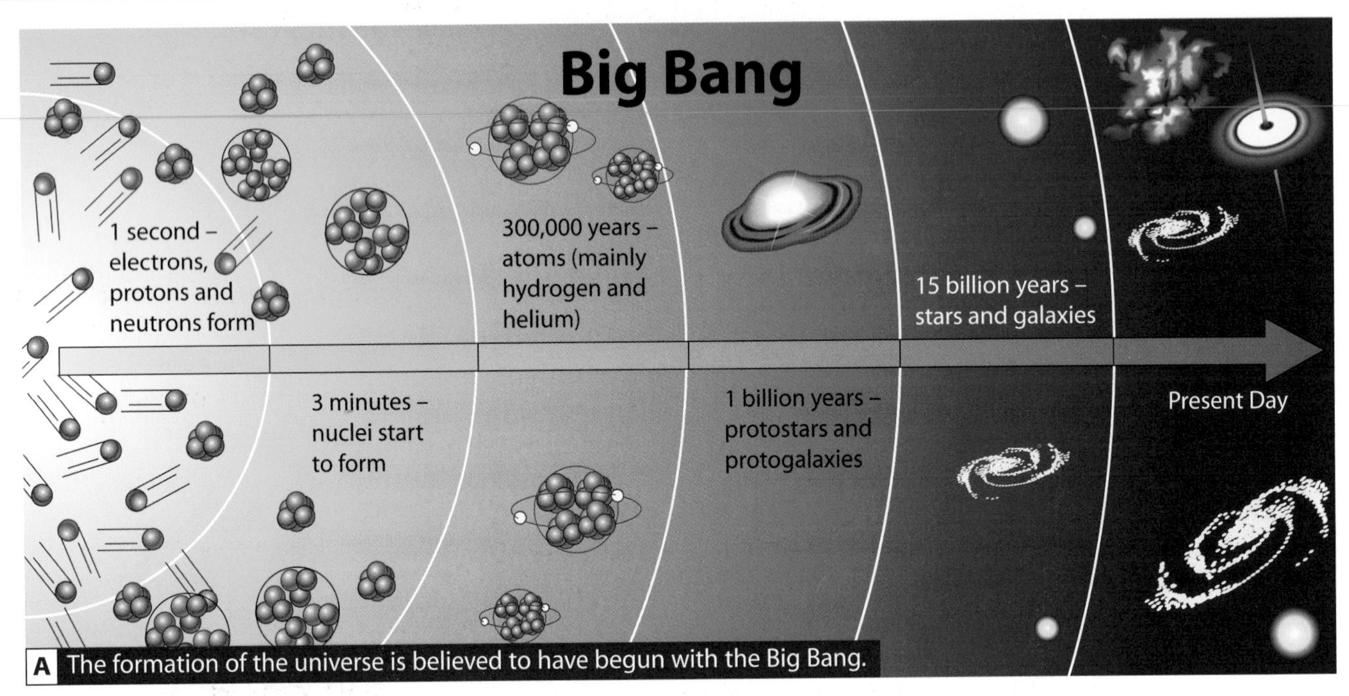

A The formation of the universe is believed to have begun with the Big Bang.

There are several theories of the universe. The **Big Bang** theory states that an enormous amount of concentrated energy and matter appeared suddenly. Nothing existed before this moment. The universe started off very small and has expanded ever since. Gravity caused stars to form by attracting together hydrogen and helium atoms. Gravity is also gradually slowing down the expansion. The future is uncertain. The universe may carry on expanding, stop expanding or it might contract and eventually implode – the big crunch.

1 According to the Big Bang theory, what was there before the Big Bang?

2 Explain why gravity is responsible for creating stars.

Big Bang followed by expansion of the universe

contraction of the universe followed by a big crunch

a new Big Bang marks the birth of the next universe, and the cycle repeats

B The oscillating theory of the universe.

Some people find it difficult to visualise that there was nothing before the Big Bang. The **oscillating theory** of the universe suggests that the creation of this universe followed the death of the last universe. The cycle repeats itself, with a series of universes, one after the other.

3 Explain what is meant by an oscillating universe.

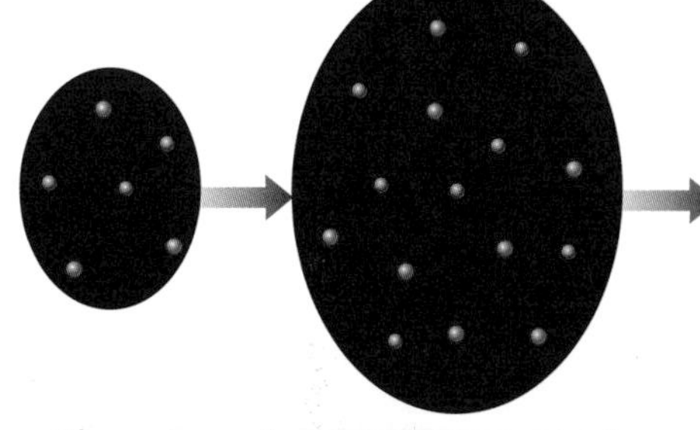

The universe is expanding constantly and a supply of hydrogen and helium atoms fills in the spaces.

C The steady-state theory of the universe.

In the **steady-state theory** the appearance of the universe has never changed and never will. It assumes that the universe is expanding but does not accept that the matter in the universe occupies an ever-increasing volume. It states that as the universe expands a supply of hydrogen and helium atoms fills in the spaces. The overall result of this is that the same amount of matter would occupy a particular volume at any time.

4 Density measures how much matter there is in a particular volume. Explain why the density of the universe remains roughly constant in the steady-state theory.

H Edwin Hubble showed that distant galaxies were moving away from us and each other. The colour of light from these galaxies was shifted towards the red end of the visible spectrum. This is called **red shift**. For instance, you might expect to see the colour blue but instead see yellow, which is nearer the red end of the spectrum than blue. Red shift happens if a light-emitting object is moving away from you. The further a galaxy is away from us, the faster it is moving. The light from very distant galaxies is red-shifted more. Hubble's observations are evidence for an expanding universe.

Penzias and Wilson were collecting radio signals. They noticed that they were picking up microwaves which were coming equally from everywhere in space. Microwave energy is what is left as the universe cools down from once being very hot. It is like going into a room a long time after the heating has been turned off. There will still be a little warmth left.

The black lines can be seen in the light from the Sun.

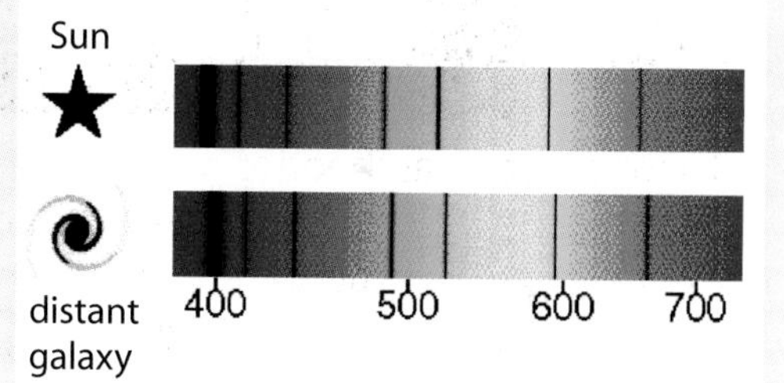

When a distant galaxy is moving away, the black lines in its spectrum shift towards the red end.

D Red shift.

5 State two differences between the Big Bang and the steady-state theories of the universe.

6 Explain whether the oscillating universe is an extension of the Big Bang or the steady-state theory.

7 Describe and compare two theories of the universe, explaining how each addresses its origin, current state and fate.

Higher Questions

Endless discoveries

H ***By the end of these two pages you should be able to:***

- recognise scientific questions which remain unanswered, such as the nature of the 'dark matter' that makes up much of the universe's mass
- discuss the benefits to society, economics and technology of exploring the universe.

Have you ever wondered:

Do physicists really have no idea what most of the universe is made from?

What we know about the universe is based largely on the light that **telescopes** have collected. We have theories to suggest how the universe is evolving, how a star is born, how galaxies and planets formed and whether the universe might end. These theories aren't necessarily correct. Observations may suggest that a theory is wrong. People like to think and suggest possible alternatives. The steady-state and Big Bang theories of the universe are good examples of this.

The Big Bang theory suggests that the future state of the universe depends on the strength of gravity. If there is enough mass then the universe may start to contract and end with a big crunch.

The current evidence suggests there is not enough matter for this to happen. There is a debate about the amount of hidden or **dark matter**, which is not easily observable. Some particles are difficult to detect but could increase the current estimates of matter. We do not have an answer yet.

1 Describe another example of when the force of gravity depends on the amount of mass.

The USA spends about $20 billion a year on space exploration. Politicians have to decide how much to spend each year. Although this is a massive amount of money it is just 0.8% of the USA's total budget and apart from engaging with front-line discovery there are many useful spinoffs.

Flat-screen technology is becoming more common. This technology was originally developed by NASA.

A Flat-screen television screen technology was developed at NASA.

2 NASA has investigated various high-efficiency food products for astronauts. They developed an oil containing two essential fatty acids found in human milk but not in most baby formulas. Suggest how this development could alter the argument that breast-feeding is best for a baby.

B This material was developed by NASA to radiate or reflect heat and keep objects cool.

NASA developed a lightweight material which was particularly good at reflecting and radiating heat. This material is now used for insulating houses. It will help to keep homes cool in hot summers and warm in cold winters. It will cut heating bills and mean we burn less fossil fuel.

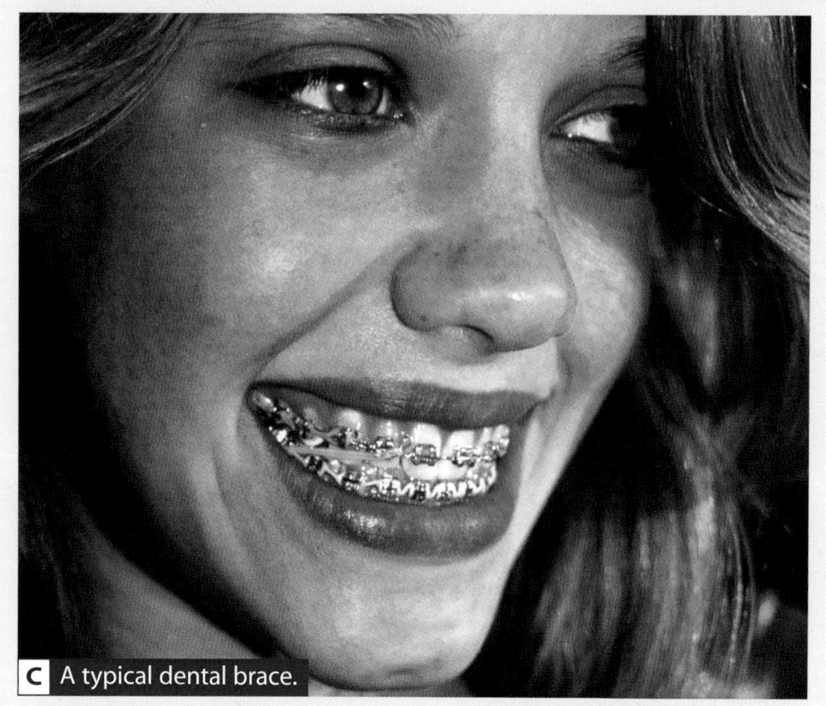

C A typical dental brace.

Dental braces have a wire which needs to be tightened regularly to hold teeth in the right positions. NASA developed a new alloy which 'remembers' the shape it needs to become. It holds teeth in the correct position without regular adjustment.

But perhaps the most important reason for exploring space is because we are curious and we want to find out more.

3 Explain why the argument for an oscillating universe is dependent on the amount of dark matter.

4 Give three reasons for exploring space.

5 Suggest one way in which the exploration of space has influenced your life.

6 Write down four implications of the force of gravity on the universe.

Higher Questions

Questions

Multiple choice questions

1 Which of the following objects is closest to Earth?
A the Sun
B Pluto
C Saturn
D Mars

2 The Milky Way is
A another name for the universe.
B another name for the solar system.
C the galaxy which the Sun is part of.
D the name of a space shuttle.

3 Put these in order of their size (largest first).
A galaxy, universe, Earth, Sun
B universe, galaxy, Sun, Earth
C universe, Sun, galaxy, Earth
D galaxy, Sun, Earth, universe

4 The Moon has no atmosphere. This means that
A the Moon has no gravity.
B you would feel weightless on the Moon.
C you would have no mass on the Moon.
D you would need breathing equipment on the Moon.

5 The Apollo missions to the Moon took about 5 days to go there and back. How long would it take to get to Mars and back at roughly the same speed? Mars, at its closest, is about 150 times further than the Moon.
A 2 years
B 1 year
C 3 months
D 40 years

6 Which of the following statements correctly describes the reason why fuel makes a rocket move?
A The fuel has to heat the air under the rocket.
B The fuel pushes the rocket backwards.
C The fuel pushes backwards which creates a force on the rocket forwards.
D The fuel has to be exploded.

7 The acceleration of free-fall on Earth is 9.8 m/s^2. The weight of a mass of 3 kg is
A 3 N.
B 3 kg.
C 27 N.
D 27 kg/N.

8 Which of the following usually has an eccentric orbit around the Sun?
A the Moon
B Mars
C a comet
D a galaxy

9 The reason the Earth orbits around the Sun is
A due to the atmosphere.
B because planets always go in a circle.
C due to gravity.
D because of fusion.

10 The force acting on an object of mass 2 kg is 300 N. The acceleration is
A 600 m/s^2.
B 150 m/s^2.
C 600 N.
D 0.007 m/s^2.

11 Which of the following was not a robotic spacecraft?
A Galileo
B Voyager
C Apollo
D Viking

12 The temperature, in °C, of deep space is about
A 0
B 100
C 3
D –270

13 The theory of the universe which suggests that it will always look largely the same is called
A the Big Bang.
B oscillating.
C the big crunch.
D the steady state.

14 Which of the following correctly describes the stages in the evolution of the Sun
A nebula, red giant, white dwarf
B nebula, white dwarf, red giant
C white dwarf, red giant, nebula
D red giant, white dwarf, nebula

15 Which of the following does a spacecraft not have to provide?
A air
B water
C gravity
D temperature control

Short-answer questions

1 Write four sentences describing the relationship between stars, galaxies and the universe.

2 Write a paragraph explaining the needs of astronauts on a trip to Mars.

3 Use four sentences to explain the difference between mass and weight.

4 Compare three alternative theories to explain the past, present and fate of the universe.

H 5 Write a paragraph explaining the evidence for the Big Bang theory of the universe.

Glossary

acceleration Change of velocity.
action Forces come in pairs. One of the forces is usually called the action.
asteroid A solid rock that can be anything from about 10 to 500 km across.
atmosphere The gases surrounding a planet.
Big Bang One of the theories of how the universe came into existence and time began.
black hole An object with such strong gravity that it prevents the escape of light.
comet A ball of ice and dust that goes round the Sun in an eccentric orbit.
dark matter Material which is very difficult to detect in space.
extraterrestrial Not from Earth.
galaxy A local collection of stars. Our Sun is part of the Milky Way galaxy.
gravitational field The area in which a mass experiences a force.
gravity A force of attraction which every object exerts on every other object.
interplanetary The area of space in which the effects of planets are negligible.
mass A measure of how much of something there is. Measured in kg.
moon A naturally formed object which orbits a planet.
nebula A collection of gas in space, usually hydrogen.
orbit Movement around a point.
oscillating theory Suggests that the creation of this universe followed the death of a previous universe in a repeating process.
planet An object consisting of solid or gas which orbits a star.
radiation Something which moves out from a point in all directions.
reaction The opposite force to an action.
red giant A very large star, fusing helium into carbon atoms.
red shift Waves emitted by objects which are moving away have their frequencies decreased and their wavelengths increased.
SETI The Search for Extraterrestrial Intelligence, a NASA-sponsored organisation that listens for possible signals from life forms on other planets.
solar system An area of space in which objects are influenced by the Sun's gravity.
star A large ball of hydrogen and helium gas. Produces an enormous amount of energy. Our Sun is our nearest star.
steady-state theory Suggests that the universe expands but because matter is added its appearance does not change.
stellar Associated with a star.
Sun The nearest star to Earth.
supernova An enormous explosion following the end of a giant star.
telescope A device to detect and locate objects in space.
temperature A scale that describes whether something is cold or hot.
universe Includes all the galaxies. All of space.
velocity speed in a particular direction.
weight The force due to gravity measured in newtons (N). Weight = mass × gravitational field strength.
weightlessness The absence of a gravitational field.
white dwarf No longer an active star but still hot and small.

relative atomic mass — 1
H
hydrogen
atomic number — 1

Period	Group 1	2											3	4	5	6	7	0
1								1 H hydrogen 1										4 He helium 2
2	7 Li lithium 3	9 Be beryllium 4											11 B boron 5	12 C carbon 6	14 N nitrogen 7	16 O oxygen 8	19 F fluorine 9	20 Ne neon 10
3	23 Na sodium 11	24 Mg magnesium 12											27 Al aluminium 13	28 Si silicon 14	31 P phosphorus 15	32 S sulphur 16	35.5 Cl chlorine 17	40 Ar argon 18
4	39 K potassium 19	40 Ca calcium 20	45 Sc scandium 21	48 Ti titanium 22	51 V vanadium 23	52 Cr chromium 24	55 Mn manganese 25	56 Fe iron 26	59 Co cobalt 27	59 Ni nickel 28	63.5 Cu copper 29	65 Zn zinc 30	70 Ga gallium 31	73 Ge germanium 32	75 As arsenic 33	79 Se selenium 34	80 Br bromine 35	84 Kr krypton 36
5	85 Rb rubidium 37	88 Sr strontium 38	89 Y yttrium 39	91 Zr zirconium 40	93 Nb niobium 41	96 Mo molybdenum 42	(98) Tc technetium 43	101 Ru ruthenium 44	103 Rh rhodium 45	106 Pd palladium 46	108 Ag silver 47	112 Cd cadmium 48	115 In indium 49	119 Sn tin 50	122 Sb antimony 51	128 Te tellurium 52	127 I iodine 53	131 Xe xenon 54
6	133 Cs caesium 55	137 Ba barium 56	139 La lanthanum 57	178 Hf hafnium 72	181 Ta tantalum 73	184 W tungsten 74	186 Re rhenium 75	190 Os osmium 76	192 Ir iridium 77	195 Pt platinum 78	197 Au gold 79	201 Hg mercury 80	204 Tl thallium 81	207 Pb lead 82	209 Bi bismuth 83	(209) Po polonium 84	(210) At astatine 85	(222) Rn radon 86
7	(223) Fr francium 87	(226) Ra radium 88	(227) Ac actinium 89	(261) Rf rutherfordium 104	(262) Db dubnium 105	(266) Sg seaborgium 106	(264) Bh bohrium 107	(277) Hs hassium 108	(268) Mt meitnerium 109	(271) Ds darmstadtium 110	(272) Rg roentgenium 111							

Index

absorption 260
AC *see* alternating current
acceleration 275, 283
accommodation reflex 66, 69, 71
acid rain 161
acids 147
action and reaction 282
adaptation 25
addiction 95
adenoviruses 49
adhesives 181
adrenaline 72
advertising 197
AIDS 83
air 170–1
alcohol 95, 99, 186–91
alkali metals 107, 114–15
alkaline 115
alleles 40–1
alloys 108, 179
alternating current (AC) 203
ammeter 205
ammonia 146, 171
amplification 267
amplitude 250
analogue systems 266
antibiotics 44, 85
antibodies 51, 92–3
antigens 93
appliances 232–3, 237
argon 117
artificial insemination 77
artificial sweeteners 145
asexual reproduction 34–5
asteroids 276, 278–9
astronauts 280–1
atmosphere 158–9
atomic number 113
atoms 112–13

bacteria 35, 73, 82, 145
baking powder 140
barbiturates 95
barriers 88–9
bases 46
batteries 204, 212–13
BCG vaccination 84, 92–3
beer 186–7
bias 161
Big Bang 290
big crunch 290, 292
binge drinking 189, 190
bio-fuels 165
bioaccumulation 21
biological control 21
biomass 17
black holes 289
blink reflex 71
blood 45, 73, 90, 157
bonds 118
brain 58, 60, 62–3
brain tumours 63
breathability 180
breeding 20
bromine 118, 120
bronchitis 101

caffeine 94
calcium carbonate 140
cancer 47, 49, 101, 144, 265
  see also tumours
cannabis 97
capacity 212
carbohydrates 148
carbon 135
carbon dioxide 14, 141, 158–9
carbon fibres 179
carbon monoxide 101, 156–7
carriers 44–5
caustic soda 149
cells (batteries) 204, 212–13
cells (plants and animals) 34
central nervous system (CNS) 60–1
CF *see* cystic fibrosis
characteristics 23, 36
chemical analysis 110, 124
chemical barriers 88
chemical reactions 106
chemical signals 57–80
chemical symbols 106–7
chlorine 118–19, 146, 173
chromosomes 36
cilia 89
circuits 203
circulatory system 90
citric acid 147
classification 28–9
climate 158–61
clones 34, 50–1
CNS *see* central nervous system
combustion 130, 154, 156–7
comets 276, 279
competition 10–11
complete combustion 156
composite materials 179
compounds 106
computer models 13, 161
computers 211, 218–19
conduction 108–9, 178
cones 66
conservation 9
contraception 74–5
conventional current 205
copper 109
cornea 66, 71
corrosion 108
crude oil 154, 168–9
crust 130, 134
current 203, 205–9, 235
cystic fibrosis (CF) 37, 42, 44–5, 47, 49

dark matter 292
Darwin, Charles 24–7
DC *see* direct current
decomposition 130
defibrillators 245
deforestation 15
dehydration 148, 189
density 109, 253
desalination 167, 173
desertification 15
designer babies 52–3
diabetes 72–3
diatomic molecules 118
digital information 266–9
dilute acids 147
dinosaurs 25
direct current (DC) 204–5
Directly Observed Treatment, Short
  Course (DOTS) 85
disease 82, 86–7
displacement reactions 120–1
distillation 168, 187
DNA 27, 34, 36–7, 46–7, 264
DNA fingerprints 37
DNA sequencing 48
Dolly the sheep 50
dominant alleles 40–1
DOTS (Directly Observed Treatment,
  Short Course) 85
downward delivery 141
drink-driving 191
drugs 84–5, 94–101
dry-cell batteries 204, 213
ducking reflex 69
ductile 109
dynamo effect 206–7

E. coli 35
ears 59, 65
Earth 153–76, 287
earth wire 232
earthquakes 254–5
ECG *see* electrocardiogram
ecosystems 16
effectors 61
efficiency 236–7, 240–1, 243
electrical signals 57–80

electricity 135, 201–48
electricity meters 238–9
electrocardiogram (ECG) 245
electrolysis 172
electromagnetic induction 206–7
electromagnetic spectrum 257–61, 264–5
electromagnets 221
electrons 112, 205
elements 106–7, 123
embryo screening 52–3
emission 259
emphysema 101
emulsifiers 194–5
emulsions 194–5
endocrine system 72
endothermic reactions 115
energy 15, 18–19, 154–5, 236–7, 240–1
environment 9–32, 34, 42–3
epicentre 255
epilepsy 63
estimating populations 12–13
ethanoic acid 194
ethanol 155, 165, 188
ethical issues 51, 53, 77
Euglena 29
evolution 24, 26–7, 39
excess 136
exothermic reactions 115
extinction 25
eyes 64, 66–7, 89

factory farming 20
fermentation 186–7
fertilisation 38, 75
fertility 76–7
fireworks 139
flame tests 124–5
flammable 133
flavourings 145
fluorescence 261
food 42–3, 144–5
food chains 16–19
food pyramids 16–17
food webs 16–17
forces 282–3
foreign bodies 93
forensic science 37, 124–5
formulae 111
fossil fuels 158–9, 162–3, 226
fossils 26
fractional distillation 168, 170–1
fractionating column 168
frequency (waves) 250, 256–7
fuels 154–5
fuses 232–3

galaxy 276
gametes 38
gamma rays 257, 265
gas syringe 131
gaseous exchange 89
gases 146–7
gene therapy 47, 48–9
generations 40–1
genes 34, 36–7
genetic engineering 23, 51, 73
genetically modified (GM) organisms 23, 73
genetics 33–56
glands 72
global warming 14, 158–9
glucagon 72
glucose 73
GM *see* genetically modified
gold 109
Gore-tex™ 180
grand mal 63
gravity 274–5, 278, 281
growth 42–3

halogens 107, 118–21
hazard symbols 139
helium 116
heroin 97
HGP *see* Human Genome Project
HIV 83, 97
hormones 72–7
Human Genome Project (HGP) 46–7
hydration 148
hydrocarbons 155, 168
hydrogen 132–3, 164, 172
hydrogen peroxide 130
hydrophilic/hydrophobic 195

ICT 211
ignition 169
immune system 86
impulses 59, 60, 62
in-vitro fertilisation (IVF) 76–7
incomplete combustion 156–7
inert gases 116
infection 86
infertility 76
inflammation 91
infrared radiation 65, 259, 264
inheritance 44–5
insoluble substances 110, 132, 138
insulation 240
insulin 72, 73
intelligent packaging 192–3
interdependency 11
interplanetary space 280
inter-species competition 10
intra-species competition 10
involuntary responses 68
ionisation 265
iris 67, 69, 71, 258
iron 108
IVF see in-vitro fertilisation

Jupiter 285

KevlarTM 180, 281
kingdoms 28

lamps 210
LDR *see* light-dependent resistor
lecithin 194
light bulbs 236
light-dependent resistor (LDR) 214–15, 217
liposomes 49
liquefication 170–1
liquids 146–7
lithium 115
liver 96, 99
longitudinal waves 251
lubrication 169
Lycra™ 178
lysozyme 89

Maglev train 221
magnesium 131
magnets/magnetic fields 206–7, 220–1, 230–1
malleable 109
Mars 284
mass 274–5
materials 178–81
mayonnaise 194
medicine 244–5
Mendel, Gregor 40–1
Mendeleev, Dmitri 122–3
menstrual cycle 75
metals 132, 134–5, 143, 167
methane 156
microbes 86
microorganisms 83, 86–9
microporous membrane 181
microwaves 258, 261, 262–3, 291
Milky Way 276
MMR vaccination 93
mobile phones 203, 212, 219, 258, 261, 262–3
molecules 118
Moon 274–5, 276
morphine 97
Morse Code 266–7
moths 39
motor neurones 61, 70
motors 230–1, 234
mucus 37, 44, 89
muscles 59
music technology 268–9
mutation 264, 265

nanobots 184–5
nanocomposites 184
nanoparticles 182–5
nanotechnology 183

nanotubes 184
NASA 284–5, 292–3
National Grid 228–9
natural selection 24–5, 39
nature and nurture 42–3
nebula 288
neon 117
nervous system 60–1
neurones 60, 94
neutralisation 136–7
neutrons 112
nicotine 100–1
nitric acid 171
nitrogen 171
noble gases 107, 116–17
non-specific immune system 91
nose 65, 89
nucleus 34, 112
nutrients 42–3

oestrogen 72, 75
oil 164
opiates 97
optical fibres 267
orbits 276–7
ores 134–5
organic farming 20–1
organisms 11, 22–3, 86
oscillating theory 290
oscillation 250
overdose 96
ovulation 75, 77
oxidising agents 139
oxygen and oxidation 130–1, 134–5, 171
ozone layer 265

P waves 255
packaging 192–3
pain sensors 70
painkillers 95–7
pancreas 73
paper 167
paracetamol 95, 96
Parkinson's disease 63
particles 112
particulates 163
pathogens 86
pathways 59
pay-back time 240
periodic table 106–7, 113, 122–3
photochemical smog 163
photosynthesis 160
photovoltaic cells *see* solar cells
physical barriers 88–9
pituitary gland 72
placenta 87
planets 276–7
plants 34–5, 40–1, 43
plasma 73
plaster of Paris 148
plastics 167
populations 10, 12–13
potassium 115
power 234–5
precautionary principle 162
precipitation 110, 138
predators 11, 21
pregnancy 74
prey 11
primary consumers 16
primates 28
producers 16
progesterone 72, 75
properties 108–9, 196–7
Protoctista 28
protons 112
pupil 67
pus 91
pyramids of biomass 17
pyramids of number 17

quadrat 13
quantitative methods 17

radiation 158, 178, 259, 281, 285
radio waves 252, 258, 285
rancid 193
rate (power) 234
raw materials 143
RCCBs *see* residual current circuit breakers
reaction times 58–9, 94
reactions and reactivity 114–15, 129–52
reactivity series 132–3, 143
receptors 59, 61, 64
recessive alleles 40–1
rechargeable batteries 204, 213
recycling 15, 162, 166–7, 213
red blood cells 45, 73
red giants 288
red shift 291
reduction 134–5
reflection 252–3
reflex arc 70–1
reflexes 68–9
refraction 254
relay neurones 70
renewable resources 226–7
reproduction 25, 34–5, 38–9
residual current circuit breakers (RCCBs) 233
residue 155
resistance (drugs) 85
resistance (electricity) 208–11, 216–17, 233
resistors 209
retina 66
rods 66
rough grazers 19
rust 108

S waves 255
salts 120–1, 132–3, 138–9, 145, 149, 172
sampling populations 12–13
saviour siblings 52
scanning 253
scavengers 11
SCD *see* sickle-cell disease
sea water 172–3
Search for Extraterrestrial Intelligence (SETI) 286–7
secondary consumers 16
sedatives 95
sediment 110
sedimentary rocks 26
seismic waves 254
selective breeding 22
sense organs 59, 64–5
sensors 214, 285
sensory neurones 61
Serengeti-Mara 16–17
SETI *see* Search for Extraterrestrial Intelligence
sexual reproduction 38–9
shape-memory alloys 179
sickle-cell disease (SCD) 45
silver 109
skin 88, 91
smart materials 179
smart meters 239
smog 162
sodium 115
sodium chloride 145, 149
sodium hydroxide 149, 173
solar cells 204, 227, 242–3
solar power 227
solar system 276
solids 148–9
soluble substances 110, 131, 138
solutions 110
solvents 98–9
soot 154, 163
space 273–96
space probes 284–5
space travel 280–1
species 10
specific immune system 93
speed (waves) 250, 252
spinal cord 60
stable oxides 135
stars 288–9
steady-state theory 291
steel 108
stem cells 48, 51
sterilisation 119
stimulants 94–5
stimuli 58
stroke 63

sugar 148, 186
sulphuric acid 148
Sun 276–7, 288
sunblock 183
superconductors 221
supernova 289
suspension 110
sustainable development 9, 167
synapses 61, 70–1, 94

table salt 145
target organs 72
TB *see* tuberculosis
technology 220–1, 244–5
Teflon™ 181
telecommunications 202–3
telescopes 292
temperature 280
testosterone 72, 74
thermal decomposition 142–3
thermistors 214–15, 217
Thinsulate™ 178
thyroxin 72
tobacco 97, 100–1
tolerance 95
tongue 65
total internal reflection 267
toxic 119, 156
transgenic organisms 50–1
transistors 218
transition metals 107, 108, 110
transmission 87
transplants 51
transverse waves 250, 251
trends 115
tsunamis 255
tuberculosis (TB) 82–7, 90–1
tumours 63, 263

ultrasound 65, 253
ultraviolet light 261, 265
universe 276–7, 284–5, 290–3
upward delivery 133

vaccination 84, 92–3
vacuum 257
vanilla 145
variation 28–9, 38–9
vector-borne disease 87
vehicle-borne disease 87
velocity 283
viral infection 97
viruses 49
viscosity 169
voltage 206–9, 235
voluntary responses 68

water 147
wave equation 256–7
wavelength 250, 256–7
waves 249–72
weight 274–5
weightlessness 281
white blood cells 48, 73, 90–1
white dwarves 288
wind power 227
wine 186–7
wires 230–3

X-rays 257, 260, 264

zinc 133